Bituminous mixtures in road construction

Edited by Dr Robert N. Hunter

⌐ŢI Thomas Telford, London

Published by Thomas Telford Services Ltd, Thomas Telford House, 1 Heron Quay, London E14 4JD

First published 1994
Reprinted 1995

Distributors for Thomas Telford books are
USA: American Society of Civil Engineers, Publications Sales Department, 345 East 47th Street, New York, NY 10017-2398
Japan: Maruzen Co Ltd, Book Department, 3–10 Nihonbashi 2-chome, Chuo-ku, Tokyo 103
Australia: DA Books and Journals, 11 Station Street, Mitcham 2131, Victoria

A catalogue record for this book is available from the British Library

Classification
Availability: Unrestricted
Content: Guidance based on best current practice
Status: Refereed
User: Civil engineers in road construction

ISBN: 0 7277 1683 2

Typeset in Great Britain by MHL Typesetting Ltd, Coventry

Printed and bound in Great Britain by Redwood Books, Trowbridge, Wilts.

Foreword

Road-making and maintenance form a significant part of Britain's economy. Nearly three people in every 100 earn their living serving this vital part of our infrastructure. Around 10% of the cost of producing goods and services arise in road transport. For Britain's more remote and peripheral areas, our roads provide the essential links to major markets.

The road network allows these communities to participate in the prosperity that greater trade can bring. Today, the national roads programmes in England, Scotland, and Wales are responding well to the growth in traffic, in particular on our trade routes. Maintenance expenditure on all Britain's extensive network of roads continues at a high level.

The pressure on the available resources is nevertheless immense in the face of rising traffic. Road engineers need to get the most from every penny spent. Initiatives such as the Citizen's Charter also mean that the public, rightly, is looking for rising levels of service from our roads. The public, of course, has limited interest in the details of how roads are built. Their main concern is to pay as little as they need to as taxpayers or through the toll booth. They do not want to be held up or irritated by the presence of road works. They expect a reasonable ride and relatively consistent response from the road surface when they brake. They do not like spray, and those living close to roads do not want to be bothered with road surface noise. New methods and materials are available which help meet the demands for quicker construction of quieter and more durable roads.

This book presents the best of British expertise. Each chapter is contributed by a leading figure in the particular field covered, and fills an important gap for road engineers. Until this publication, a comprehensive coverage of early 1990s state-of-the art was simply not available. It provides up-to-the-minute thinking in materials specification, test methods and harmonization of standards. Its appeal will be wide.

The particular contribution this book makes is in its comprehensive coverage of all aspects of fully flexible road construction from foundation design through to surface treatment. It provides an excellent reference for materials technologists, graduate engineers and practising engineers. The coverage of pavement design is a particular pleasure to me. This caters for the needs of all design engineers on both new works and maintenance and not just the engineer working on new major inter-urban highways. It also complements the Scottish Office and Scottish County Surveyors' Society initiative on *Fitting roads*, which similarly aims to provide guidance on the design of more minor roads which form the vast majority by length of the road network.

Education and training is essential to our future. Materials technologists and graduate engineers will welcome this comprehensive guide to all they need to know about the 'black' art. Degree courses cannot cover all aspects of materials technology in their crowded syllabus and I am sure this book will become a reference work.

The latest advances in binder technology are well covered. Better binders reduce costs and increase durability. They enhance road safety by extending the use of rapid cost-effective treatments to maintain skid resistance, which in turn means reduced delays at roadworks. Modified binders mean more recycled materials can be used, particularly where local aggregates are scarce and environmental constraints are pressing.

All users of this book will be grateful for the time and effort of those who have contributed and to its editor. I am particularly pleased at the significant contribution that has been made from Scotland and commend the book to all practising and aspiring road engineers.

John Dawson
Director of Roads and Chief Road Engineer
The Scottish Office Industry Department

Preface

I was the Engineer's Representative on a Trunk Road re-alignment contract in 1979. Shortly after laying, the hot-rolled asphalt wearing course suffered the loss of a substantial proportion of the pre-coated chippings. I thought that there would be some texts and papers on the subject to provide advice on this failure. Unfortunately there were very few relevant publications. I soon realised that the whole subject area of bituminous mixtures technology was poorly served by documentation. The situation has altered little since that time, so I was delighted to be asked to edit a definitive treatise on the subject.

This book covers all the main areas of bituminous mixture technology and has contributions by leaders in their fields. It is the first time that much of this knowledge has been published. The reader is advised to check the index carefully since many aspects are covered in part in different chapters.

This book was written to appeal to the practising engineer, surfacing contractors and undergraduates. I hope you find it as interesting to read as I found it to edit.

Robert N. Hunter, February 1994

Biographies of contributors

Robert N. Hunter BSc MSc PhD CEng MIHT MICE MIAT
Robert Hunter has a BSc in Civil Engineering from Heriot-Watt University. He has wide experience on site and it was whilst supervising major surfacing operations that he became interested in bituminous mixtures. His first higher degree involved an examination of the fretting of chipped hot-rolled asphalt wearing course. His second higher degree produced a computerized prediction of temperatures within unchipped and chipped bituminous mixtures and, using the derived results, he was able to explain a number of phenomena associated with the laying of these materials. He has had papers published on subjects associated with bituminous mixtures and has given a number of papers to meetings of the Institute of Asphalt Technology, the Institution of Highways and Transportation and the Society of Chemical Industries.

Derek Fordyce BSc MSc MIHT MIAT
Derek Fordyce was born and educated in Edinburgh. He was indentured with Blyth and Blyth, Civil and Structural Engineers from 1967 until he graduated with an Honours degree in Civil Engineering from Heriot-Watt University in 1970. He continued to work for Blyth and Blyth as a Graduate Engineer until 1972. From 1973 until 1980 he was a lecturer in the Department of Civil Engineering at Napier College, now Napier University, in Edinburgh. He joined the Department of Civil Engineering at Heriot-Watt University in 1980. He received an MSc for research into the ageing of petroleum bitumen in 1982 and was made a Senior Lecturer in 1992.

Eric O'Donnell MPhil IEng
Eric O'Donnell is employed by Frank Saynor and Associates as Laboratory Manager of both Edinburgh and Glasgow offices. He has worked exclusively in the field of material science for over 15 years. This experience covers all construction materials with recent research concentrated on bituminous mixtures.

In 1992 Eric gained a higher degree at Heriot-Watt University. As well as continuing research work he is currently involved in CEN committee work and a European open learning scheme for Highway Technicians.

Professor Stephen Brown DSc FEng
Professor Stephen Brown is Head of the Civil Engineering Department at the University of Nottingham, where he has spent most of his career. He has been

involved in research, teaching and consulting in the general field of pavement engineering for 30 years. His work has embraced material properties, pavement design and evaluation. In addition to leading the Department's Pavement and Geotechnical Engineering Research Group, he is a director of SWK Pavement Engineering Ltd., a consulting firm which evolved from his research activities. Professor Brown's research has been supported by a range of government and industrial organizations both in the UK and the USA. He has travelled widely and spoken at major conferences in various parts of the world. In addition to publishing over 130 technical papers, he has contributed to books and pioneered the teaching of pavement design principles based on mechanistic concepts.

Ian D. Walsh BSc(Eng) DMS CEng MICE MIHT MBIM MIAT

Since graduating, Ian Walsh has spent most of his career in highway engineering with contractors, a road construction unit, District and County Councils. He is now Head of Laboratory for Kent County Council and manages a staff of fifty. The laboratory carries out geotechnical fieldwork and design to BS 5750 : Part 1, evaluates, tests and provides advice on materials to NAMAS requirements. He is a member of a number of BSI/CEN Committees on QA and road materials and characteristics.

As Quality Assurance Manager for the Highways Department he introduced a Quality System for Highway Design and Construction which achieved third party certification with Lloyds Register QA in February 1992.

John P. Moore BA(Hons) MIAT

John Moore is the Technical Director of the ACP group of companies, a leading manufacturer of asphalt and tarmacadam plants. He is responsible for Engineering, Estimating and Technical Publications Departments and undertakes extensive overseas business travel. He has an Honours degree in mathematics and is an Associate of the Institute of Quarrying with over 25 years of experience gained in the quarrying industry. Since 1987 he has been engaged in the draft committee for the CEN European Standards for Asphalt Plants.

Bob McLellan BSc(Hons) CEng MICE MAPM MIHT MIAT ACIA

Bob McLellan graduated with a BSc(Hons) degree in Civil Engineering at the University of Strathclyde, Glasgow in 1982. Subsequently, he began employment with Lothian Regional Council Department of Highways and held a number of appointments including Works Engineer, Works Manager and Regional Surfacing Manager. He is presently Assistant Director (Network) in Lothian Transportation.

Bob is currently undertaking a PhD in the use of project management procedures by public and private contracts in the Construction Management Division of the University of Strathclyde, Glasgow.

Gerry Hindley FIAT MIHT AMIQ MBAE

Gerry Hindley is a Bituminous Pavement Construction Consultant with Gerald Hindley Associates. He has 30 years' experience in the production and laying of bituminous mixtures which was gained whilst engaged in Senior Management appointments with two of the UK's major construction companies. Gerry specialized in major roadworks contracts from green field sites to motorway reconstruction contracts.

Gerry is currently a member of the National Council of the Institute of Asphalt Technology with a particular interest in training. He is also a member of the recently formed British Academy of Experts specializing in bituminous pavement issues.

John Richardson BSc CChem MRSC CEng MInstE

John Richardson is Quality Manager with Wimpey Minerals and is responsible for the establishment and maintenance of Quality Management Systems and for the introduction of other quality initiatives in the operating companies of this division of George Wimpey plc. The range of business activities include quarry products, asphalt surfacing, waste management and open cast coal mining.

John graduated from Strathclyde University with an Honours degree in Applied Chemistry in 1973. After a short period with the Water Board in the South of Scotland, he joined Wimpey Asphalt Ltd as a Trainee Quality Engineer. He subsequently became involved in research and development, quality control and overseas contracting before being appointed Quality Manager.

John is a committee member of the Construction Materials Group of the Society of Chemical Industries and a representative on the ACMA Technical Working Panel. He is also the BACMI representative on the BSI committee B/510/1 for Sampling and Examination of Bituminous Mixtures, and a BSI delegate to the CEN TC227/WG1 Task Group 1 on Test Methods for Bituminous Mixtures.

Terry Hoban CEng MIMechE MIAT MInstE

Terry Hoban was educated at Woolwich Polytechnic and Kings College London. He served an apprenticeship with the General Electric Company and, after a period in design, he was appointed Senior Research Engineer in the Turbine Development Department. Terry Hoban joined Shell in 1963 and worked as a Development Engineer in the fuel oil and liquefied petroleum gas markets before moving to the bitumen market in 1974. He was appointed Manager of the Technical Department in 1980, where his responsibilities include the preparation and implementation of Research and Development programmes and the provision of technical services to customers. He is a member of the Technical Committee of the Refined Bitumen Association in the UK, and represents the bitumen supply industry on a number of British Standards Committees. He is a member of the CEN committee concerned with the specifications for modified bitumen binders and has presented papers to a number of national and international conferences including Eurobitume in 1981.

Tony Pakenham MIAT

Tony Pakenham came from a mechanical engineering background and eventually joined Blaw-Knox in 1960. His first year was involved in product manufacture and subsequently in product support in the field. As a Sales Engineer he was initially attached to the Blaw-Knox Concrete Construction Plant Division. After a period as a Regional Manager he was appointed Sales Manager in 1975. In recent years he has been totally involved in the 'black top' industry and is a long standing member of the Institute of Asphalt Technology. He was elected to the Council in 1980 and elected President in 1993.

Jeff Farrington CEng MICE MIHT MIAT DMS

Jeff Farrington is a sharp-end Civil Engineer who has been involved in various aspects of road building and re-building for most of his working life. He is particularly interested in bituminous mixtures and is very aware that most Engineers have little knowledge of most aspects of both specifying and using these materials. He has served on British and European black-top specification committees.

Acknowledgements

The front cover photograph was kindly supplied by INROADS, the contracting division of the Transportation Department of Lothian Regional Council.

Robert Hunter would like to thank the Transportation Department of Lothian Regional Council for giving him access to its staff, Jack Edgar for his comments on a couple of chapters, Ross Nicolson for providing technical information at short notice, Joan Morrell for her typing skills, all the companies and individuals who provided photographs, the contributors for tolerating my nagging mostly with good grace, Russell Hunter for being a very special little boy, scanning some photographs and doing odd bits of typing and Jeff Farrington for his excellent advice.

Terry Hoban would like to thank David Whiteoak of Shell International Chemical Company (SICC) and his colleagues at Shell Bitumen.

John Moore would like to thank Kathy Dunne for the excellent typing, Nigel Moreton for helping (advice only) with the geological content and reading the script, Paul Bolley and Martin Toon for drawing the flow diagrams and Roland Slack for advice on the use of marine aggregates in hot bituminous mixtures and the storage of mixed materials.

Gerry Hindley would like to acknowledge the help of Fred Mason, a superb surfacing foreman with Tarmac Roadstone, and Martin Southall, Contracts Manager with Tarmac Roadstone and Wrekin Construction.

John Richardson would like to express his gratitude to John F. Hills for commenting on his first draft and Chris A.R. Harris for inspiring the greater use of statistical analysis.

Jeff Farrington would like to acknowledge the assistance of his typist and his colleagues in European specification work.

Bob McLellan provided some material for chapter 6.

Contents

1. Bituminous pavement materials: their composition and specification

1.1. Introduction

Bituminous mixtures evolved from dry stone mixtures. Telford and Macadam produced individual dry-bound pavement structures which were subsequently sprayed with a sealing tar blend to bind the surface together. These composite mixtures relied on the interlock of the stone for their strength. However, the advent of powered motor transport quickly highlighted weaknesses in their performance. The first truly bituminous mixtures were produced in the 1870s in Paris, and first used in the UK around the turn of the century, although they were not extensively available until the 1930s.

Although highway technology has changed markedly since, principal road layers still consist of a mixture of stone, in the form of graded aggregate, and binder, in the form of a petroleum bitumen or a derivative of petroleum bitumen. Engineers today have a wide choice of materials available for road pavement construction.

This chapter gives an overview of the various materials used in road pavement construction in the UK. It also provides a basic understanding of the nature and properties of the material ingredients which form bituminous mixtures used for pavement construction in the UK, and offers guidance for the selection of those ingredients.

1.2. Bituminous materials available in the UK

The various layers which together form a road are called a pavement. Fig. 1.1 shows the layers which make up a typical flexible, or flexible-composite construction in the UK.

The wearing course provides the running surface for traffic and, as such, has to be laid to close tolerances to ensure an even profile for user comfort. The wearing course, or the basecourse, also have the function of sealing the construction against the ingress of water. Water permeating a material reduces its mechanical strength. Water permeating a pavement structure will reduce its serviceable life.

Most major carriageways have a hot-rolled asphalt wearing course. Bitumen-coated stone chippings are partly embedded into the surface of the layer. Hot-rolled asphalts used for road pavement wearing course layers are impermeable. The chippings are spread onto the surface of the wearing course immediately after laying and are embedded during layer compaction. The compaction process is by rolling. The coating of the stone chippings is to ensure adhesion

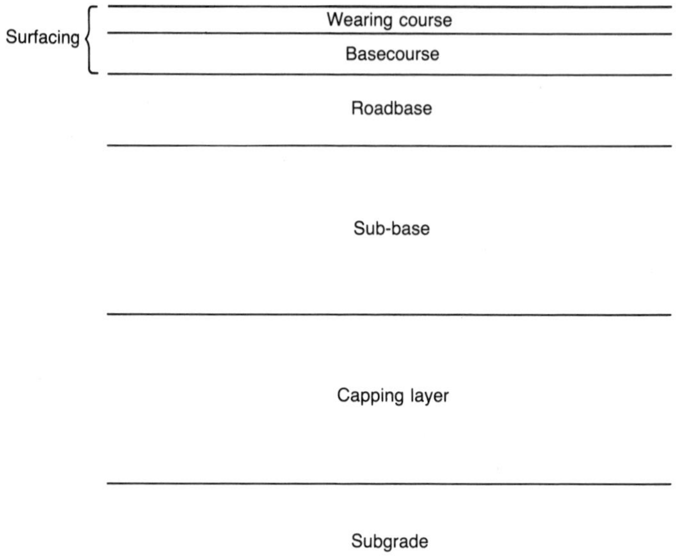

Surfacing {

| Wearing course |
| Basecourse |

Roadbase

Sub-base

Capping layer

Subgrade

Fig. 1.1. Layers in a typical UK road

of the chippings to the hot-rolled asphalt. The surface of the stone chippings provides the principal contact with vehicle tyres and is the source of the basic frictional properties of the surface. The embedment of the stone chippings is to ensure the running surface has a texture for the maintenance of frictional properties in wet surface conditions.

The wearing course of a road pavement may also be a bituminous macadam. Dense bituminous macadams provide a surface texture as a result of the stone content of the material, and when compacted they can have a low permeability. Pervious macadam (porous asphalt) wearing course mixtures are increasingly being specified for major routes. Pervious macadams allow drainage through the layer reducing traffic spray. The main advantage of pervious macadams is a reduction in road noise when compared with a conventional bituminous or concrete running surface.

The purpose of the basecourse to a road pavement is to provide a well-shaped surface upon which the wearing course can be laid. The basecourse is normally a dense bitumen macadam but it may also be a hot-rolled asphalt. The wearing course and the basecourse together form the surfacing layers.

The roadbase is the main structural layer within a road pavement. The thickness of the roadbase is determined by the cumulative loading from the traffic over the design life of the carriageway. Roadbases with flexible pavement constructions may be formed from bituminous mixtures, unbound aggregate mixtures, selected hardcore, or stabilized soil or gravel. Roadbases with flexible-composite pavement constructions are formed with cement-bound materials.

The primary function of the sub-base is to provide a surface of uniform strength for the construction of the roadbase and surfacing layers. The sub-base also provides a running surface for site traffic and it may act as a drainage layer for the construction to intercept rising water and transmit it sideways to the verges. Capping layers form a sub-structure to sub-bases where the subgrade is particularly weak. The combined thickness of the sub-base and capping layer varies with the strength of the subgrade.

The functions of all the pavement layers are covered in detail in chapters 2 and 3.

1.2.1. Composition of bituminous mixtures

Bituminous mixtures are a combination of mineral aggregates or slag aggregates, filler and bitumen or a bitumen-based binder. Aggregates, mineral or slag, are stored in a quarry as single sizes with material greater than 6.3 mm. The standard sizes are 50 mm, 37.5 mm, 28 mm, 20 mm, 14 mm and 10 mm. Aggregate greater than 6.3 mm is described as coarse aggregate. Aggregate of size 6.3 mm–75 μm is described as crushed rock fine aggregate, or dust, or sand fine aggregate when it is excavated from a natural deposit. The fine aggregate fraction is stored as one material. Filler is material which, substantially, is less than 75 μm.

The mechanical properties of bituminous mixtures result from friction and cohesion. The friction comes from the interlock of the aggregate and depends on the maximum aggregate fraction size used to form a blend, and on the grading of the aggregate fractions which are blended together. The cohesion comes from the properties of the bitumen or bitumen-based binder. The flow or rheological properties of the binder are used to enable a mixture to be adequately compacted.

There is a range of mixtures formed from the mixing of coarse and fine aggregate, filler and a bitumen or bitumen-based binder. The mixtures are the result of a specific blend of aggregate fractions. Table 1.1 shows typical mixtures. The headings are a convenient categorization of bituminous mixtures. Coated stone is an open-graded material resulting from a limited mass proportion of fine aggregate. It is used for surfacing layers and has a reduced

Table 1.1. Typical compositions of bituminous mixtures

		Coated stone	Continuously-graded	Gap-graded	Mastic
Coarse aggregate	% wt	86.0	52.0	30.0	30.0
Fine aggregate	% wt	7.0	38.0	53.0	26.0
Filler	% wt	3.0	5.0	9.0	32.0
Bitumen	% wt	4.0	5.0	8.0	12.0
Coarse aggregate	% vol	64.5	44.1	25.7	27.5
Fine aggregate	% vol	5.1	32.2	46.0	18.9
Filler	% vol	2.1	4.2	7.8	27.0
Bitumen	% vol	8.3	11.5	17.5	26.6
Void content	% vol	20.0	8.0	3.0	<1.0
Grade of bitumen	pen	100–300	100–200	35–100	15–25

Table 1.2. Grading ranges for common bituminous mixtures

Sieve size	Percentage by weight passing each sieve			
	Bitumen macadams		Asphalts	
	20 mm open-graded crushed rock basecourse	20 mm dense crushed rock basecourse	Column 2/2 50/14 roadbase, basecourse and regulating course	Column 3/2 30/14 type F wearing course
50 mm				
37.5 mm				
28 mm	100	100		
20 mm	95–100	95–100	100	100
14 mm	50–80	65–85	90–100	85–100
10 mm		52–72	65–100	60–90
6.3 mm	15–35	39–55		
3.35 mm	10–25	32–46		
2.36 mm			35–55	60–72
600 μm			15–55	45–72
300 μm		7–21		
212 μm			5–30	15–50
75 μm	0–9	2–9	2–9	8–12

resistance to internal movement compared with dense or continuously-graded mixtures. Such mixtures are used on minor routes and parking areas where there may be a limited amount of compaction applied to the material, and therefore ease of compaction is required. Continuously-graded materials include dense bitumen macadams (DBMs) commonly used for roadbases and basecourses and for wearing courses on minor roads. Gap-graded materials include hot-rolled asphalt wearing courses. Mastics are materials which have high bitumen and filler contents and zero air voids. Typical grading ranges are shown in Table 1.2. Type F wearing course asphalts are specified for fine sands. Type C are specified for coarser sands. There are other requirements in the standards but Table 1.2 illustrates the difference between bitumen macadams and asphalts.

Since the proportion of most stone sizes has a range of acceptability the limits can be shown graphically. (See Fig. 1.2).

Grading envelopes similar to those in Fig. 1.2 are used to plot the results from mix analysis and so ascertain conformance.

1.2.2. Bituminous macadams

The composition of bituminous macadams is specified in BS 4987: Part 1.[1] They are specified by three characteristics: their function in the pavement construction i.e. roadbase, basecourse or wearing course; the nominal maximum size of the aggregate used; and the grading of the aggregate blend, described as open-graded, close-graded or dense. Nominal maximum in relation to aggregate size means that there may be a small percentage of material larger than that size. For example, with a 40 mm DBM, 5% by weight of the stone can range between 37.5–50 mm. Here the actual sieve size is 37.5 mm, but

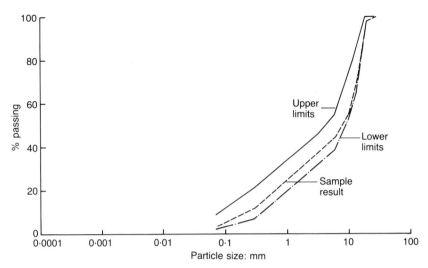

Fig. 1.2. Grading envelope for a 20 mm bitumen macadam. From BS 598: Part 102: Compositional analyis

is rounded off to 40 mm for convenience.

The selection of a nominal maximum size of aggregate used with a bituminous macadam is judged by the compacted thickness of a layer, or part-layer where it is thicker than 200 mm. The rule of thumb is that a compacted layer thickness should not be less than 2.5 times the nominal maximum size of aggregate. For example, a 40 mm DBM would form a layer no less than 100 mm in compacted thickness. Where layer thicknesses are less than 1.5 times the maximum nominal size of aggregate in a mixture, crushing of the larger aggregate will occur with a consequent loss in the mechanical properties of the layer. Practically, the maximum compacted thickness of a material is 200 mm. Layers which have to be greater than 200 mm in thickness are laid in two or more sub-layers.

Increasing the mass proportion of finer fractions within a bituminous macadam reduces the voidage within the aggregate structure, or reduces the voids in the mineral aggregate (VMA). Increasing the mass proportion of finer fractions therefore 'closes up' the mixture. A close-graded macadam therefore contains a higher percentage of finer fraction sizes. This reduces the permeability of a bound mixture and increases the internal friction and resistance to internal movement of a bound mixture for a given nominal maximum size of aggregate. The larger the nominal maximum size of aggregate the greater the internal friction.

More continuous grading reduces permeability and increases internal friction, resulting in layer stiffness (which is measured by its elastic modulus). This relates to resistance to flexing under traffic loading. Increasing internal friction increases the amount of compactive effort needed during the placement of a layer of material. Compaction temperatures are more critical with DBM mixtures compared with open-graded mixtures. Pervious macadam mixtures are open-graded mixtures. These wearing course mixtures are used for their permeability to allow the free passage of moisture to reduce spray, and free passage of air to reduce tyre noise.

The mechanical properties of bituminous macadams result more from internal friction as a consequence of aggregate interlock than from the cohesion of the binder. Bitumen macadams have a high resistance to internal movement and are therefore specified where high loading or long-term loading is applied, such as with car parks. Smaller nominal maximum size of aggregate is specified with surfacing material to provide a quality of surface finish in terms of texture, appearance and evenness. Bitumen macadams have a relatively low binder content and therefore may be more susceptible to fatigue cracking than a hot-rolled asphalt. Fatigue cracking happens when a layer is repeatedly flexed by traffic loading, but the repeated strain is less than that necessary to cause material fracture with a single applied load. The full range of bituminous macadams which can be specified with pavement construction is given in BS 4987: Part 1.[1]

1.2.3.   Hot-rolled asphalt
The composition of hot-rolled asphalts is specified in BS 594: Part 1.[2] They are mainly fine aggregate, filler and binder (referred to as the asphalt mortar) mixtures to which a single size coarse aggregate is added, and are specified in terms of three characteristics

(a)   their function in the pavement construction i.e. roadbase, basecourse, regulating course or wearing course
(b)   the nominal content by mass of the single size coarse aggregate
(c)   the nominal size of the coarse aggregate.

For example, a wearing course hot-rolled asphalt can be specified as a 30/14. This means 30%, by mass, of 14 mm single size coarse aggregate is contained in the fine aggregate, filler and binder mortar. Regulating courses are a specified use of hot-rolled asphalts to remove variations in the profile of an existing carriageway before the placement of a wearing course.

The same relationship between the nominal maximum size of contained aggregate and compacted layer thickness applies to hot-rolled asphalt. A compacted layer thickness should not be less than 2.5 times the nominal maximum size of contained aggregate. A wearing course hot-rolled asphalt is normally no less than 40 mm in thickness. With such layers a 14 mm stone is used as the coarse aggregate fraction.

The mechanical properties of hot-rolled asphalt come more from the cohesion of the binder within the fine-graded aggregate mortar. This means that hot-rolled asphalts can be less resistant to internal movement, but can have better fatigue cracking resistance than bituminous macadams. Hot-rolled asphalt mortars require a relatively high binder content and are therefore more expensive than bituminous macadams. The volume of a hot-rolled asphalt mortar is increased by adding a coarse aggregate. Where the amount of coarse aggregate is less than 40% by mass, the coarse aggregate does not interlock and the mechanical performance of the mixture relies mainly on the properties of the mortar. Where the coarse aggregate content exceeds 40% by mass of the mixture, aggregate interlock occurs. This reduces the binder content and, therefore, the cost of the mixture, but it also reduces the ability to produce an even surface.

An included coarse aggregate content of more than 40% by mass also reduces the impermeability of a layer, but it enhances the resistance to internal movement of the mixture. Hot-rolled asphalt roadbase mixtures consequently specify either 50% or 60% including coarse aggregate by mass of the mixture. Wearing course mixtures usually specify 30% by mass, including coarse aggregate. However, where heavy vehicles are parking on such a surface it is common to specify a 40% coarse aggregate asphalt; where channelled traffic is using a route, a 55% coarse aggregate content mixture can be specified. Where a 55% coarse aggregate content hot-rolled asphalt mixture is specified, chippings are not embedded into the surface as the material naturally creates a surface texture.

1.2.4. Materials and layer thicknesses
On major routes, the surfacing will be specified as a 40 mm wearing course and 60 mm basecourse. The wearing course will be a 30/14 hot-rolled asphalt. The basecourse will be a 20 mm DBM, or 50/14, or 50/20 hot-rolled asphalt.

Roadbase thicknesses vary with traffic loading. Thicknesses lie in the range 100−300 mm in flexible constructions. Hot-rolled asphalts and macadams may be specified. The nominal maximum size of aggregate will relate to the compacted layer thickness, to a maximum of 40 mm. With flexible-composite constructions the roadbase thickness can vary between 150 mm and 240 mm. The surfacing with thicker cement-bound roadbases with flexible-composite constructions increases from 100 mm to 150 mm. The increased thickness is included in the basecourse layer.

On extremely heavily-trafficked routes, the roadbase may be a composite construction consisting of an upper roadbase of DBM and a lower roadbase of hot-rolled asphalt. This construction optimizes the application of the two materials. At the bottom of the roadbase is a material more resistant to fatigue, and at the top is a material more resistant to internal movement possible under the higher cumulative vertical stressing. Fig. 1.3 is a family tree of the variants of bituminous mixtures.

1.2.5. Alternative bituminous mixtures

Heavy duty macadam (HDM) and DBM 50
These materials are listed in the *Specification for highway works*.[3] They are commonly used in roadbases and basecourses where the pavement structure is expected to sustain very high traffic volumes. They may also be used to decrease overall construction thickness. These mixtures are stronger and stiffer than conventional roadbase and basecourse materials but are more expensive. They are all arguably more difficult to lay as they require higher laying temperatures. In appropriate circumstances stiffer bitumens and mixtures have considerable merit.

HDM is a development of DBM roadbase to BS 4987: Part 1,[1] with 3% additional filler and 50 pen bitumen as opposed to the normal 100 or 200 pen. DBM 50 is a dense bitumen macadam roadbase material mixed with 50 pen bitumen. Both materials show an increase in stiffness modulus over

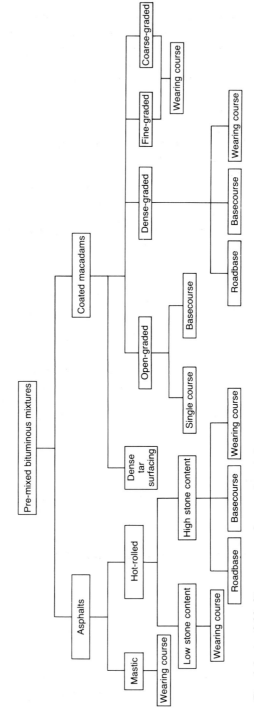

Fig. 1.3. Available bitumen macadams and asphalts

conventional DBM by factors of 2.2 and 1.7 respectively. The increase in stiffness, or stiffness modulus, is caused by an increase in binder stiffness and a reduced air voids content of the compacted layer. The higher loadbearing capacity can therefore reduce the pavement thickness or increase its life expectancy compared with the same thickness of conventional material.

Mastic asphalt

Mastic asphalt is a dense, voidless material consisting of a hard bitumen, or a bitumen/refined Trinidad Lake Asphalt (TLA) binder in the range 12−30 pen mixed with limestone mineral matter. Blocks of the material are melted on the laying site, coarse aggregate is added and the mix usually hand-laid with wooden floats. It is impermeable and used for heavy traffic applications, where its resistance to deformation and fatigue and its waterproofing qualities are particularly appropriate, e.g. bus lay-bys, bridge decks and multi-storey car parks and other uses where the improved performance justifies the extra cost.

Grouted macadam

Grouted macadam is a two-stage process which consists of a predominantly single-sized coated stone bound with a relatively hard binder of 100−200 pen. The void content is maintained in the range 20−25%. After cooling to ambient temperature, a cementitious grout of cement/sand/water and polymer modifiers is vibrated into the remaining voids and allowed to cure. It is used in areas where high resistance to point loading is required or where chemical and fuel spillage is likely e.g. container parks, docks, airport aprons or garage forecourts. It has also been used experimentally to resist the formation of wheel track rutting in contraflow bus lanes. Some work using grouted macadams in utilities reinstatements is currently underway. This material performs best in thick-lift constructions.

1.3. Aggregates

Aggregates are the fundamental component in bituminous mixtures. They must have certain properties to withstand the stresses imposed in the road surface and sub-surface layers. Aggregate properties are a function of the properties of the constituent particles and also the manner in which they are bound together. This section explains how these properties are defined, what they should be, and provides typical examples for reference purposes.

1.3.1. Basic petrography

Petrography is the science of the origin, chemical and mineral composition and structure, and alteration of rocks.

The vast majority of aggregates used for bituminous mixtures in the UK today are derived from natural sources. These can be blasted and quarried from single lithology sources, dredged or excavated from pits, estuaries and rivers. Artificial aggregates from the by-products of steel manufacture or the remains of spent fuel can provide suitable alternatives to natural sources.

Natural aggregates

Rock is formed as a result of three fundamental processes.

(a)   The slow cooling of massive forms of magma which have reached the earth's surface (extrusive rocks) or the slow cooling of magma below the earth's surface (intrusive rocks) cause distillation of the minerals within the magma to form a variety of rocks described as igneous.

(b)   The products of the disintegration of any rock where the process causes the deposition of these materials into stratified layers which are consolidated or cemented by chemical action into a new rock, form a variety of rocks described as sedimentary. Sedimentary rocks can also be formed by the chemical precipitation of minerals which were dissolved in water.

(c)   Igneous or sedimentary rocks which are subjected to extremely high temperatures result in the re-crystallization of the constituent materials to produce a new rock, form a variety of rocks described as metamorphic.

Geologists have identified hundreds of different rock types which may be classified into one of these three groups. Each geological grouping yields rocks of different properties in terms of their road-making properties. The principal rock types quarried for road-making in the UK are shown in Fig. 1.4.

The British Standards for macadams and asphalts[1,2] state that aggregates shall be 'clean, hard and durable'. The Scottish Office Roads Directorate produces a broad-based list of acceptable rock types[4] which is more helpful to the practising engineer, although there is a danger that it may unnecessarily restrict the use of alternative available materials (Table 1.3). Many aggregates have local or traditional names and this can cause confusion. Such names should not be used in specifications or contract documents. O'Flaherty[6] has listed some traditional names against their designated British Standard names (Table 1.4).

Table 1.3. *Aggregates in flexible road pavements*

Definitions: crushed rock	
Gravel (crushed or uncrushed) Crushed slag Slags must comply with BS 1047[5]	
Crushed rock/gravel must come from the following groups	Granite Basalt Gabbro Porphyry Quartzite Hornfels Gritstone Limestone

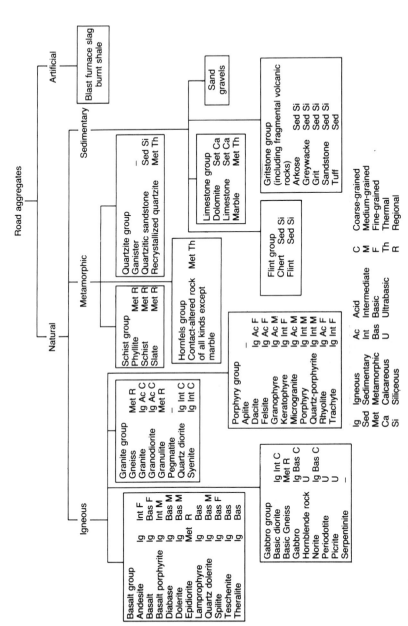

Fig. 1.4. Principal types of rock quarried for making roadstone

Table 1.4. Some common traditional rock names

Traditional name*	Appropriate group in British Standard classification for roadstone
Clinkstone	Porphyry (rarely basalt)
Cornstone	Limestone
Elvan (blue elvan)	Porphyry (basalt)
Flagstone	Gritstone
Freestone	Gritstone or limestone
Greenstone	Basalt
Hassock	Gritstone
Hornstone	Flint
Pennant	Gritstone
Rag (stone)	Limestone (rarely gritstone)
Toadstone	Basalt
Trap (rock)	Basalt
Whin (stone)	Basalt

* As the traditional names are often applied loosely, the information in this Table cannot be precise.

Artificial aggregates

Artificial aggregates are mainly made from air-cooled, blast-furnace slag which is a by-product of steel manufacture. Their use is restricted in the UK because of limited availability. Its inherent variability and, in some cases, its excessive bitumen demand make it relatively unattractive as an alternative to natural rock. However, attempts to produce more consistent materials in recent years will make their use more commonplace.

Pulverized fuel ash is rarely used as a filler replacement material, but recent work indicates that there is scope for its further use.

1.3.2. Properties of roadstone aggregates

It is not unusual for aggregate requirements to be specified using a general statement e.g. 'roadstone aggregates must be clean, hard and durable'. The standard UK specification[3] offers additional guidance in the form of acceptability limits for key properties. In reality, the use of aggregates is complicated by the fact that they have many different and widely varying properties. Another problem is that roadstone aggregates have to possess many conflicting properties, such as the retention of a highly textured surface and also high abrasion resistance. To help the engineer make the right choice, a suite of tests is available to allow properties to be ascertained.

Cleanliness of roadstone aggregates, particularly in the coarse fractions, is important for good bitumen adhesion. Fine, particulate matter on the surface of the stone can interfere with the coating process which can cause incomplete bitumen films or weak bonding to the surface of the stone. This could lead to stripping of the bitumen film and consequently failure of the bituminous mixture. The specification[3] relating to constituent aggregates specifies maximum dust contents. These range from 1−5% in coarse aggregate and up to 8% in fine aggregates.

Hardness of a roadstone aggregate relates to mechanical properties within the rock. All hardness tests are performed on the coarse aggregate fraction only. Characteristics such as crushing, impact, abrasion and polishing are important. Aggregates must meet certain criteria related to these parameters and details of these are given in Table 1.5.

Durability of aggregates is a difficult concept to define. Any prediction of the effective life of rock can only be speculative. However, using advanced petrographic and testing techniques, it is possible, with experience, to make a considered judgement regarding durability. Important factors which affect durability are the aggregate porosity, its chemical composition and the environmental conditions under which it will operate.

1.3.3. Testing of aggregates

Since aggregate petrographical properties vary significantly from source to source, it follows that their engineering properties will also differ. It is therefore important to be able to define the variation by means of standard tests. This allows the engineer to make judgements about the acceptability of particular aggregates. Aggregate tests in the UK are usually stipulated by reference to BS 812.[7-24] This standard details procedures for each type of test which may be carried out on an aggregate. Tests can be grouped into headings of physical, mechanical, chemical and petrographic. Miscellaneous tests such as stripping and resistance to weathering are also specified.

(a) Physical property tests traditionally encompass characteristics such as shape, water absorption, voidage, gradation and relative density tests. These tests are useful in classifying aggregates. For example, it is possible to describe flakiness, elongation and angularity and these characteristics will indicate the degree of interlock of the particulate material which can be expected. However, many of these tests are only applied to the coarse aggregate fraction.

(b) Water absorption can indicate bitumen absorption into the aggregate. However, it is commonly used to assess durability on the basis that high water absorption usually means potentially low durability.

(c) Bulk density can indicate voidage in the uncompacted/compacted mineral aggregate. This is traditionally used to indicate bulking factors for the transportation of materials although it can be used to assess the compactibility of aggregates.

(d) Gradation tests are used to define the particle size distribution of the aggregates, and are important because they are commonly used to assess compliance. Knowledge of the aggregate gradation and how to manipulate it is a prerequisite for bituminous mix design.

(e) Relative density tests determine the unit weight of aggregates to be ascertained relative to water. The results of these tests can be used in design calculations to assess the theoretical density of bituminous mixtures and thus the coverage i.e. the number of square metres per tonne, for a known thickness. Minerals can be identified by their relative densities. It is not unusual for rocks to be apportioned an arbitrary relative density e.g. granite is traditionally given a relative density of 2.65 and this is a useful means of classifying aggregates.

Table 1.5. Typical values possible for roadstone aggregates in relation to their geological classification*

	Rock types	Mechanical				Physical		Weathering		Stripping
	Test	ACV	AAV	AIV	PSV	RD	WA	S	FT	
Igneous	Basalt range	14 (15–39)	8 (3–15)	27 (17–33)	61 (37–74)	2.71 (2.6–3.4)	0.7 (0.2–1.8)	Low to high	Low to high	No
	Porphyry range	14 (9–29)	4 (2–9)	14 (9–23)	58 (45–73)	2.73 (2.6–2.9)	0.6 (0.4–1.1)	Medium	Low	No
Metamorphic	Granite range	20 (9–35)	5 (3–9)	19 (9–35)	55 (47–72)	2.69 (2.6–3.0)	0.4 (0.2–2.9)	Low	Low	Yes
	Quartzite range	16 (9–25)	3 (2–6)	21 (11–33)	60 (47–69)	2.62 (2.6–2.7)	0.7 (0.3–1.3)	Low	Low	Yes
Sedimentary	Gritstone range	17 (7–29)	7 (2–16)	19 (9–35)	74 (62–84)	2.69 (2.6–2.9)	0.6 (0.1–1.6)	Low to high	Medium	No
	Limestone range	24 (11–37)	14 (7–26)	23 (17–33)	45 (32–77)	2.66 (2.5–2.8)	1.0 (0.2–2.9)	Low to high	Low to high	No
Pits	Gravels range	20 (18–25)	7 (5–10)	15 (10–20)	50 (45–58)	2.65 (2.6–2.9)	1.5 (0.9–2.0)	Low to high	Low to high	Yes
Artificial	Slag range	28 (15–39)	8 (3–15)	27 (17–33)	61 (37–74)	2.71 (2.6–3.2)	0.7 (0.2–2.6)	Low to high	Low to high	No

*ACV = aggregate crushing value
AAV = aggregate abrasion value
AIV = aggregate impact value

F = freeze thaw
PSV = polished stone value
RD = relative density

S = soundness
WA = water absorption

(f)   Normally, the mechanical properties of an aggregate relate to its ability to withstand stresses induced by wear and tear, impact and crushing. However, any assessment of mechanical properties should also include the determination of the ability of the aggregate to withstand the dynamic action of freeze/thaw cycles. These tests are often used to categorize the aggregate in qualitative terms and in relation to its use in each of the pavement layers. For example, an aggregate may be excluded from bituminous mixtures if its 10% fines value (TFV) is below 140 kN.

(g)   Polishing and abrasion resistance values are important for the running surface of the road as this is where the forces which cause aggregate smoothing and wear occur. Aggregates used in pre-coated chippings must meet certain minimum standards related to these properties to ensure that they will provide adequate skidding resistance in the running surface. Abrasion is loosely linked with the hardness of an aggregate.

(h)   Impact resistance is the measure of the resistance of a particular aggregate to rapid loading. Tests such as the aggregate impact value (AIV) test and the Los Angeles abrasion test are commonly used for such assessments.

(i)   The toughness of an aggregate is defined as its ability to resist slow loading. Aggregate crushing and 10% fines values are commonly used as a measure of toughness.

(j)   Chemical tests are commonly used to identify the presence of compounds in the aggregate which may be unstable or potentially deleterious to the coated stone.

(k)   Slag can be unsound and it can break down in the presence of water. Some specifications limit sulphate content in coarse aggregate fractions.

(l)   Stripping tests are notoriously unreliable. They attempt to assess the likelihood of the bitumen film retracting or stripping from the stone. There are many versions, and the best ones use mechanical immersion methods.

(m)   Resistance to weathering or durability tests attempt to predict the ability of aggregate to withstand the repeated action of freezing and thawing or the effect of water. Commonly, magnesium sulphate soundness or freeze/thaw tests are used to evaluate this property.

(n)   Petrographic tests using microscopy techniques are extensively used to assist the engineer to make considered judgements about the properties of aggregates. As knowledge of aggregates increases and sources of supply dwindle, this approach will become increasingly useful.

1.3.4.   Importance of shape, grading and consistency

The aggregate framework or skeleton which forms when bituminous mixtures are compacted determines the stability of these materials. Recent research[25] confirms that it is the most important factor in determining the success of the material when laid. Designing this framework to achieve maximum interlock through shape and gradation while allowing sufficient room for mortar is the key to success in bituminous mixture technology. Aggregates which have a large number of crushed faces of random shape are likely to have good interlock characteristics. Also, since gradation is important, effective blending of aggregates is fundamental. Above all, once the shape, blend and gradation has been set, consistency is essential to achieve the predicted performance in the laid material.

1.3.5. Selection of aggregates for bituminous mixtures

Fig. 1.5 shows the processes an engineer would typically follow when deciding which aggregates to use as roadstone.

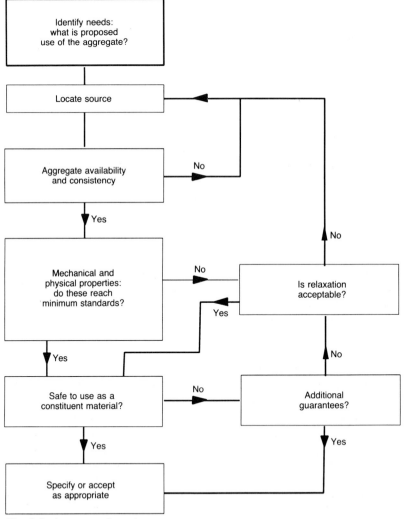

Fig. 1.5. Aggregate flow chart

Table 1.5 gives examples of test result values which can be expected from roadstone aggregates in relation to their geological classification.

Table 1.6 gives typical specification limits which apply to roadstone aggregates in relation to their use in the pavement.

Most production sources will have a data bank of documented information on the history of the quality of the rock from the source. This should be accessed to evaluate the properties of the aggregates. The information for key properties, particularly mechanical properties, should be provided through a testing authority which is independent of those providing the data. The data reviewed should be taken over a number of years to assess fluctuations in the quarried materials. If there is any parameter which requires confirmation, this should be independently checked before approval is granted. If these steps are undertaken before material production time and money will not be wasted.

1.3.6. Summary

Many factors are relevant when specifying aggregates for use as roadstone. These parameters relate to the stresses which will operate within the pavement and the quality of the rock used, and are

- (a) the type of pavement e.g. highway, runway, hardstanding
- (b) the type of bituminous mixture e.g. macadam, hot-rolled asphalt, porous asphalt
- (c) the expected intensity of traffic and the loadings imparted
- (d) the location in the pavement structure i.e. running surface, basecourse, roadbase.

Factors which relate to the aggregate are

- (a) the proven quality of the rock from the source
- (b) the consistency of the properties and supply
- (c) the availability of the required quantities of supply.

As the industry moves into using end-product specifications for bituminous mixtures, the emphasis is likely to be placed on maximizing the use of available aggregates. This may influence the quality of the constituents which can be used in the bituminous mixture.

1.4. Binders

1.4.1. The nature of petroleum bitumen

Petroleum bitumen is a solution of complex chemical structures. Three generalized forms of structure make up a petroleum bitumen.

- (a) A chain type form is the physical form of oils. Oils are liquids. There is no surface charge along a chain and they are described as being non-polar. Chains will associate through intertwining (Fig. 1.6), and movement (or flow) is through the relative movement of the chains. One chain has no preference for another chain. When a shearing action is stopped, the chains form a new but stable intertwined system. The viscosity (or resistance to flow) will depend on the complexity of the chains.

Table 1.6. Typical specification limits applied to roadstone aggregates in relation to their proposed use in the pavement

Surfacing material	Flakiness index, % max		Aggregate crushing value, % max		10% fines value, kN		AIV, max		Absorption, % max	Aggregate soundness, % max	Stability	Sulphur content, % max	Bulk density, kg/m³, min	Absorption, % max	Stripping test
	Crushed rock or slag	Gravel	Crushed rock or slag	Gravel	Crushed rock or gravel	Slag	Crushed rock or gravel	Slag							
Marshall asphalt	30	30	30	25					2	18‡	—	—	—	—	Not greater than 6 particles from a 150 particle test sample shall indicate evidence of stripping
Marshall DTS	30	30	30	25					2	18‡	—	—	—	—	Not greater than 6 particles from a 150 particle test sample shall indicate evidence of stripping
Rolled asphalt wearing course	30	30	30	25	140	85	30	35	2	18‡ 25§	Requirements to Appx (A) of BS 1047	2%	1120†	4	Not greater than 6 particles from a 150 particle test sample shall indicate evidence of stripping
Macadam basecourse	30	30	30	25	140	85	30	35	2	25§	Requirements to Appx (A) of BS 1047	2%	1120†	4	Not greater than 6 particles from a 150 particle test sample shall indicate evidence of stripping
Friction	25*	—	16*						1.5	18‡	—	—	—	—	Not greater than 6 particles from a 150 particle test sample shall indicate evidence of stripping

Spanning group headers: All aggregates (crushed rock, gravels and slag) · Coarse aggregate test properties [Crushed rock and gravel (BS 812) · Slag (BS 1047) · All aggregates]

* Crushed rock only of specified groups.
† Minimum when aggregate sample used in test complies with Table 1 of BS 63 for 20 mm nominal single-sized aggregate.
‡ Overall percentage mass loss based on the mass losses of the various nominal sizes tested from any one supply source (for BS 812 losses are now expressed as a percentage retained, i.e. 18% becomes 82%).
§ When tested in accordance with BS 812 maximum loss allowed in any one fraction.

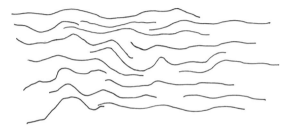

Fig. 1.6. Generalized form of oils

(*b*)   A form composed of closely-associated flat plates effectively making up an irregularly shaped solid defines the chemical structure called asphaltenes (Fig. 1.7). The plates have a strong surface charge leading to close association and these structures are described as being polar. Plates have a high resistance to shear. The whole structure acts as a solid. When extracted, they look and feel like very fine solid particles. Oils and asphaltenes do not mix.

Fig. 1.7. Generalized form of asphaltenes

(*c*)   A hybrid of the previous forms consists of plate-like structures which have chains coming from them. The plates associate with each other (Fig. 1.8). This form describes a group of chemical structures defined as resins. Resins can associate with both asphaltenes and oils. Resins allow the two other components, the oils and the asphaltenes, to form a solution.

Petroleum bitumens are a solution composed of particles of asphaltene structures surrounded by resins which exist within an oil medium (Fig. 1.9).

Asphaltene structures are described as forming the 'body' in a petroleum bitumen. As there are no physical bonds between the three chemical structures, they will act as a liquid when sheared. On application of a shearing force the system will move (flow) and on removal of the shearing force movement will cease. As a result of the complex shapes of the chemical structures and the fact that some of the structures have a surface charge, the strength of the

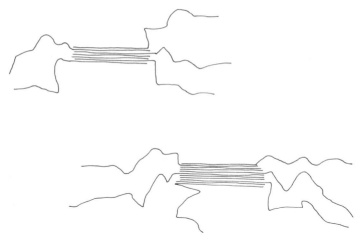

Fig. 1.8. Generalized form of resins

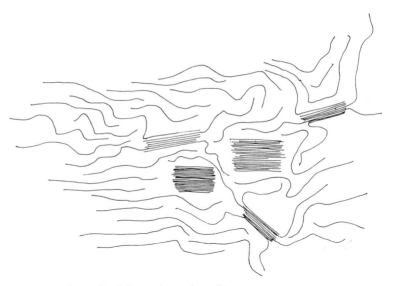

Fig. 1.9. Generalized form of petroleum bitumen

association between structures can lead to a high resistance to shear. Such materials are described as being viscous. The viscosity of bitumens will be influenced by the relative proportion of the three chemical groupings and the temperature of the material. Simply, petroleum bitumens, having a higher proportion of asphaltenes (and lower proportion of oils), will be more viscous. The higher the temperature of a bitumen, the more energy and movement the individual particles will have and the less the associated bond between particles will be. The consequence will be a lower resistance to shear, quantified by a lower viscosity.

Rheology is defined as the science of the deformation and flow of matter. Petroleum bitumens have two important rheological properties, being thermoplastic and visco-elastic. Thermoplastic means that the viscosity of petroleum bitumens reduces on heating and increases on cooling. The process is reversible. Visco-elastic means that when a force is applied, the structures making up the bitumen will distort as well as flow. Viscous flow is irrecoverable movement. Distortion is recoverable movement and describes elastic behaviour. The relative amount of viscous and elastic response that a petroleum bitumen exhibits to an applied force depends on its chemical make-up and its temperature. More viscous materials at lower temperatures will respond more elastically to a short-time applied force. The material will deform rather than flow but some flow will occur. Less viscous materials at higher temperatures, and under the action of a force applied for some time, will flow. On removal of the applied force, the material will exhibit little recovery of shape.

The responses of a material to an applied force are a function of its mechanical properties. A force acts on a material in the form of a stress as the result of the force acting over an area of contact. The response of a material is through movement. Where the movement is recoverable the degree of movement may be related to the original dimension or shape of the material. Recoverable movement is recorded as a strain. Where the movement is irrecoverable and flow occurs, the movement is recorded as rate of strain (strain per unit time).

Defining the amount of strain (or rate of strain) exhibited by a material when subjected to a stress is important. For a liquid, viscosity is the parameter which is the ratio of shearing stress to the corresponding rate of shearing strain. For elastic solids, the parameter is the elastic modulus i.e. the ratio of applied stress to corresponding strain. The term 'stiffness modulus' is used with petroleum bitumens because of their visco-elastic properties.

As part of a pavement construction petroleum bitumens will be subjected to a particular form of loading caused by vehicle wheels moving over the pavement surface. The surface load is cyclic i.e. the load builds up, peaks and then drops off. The pattern repeats itself. The cycle between load applications vary. The stress on the petroleum bitumen between the particles of aggregate within a pavement layer will build up, peak and then drop off. When a sample of petroleum bitumen is loaded in such a manner, but with the stress cycles being continuous and alternating between compression and tension, the ratio of the peak stress to peak strain defines the parameter stiffness modulus, often shortened to stiffness. A petroleum bitumen may be described as having a high stiffness. A petroleum bitumen with a high stiffness or high stiffness modulus will respond more elastically to an applied stress.

In terms of physical properties, petroleum bitumens are non-wetting liquids, i.e. they will not move across a solid surface by attraction of the liquid to the surface. This is because of the low polarity of the material as a whole. Water is a highly polar material and will, generally, wet a solid surface. If water and petroleum bitumen are together on a solid surface such as a roadstone, the water will preferentially wet the surface of the stone (Fig. 1.10). The petroleum bitumen will be 'stripped' from the surface of the stone.

The adhesion mechanism of a petroleum bitumen to a clean stone surface is mainly one of mechanical interlock with the microtexture of the stone surface.

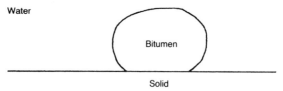

Fig. 1.10. Bitumen and water on a solid surface

The cohesion which the material provides between solid surfaces is due to the viscosity of the material.

The chemical structure of petroleum bitumens makes them unstable materials. The resinous chemical structures can break down by the loss of the branch chains to form asphaltenes. The result will be an increase in the viscosity and stiffness of the material and the material will also become less ductile. This type of breakdown is caused by heat and exposure to oxygen in the atmosphere. The degree of breakdown (oxidation) and 'hardening' of a bitumen will depend on the temperature, the time held at that temperature and the ratio of the exposed surface area to the volume of the bitumen. Critical high temperature processes for bitumen occur during mixing with an aggregate and during the storage of the resultant mixture. The effect of heat on the rate of reaction of a bitumen is that the reaction rate doubles for every 10C° above 100°C.[26] The mixing process is more critical. The temperature of the material will be high and, because of the mixing process, a high surface area to volume of material exists.

In storage, where the material temperature may be lower than during the mixing process and the voids within the material are not continuous, and the exposed surface area is therefore lower, the degree of hardening will be reduced. In service, exposure to the atmosphere at ambient temperatures will still cause a hardening of the petroleum bitumen but at a much reduced rate. In service, water will leach out some of the chemical structures within the resinous fraction of the bitumen. This causes a further hardening of the material. Fig. 1.11 shows the nature of the increase in relative viscosity of a bitumen with time, from mixing with an aggregate, transportation to site, compaction and exposure in service.[27]

1.4.2.   *Specification of petroleum bitumen*
The specification of petroleum bitumens used for road purposes[28] is based on measures of the rheological properties of the material, its volatility, the content of residual material (purity), and a measure of its electrical properties.

The most relevant specification relates to the rheological properties of the material. Petroleum bitumens are graded by two measures of viscosity: penetration and softening point temperature. Commonly, penetration alone is used to specify a petroleum bitumen.

Bitumens used for the manufacture of bituminous mixtures are generally penetration grades of 50 pen, 70 pen, 100 pen and 200 pen. The pen value is the penetration recorded in tenths of a millimetre of a standard weighted needle into a sample of bitumen at a temperature of 25°C.[29] It is an indirect measure of viscosity at one temperature. Lower pen values mean higher viscosity and harder bitumen. Rolled asphalts, which rely more on the properties

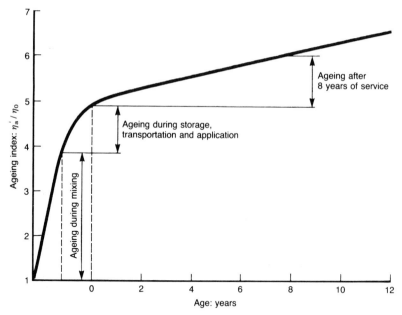

Fig. 1.11. Relative increase in viscosity of a bitumen with time from mixing with an aggregate[27]

of the bitumen for their strength, use 50 pen, 70 pen and, possibly, 35 pen grade bitumens. The latter is used where higher ambient temperatures are expected.[2] In Scotland, 70 pen may be specified during winter whereas 50 pen would be the standard specification for laying during the other seasons. Coated macadams use 100 pen and 200 pen bitumens because the graded aggregate has a greater influence on material properties.[1]

The softening point temperature differs from the penetration in that it is the temperature at which all petroleum bitumens have the same viscosity. It is an equi-viscous temperature. In penetration terms, it is the temperature at which the majority of bitumens have a penetration of 800 pen. In absolute terms it is the temperature at which all bitumens will have viscosity of about 1200 Pascal seconds (Pa.s). Whereas penetration grades vary for different bitumens, the softening point temperature is used as a common datum for comparison of the performance of bitumens at a particular operating or design temperature.

The volatility test is essentially an emissions test. It measures the loss of the oil fraction on heating. The residue test assesses the purity of material. The measure of the electrical properties of a bitumen has a relationship with the potential re-activation of the resins. Where resins are 'active', they will reduce to asphaltenes more readily on exposure to the atmosphere. In a hot-rolled asphalt running surface, the reaction of the exposed bitumen will cause the thin layer coating the fine aggregate to lose ductility and be abraded off the aggregate under tyre action. The microtexture created provides the friction required to ensure vehicle control at low speeds in wet surface conditions.

The specification tests for bitumens are mainly material consistency tests,

Table 1.7. Specification for penetration grade bitumens[28]

Property		Test method BS 2000: Part	Grade of bitumen									
			15 pen	25 pen	35 pen	40 pen HD	50 pen	70 pen	100 pen	200 pen	300 pen	450 pen
Penetration at 25°C dmm		49	15 ±5	25 ±5	35 ±7	40 ±10	50 ±10	70 ±10	100 ±20	200 ±30	300 ±45	450 ±65
Softening point °C	min max	58	63 76	57 69	52 64	58 68	47 58	44 54	41 51	33 42	30 39	25 34
Loss on heating for 5 h at 163°C (a) loss by mass % (b) drop in penetration %	max max	45	0.1 20	0.2 20	0.2 20	0.2 20	0.2 20	0.2 20	0.5 20	0.5 20	1.0 25	1.0 25
Solubility in trichloroethylene % by mass	min	47	99.5	99.5	99.5	99.5	99.5	99.5	99.5	99.5	99.5	99.5
Permittivity at 25°C and 1592 Hz min		357			2.63	2.63	2.63	2.63	2.63			

although two of the tests, penetration and softening point, are used as the basis for the assessment of material performance. Table 1.7 is an extract from BS 3690.[28] A 40 pen HD bitumen is heavy duty bitumen which has a nominal penetration of 40 tenths of a millimetre but which has an increased softening point temperature to a value of 58−68°C. This bitumen consequently has a higher penetration index (PI) value than would normally be expected. This hard grade of bitumen is, therefore, less temperature-susceptible and may be specified, for example, on heavily trafficked routes to reduce the rate at which rutting takes place in the wheel tracks. Care is required in the placement of mixtures using such bitumens as they have to be compacted at higher temperatures to ensure adequate compaction.

1.4.3. Performance of petroleum bitumens

There are three important stages in the life of petroleum bitumens as a binder in a bituminous mixture. These are the coating of the aggregate, the compaction of a layer of coated aggregate and the service performance of a layer of compacted coated aggregate. The performance of the bitumen at these three stages is viscosity or stiffness-related, both properties being temperature-dependent. The single performance relationship for a bitumen which is most useful is that of viscosity change with change in material temperature. Fig. 1.12[30] shows a bitumen test data chart (BTDC) which provides a straight line plot of change in viscosity with change in temperature. Three classes of bitumen are plotted. Bitumens used for road purposes are usually class S, which are straight-run bitumens. Class B bitumens are blown bitumens and class W bitumens are waxy bitumens. Straight-run bitumens are bitumens produced under vacuum with no modification to their chemical structure. Blown bitumens are bitumens which are air-rectified, which means that air is blown through the fluid mixture during its production. Blown bitumens have an oxidized

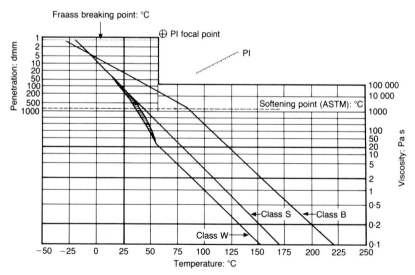

Fig. 1.12. BTDC comparing classes of bitumen[30]

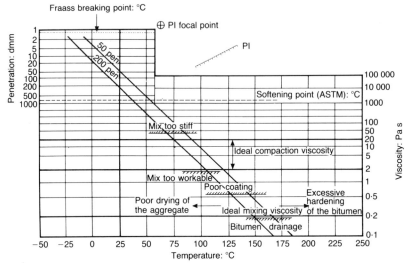

Fig. 1.13. BTDC comparing two penetration grade bitumens[31]

chemical structure. A waxy bitumen is one which has a relatively high content of wax.

Fig. 1.13 shows two penetration grade bitumens plotted on a BTDC,[30] with the three areas of importance in terms of material performance superimposed.

Generally, the steepness of the profile signifies the temperature susceptibility of a bitumen. An index used to quantify temperature susceptibility is the penetration index (PI). The PI can be calculated from an equation using either two temperatures or the pen value and softening point temperature, or it can be defined graphically, as shown at the top of the BTDC. An alternative graphical procedure along with the equation to calculate the value of PI has been published.[32] The PI for penetration grade bitumens lies in the range −0.5 to 0. Zero value is the datum. An increasing negative value represents an increasing temperature susceptibility and the profiles plot as lines of increasing steepness. An increasing positive value represents a reducing temperature susceptibility and the profiles plot as lines of reducing steepness. The PI values quoted with bitumens are those for materials which are specified and which are delivered to a quarry for the coating of aggregates.

As a result of the mixing, storage (where material is kept in a heated hopper before transportation to site), transportation and laying of a mixture the bitumen will become harder. It reduces by approximately one penetration grade. The reduction in pen value and increase in softening point temperature with oxidation may cause a consequent positive shift in the PI. The magnitude of the shift is difficult to define as it is dependent on the nature of the original source material, but it is unlikely to be more than +0.5. In service, the material may continue to change with a further shift in the penetration, softening point temperature and PI value. The changes in service are difficult to quantify as they depend on the permeability of a layer i.e. the degree to which oxygen and moisture have access to the bitumen. Fig. 1.14[33] shows a typical shift in

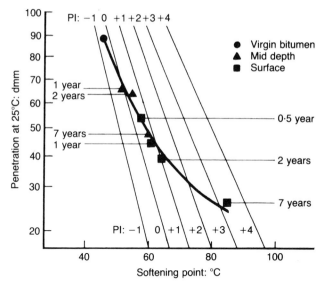

Fig. 1.14. Hardening of a binder within an asphaltic concrete wearing course[33]

penetration and PI within an asphaltic concrete wearing course. It is not the temperature of a material which has an important influence on its performance but the rheological and mechanical properties of the material at given temperatures.

Bitumens from one producer can generally be regarded as consistent and differences in performance between batches are small. The same nominal grade of bitumen from another producer may be marginally different in performance characteristics because of different sources of material and production processes.

1.4.4. Coating of aggregates

The ideal coating viscosity of a binder is around 0.2 Pa.s. Poor coating will occur above a viscosity of 0.5 Pa.s as the binder will not coat the aggregate fully and/or the coating will lack uniformity. A nominal 50 pen bitumen has a suitable mixing temperature of around 160°C while a nominal 100 pen bitumen has a suitable mixing temperature of around 150°C. A rule of thumb is that the mixing temperature is 110C° above the softening point of the bitumen.[34] The temperature of the binder during mixing with, and coating of, the aggregate will potentially harden the binder as a result of oxidation. In terms of the softening point temperature, Fig. 1.15 shows the value of the shift during a standard mixing time of approximately 30 seconds.[35] A mixing temperature of 170°C will increase the softening point of the bitumen by around 40C° when mixed for 30 seconds. A longer time in the mixer will further increase the softening point temperature.

During storage and/or transportation the bitumen will be maintained at an elevated temperature. In well-insulated lorries the heat loss will be a few degrees Celsius per hour. The effect of storage and transportation temperatures will

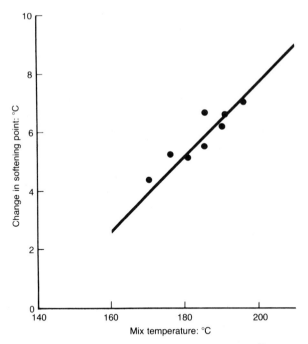

Fig. 1.15. Increase in softening point with mixing temperature[35]

depend on the time period over which the material is maintained at an elevated temperature and the degree of permeability of the mass of coated material. During storage, the nature of the atmosphere in which the material is stored will be an additional factor. In an inert atmosphere there will be little oxidation of the bitumen.

1.4.5. Compaction of a bituminous mixture

Compaction is the process of interlocking the aggregate structure making up a bituminous road material. At the same time, the bitumen is displaced from around the aggregate particles to the void space within the interlocked aggregate structure. The principal work on the material is in the displacement of the films of bitumen. The more viscous the films, the greater the energy required to fully interlock the aggregate particles. The viscosity limits for bitumen to ensure effective compaction of mixtures is approximately 2–20 Pa.s. The range of compaction temperatures for a particular bitumen relating to the viscosity limits can be defined from the BTDC.

The frictional characteristics and the shape of the aggregate and the volume of the bitumen are additional material characteristics which influence the compactability of a mixture e.g. in hot-rolled asphalts, the volume of the fine aggregate can modify the lower compaction temperature (the minimum rolling temperature) by as much as 30C°.

The upper compaction temperatures (the maximum rolling temperature) of bituminious mixtures are not currently specified. The volume of the void space or not the same quality of mix is needed for laying in a country lane or small urban road as that which is required on a heavily trafficked motorway. Many would emphatically state that standards should be uniform since materials which would give ten years good service in a motorway pavement will last for at least twice as long in less heavy trafficked roads.

Also, if clients worked to end-product type specifications, stating the performance characteristics they required from the mixtures, and assuming they were properly transported, placed and compacted, the specifications for mixtures in these two types of road would be drafted in very different terms.

Currently, in the UK at least, materials are sampled and tested and if a sample fails to comply with the specification, the material it represents in the road layer is rejected. Since all materials, including design mixes, are specified in terms of recipe gradings, it is difficult to know what difference in performance would be given by the out-of-specification material relative to those which are either just within the specification limits or even in the middle of the specification. Recent examples of fundamental testing of failed materials have shown some to be superior to those which complied with the specifications.

Clearly, binder content has a major effect on the performance of a mixture and binder non-compliance should not be treated lightly. Many thousands of square metres of pavement layers are removed each year and replaced, however, for relatively trivial instances of aggregate grading non-compliance, judged on very random sampling and testing, without the client having any idea of the effect of the non-compliance on performance. Even the most carefully-controlled production processes are subject to random fluctuations in quality and if failure is related to possible loss of life, as in some production processes, then the production control limits have to be set at levels that would ensure acceptable safety limits even if the material failed.

In the nuclear industry where these risks are high, the production control limits are also extremely high, as are the ensuing costs of production, since it is impossible to ensure a degree of failure does not occur. In the black top industry or any other industry where the risks to life resulting from non-compliance are considerably lower, it is equally impossible to ensure that all material is produced within the specification. The specified tolerance limits for acceptable material used in production control would ideally be related to a desirable performance range expected of the material. However the range is determined, no client can expect total compliance. If the client sets untenable acceptance levels then either rejection rates would be so high as to increase material costs astronomically or contractors would simply not tender if they thought the specification would be enforced. Thus the client is forced to consider acceptable quality levels. What percentage should pass is a matter yet to be debated but it is inextricably linked with production tolerance ranges.

Acceptable quality levels link the bituminous material placed on a rural road in Wales with ostensibly similar material placed on the M25, through the same EN. The production tolerances could perhaps be wider for the rural road material but, say, a constant 95% of tested samples would still be expected to comply with the range of requirements for each road.

More discussion of recipe mixtures in these terms only serves to emphasize how sensible it will be to move to specifying materials in performance terms, particularly, as in concrete, the higher material strengths can only be achieved with premium materials. At present, again using recipe mixtures, it is assumed that all dense macadams are equally fit for purpose regardless of the characteristics of the primary aggregate in the material, which in turn will determine their dynamic stiffness. This is simply not the case.

11.8.4. TG4 — quality and attestation of conformity

TG4's terms of reference are

(a)   to prepare all parts of WG1's ENs which will be required to implement the systems for attestation of conformity applying to bituminous mixtures

(b)   to consider the inter-action of test precision and specification limits and to ensure that this is taken into account where judgements are made about compliance with specification

(c)   to study all aspects of quality, conformity and attestation and to present relevant information to WG1 so that it may make decisions from a well-informed position.

Whatever level is selected, the factory production control requirements are identical. It is WG1's intention that conformity with these latter requirements will enable products to carry the CE mark and it is vital to note that any attestation requirements for the product as stated in CEN specifications are legal obligations, which will be enforceable under UK statute, through the construction products regulations, once the relevant ENs have been published.

This is an extremely significant change from current UK specification practice, as it is in other member states, and both clients and producers have realized that despite the lack of government subsidy or assistance with expense, they can ill-afford to ignore TC 227.

The two interfaces covered in WG1's ENs i.e. producer/layer and layer/purchaser need to be covered by different categories of standard. These are defined by CEN in two categories. Category A standards concern the design and execution of building and civil engineering works and their parts, or particular aspects thereof (including the basic data on one side and the application of construction products on the other), with a view to the fulfilment of the essential requirements as set out in Annex I of the Council Directive 89/106/EEC and taking account of article 2.2 of this directive.

Category B standards exclusively concern construction products subject to an attestation of conformity and marking according to articles 13, 14 and 15 of council directive 89/106/EEC. These standards may concern definitions and requirements with regard to performance and/or other properties, including durability of those characteristics that may influence the fulfilment of the essential requirements, testing, compliance criteria of a product. Those specifications that concern a family of products or several families of products are of an intermediate character and are described as horizontal (category Bh standards).

While the necessary CEN directives are still being drafted, TC 227 WG1 will produce the following set of standards, which must not conflict within a compacted aggregate structure, in relation to the volume of the bitumen, controls the upper compaction temperature of a mixture. At compaction the volume of the bitumen has increased due to thermal expansion. Where the relative volume of the bitumen is high, in relation to the void space available, the mixture will become unstable and a compacted mixture will displace rapidly around the barrels of a roller. Such a condition controls the upper compaction temperature of a mixture. The upper compaction temperature for a mixture can be ascertained using a high temperature triaxial cell.[35,36]

Lower compaction temperatures, or minimum rolling temperatures, are specified for bituminous mixtures.[37,38] The values quoted relate to bitumen viscosity. Again, the nature of the whole system has to be considered in terms of both the bitumen and the aggregate. A procedure is available to define lower compaction temperatures for whole mixes.[39,40] Below minimum rolling temperatures, a material layer will still compact but the effort required becomes greater.

There are material production and environmental factors which will influence both the upper and lower compaction temperatures. During material production, the ingredients are metered into the mixer. Individual ingredients will vary in their mass proportion in the final mixture. The effect will be that mixtures vary in their performance in relation to both placement and service performance. For example, with hot-rolled asphalts, an increase in the relative filler content will reduce the voidage within the interlocked aggregate structure. If there is no change in the bitumen content, the effect will be to raise the relative volume of bitumen. This will potentially lower the upper and lower compaction temperatures. A reduction in the relative volume of bitumen results in an increase in the upper and lower compaction temperatures. Some fine, dense mixtures are sensitive to variations in the mass percentages of the fine aggregate and filler fractions. Changing the source of filler can cause a significant effect on the placement performance of a mixture. Again, this is because of the effect on the void space within the interlocked aggregate structure.

The environment has an effect on the compaction performance of mixtures as it exerts an influence on the layer temperature and the temperature gradient within the layer. Temperature itself is not the key issue, but rather the effect of temperature on bitumen viscosity and bitumen volume.

1.4.6. Service performance

Service performance relates to the response of the pavement structure to the particular form of applied loading, which is randomly cyclic, of varying magnitude and reversible in sense (i.e. tension, compression or positive or negative shear). The visco-elastic nature of the binder influences the characteristics of a layer and a pavement structure as a whole.

The in-service performance parameter for a mixture is stiffness modulus. This is influenced by the stiffness modulus and volume of the bitumen and the volume of the aggregate. The stiffness modulus of a bitumen can be calculated using Van der Poel's nomograph[40] (Fig. 1.16). The required input parameters are the time of loading of material (which is a function of the speed of the traffic moving over the carriageway surface), the temperature difference

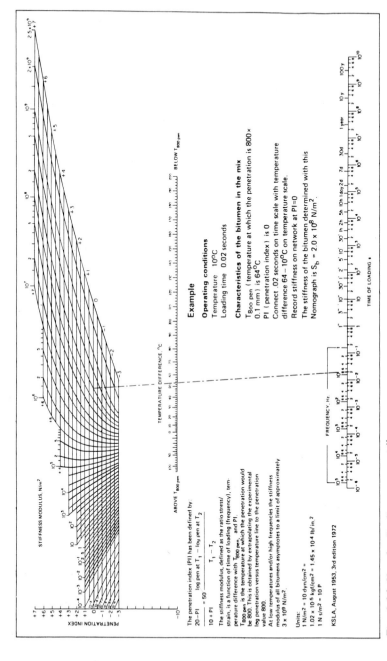

Fig. 1.16. Van der Poel's nomograph[41]

between the softening point temperature and design ambient temperature, and the PI of the bitumen. A rule of thumb for time of loading in seconds is the reciprocal of the vehicle design speed when the design speed is measured in kilometres per hour.

The nomograph for predicting the stiffness modulus of a bituminous mixture is shown in Fig. 1.17.[42] The stiffness modulus of the bitumen should relate to the properties of the bitumen in the mixed material. The properties of the bitumen within the mix will not be those of the specified binder, but those of the specified binder subject to oxidation as a consequence of the processes of mixing, transportation and laying of the bituminous mixture.

The stiffness modulus of a mixture will vary as the stiffness modulus of the bitumen changes. This will occur in practice as a result of contact between the bitumen and air. The more permeable the pavement structure, the more access oxygen and moisture have to the bitumen. Fig. 1.11 shows the typical ageing of bitumen in service.

The petroleum bitumen binder will also influence particular behaviour patterns of a material layer in service. The fatigue characteristics of a roadbase is a function of the volume of the bitumen and the softening point temperature of the bitumen. The creep properties of a bituminous material are a function of the viscosity of the bitumen. Environmental effects, which influence the properties of the bitumen, will affect the service properties of a mixture.

Controlled oxidation is important with a running surface. The bitumen will fill the microtexture of the exposed aggregate and the microtexture of the exposed surface. The permittivity test was developed to define a parameter to quantify the potential oxidation rate and, therefore, the embrittlement of petroleum bitumen. Refined Trinidad Lake Asphalt (or Trinidad Epuré) oxidizes readily on exposure to atmosphere. When petroleum bitumen was first used as a binder for wearing course hot-rolled asphalt, it did not exhibit the same ability to 'wear' and expose the fine aggregate. Although the permittivity test does appear to correlate with texture development, it is not a fundamental measure of bitumen reactivity.

Under low temperature conditions, petroleum bitumens become brittle and can fracture as a result of thermal shinkage or single stress application. The Fraas breaking point defines a temperature at which controlled fracture occurs in laboratory specimens. It measures a critical stiffness for bitumen and so, in principle, it is similar to the softening point temperature which is an equi-viscous temperature. The Fraas breaking point temperature can be defined from the BTDC. It is not a common test in the UK.

1.4.7. Modified petroleum bitumens
The rheological and mechanical properties of straight-run petroleum bitumens are not always ideal. In terms of service performance, the lower the temperature susceptibility of the bitumen, the more stable the performance of a bound layer will be over the range of ambient temperatures. Blown bitumens, with a dog-legged BTDC profile, have a reduced temperature susceptibility at service temperatures. Blown bitumens are oxidized bitumens and are therefore 'modified'. Blown bitumens are not commonly used for the production of pavement materials. Petroleum bitumens can be modified using solvents (called

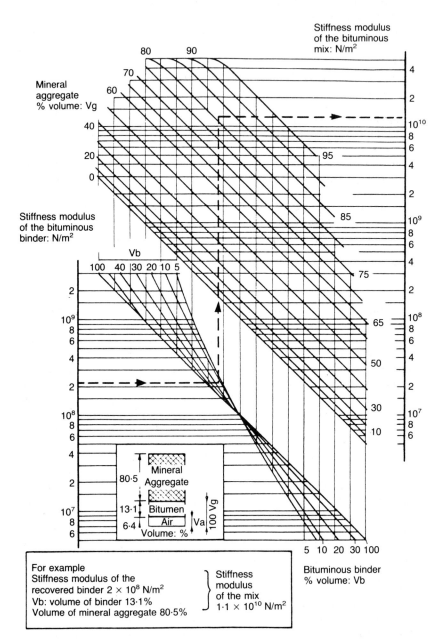

Fig. 1.17. Nomograph for predicting the stiffness modulus of bituminous mixes[42]

cut-back bitumens) forming emulsions or by modification of the internal structure of the bitumen with compatible material (thermoplastic co-polymers, rubbers and thermosetting resins) or by the addition of a material such as sulphur which can fit into the structure of the bitumen.

Cut-back bitumen
Placement of small quantities of bituminous mixtures can be difficult because of the need to maintain an elevated temperature. Maintaining small quantities of material at elevated temperatures for extended periods results in oxidation and hardening of the bitumen. Modification of the bitumen by the addition of a volatile solvent to the bitumen, such as kerosene or creosote, will reduce the viscosity of the material with less, or no, heat input. In permeable bituminous layers with access to atmosphere the solvent will evaporate with time, although this process may take several years. Cut-back bitumens are commonly used for temporary patching and for surface dressing. The base bitumen is usually a 200 pen or 300 pen grade material. As the viscosity of the cut-back binder is low, it is measured as a flow time through an orifice. The equipment used is a standard tar viscometer (STV) and viscosity is quoted in seconds. Common viscosities are 50 seconds, 100 seconds and 200 seconds. The faster the time, the less viscous the cut-back bitumen.

Bitumen emulsion
An alternative mechanism for reducing the viscosity of a petroleum bitumen is the use of an emulsion. Bitumen is immiscible in water. A stable emulsion is formed by grinding the bitumen in a mill to a particle size range less than 50 μm and adding it to water with an emulsifier. The base bitumen is usually a 100 pen, 200 pen or 300 pen grade material. The emulsifier has two chemical components. One component is a hydrocarbon chain which has an affinity for bitumen and the second component is a structure carrying a surface charge which has an affinity for water. The structure carrying the surface charge may be cationic (positively charged) or anionic (negatively charged). Emulsions are designated K types i.e. cationic, or A types i.e. anionic. Cationic emulsions are more commonly used to coat roadstone. Limestones have a weak negative surface charge and this produces a strong electrostatic bond between the bitumen and the stone surface displacing surface water. The coating of the surface of aggregate particles is described as the breaking of an emulsion i.e. the separation of the two elements, namely the water and the bitumen. Breaking has two stages, flocculation followed by coalescence. Flocculation is the build-up of droplets of bitumen around the surface of the aggregate particles. Coalescence is the amalgamation of the individual particles to form a continuous film.

The speed of initial break, the flocculation of the bitumen droplets around the surface of the aggregate particles, can be controlled chiefly by the amount of emulsifier used. Increasing the amount of emulsifier slows the rate of flocculation. A scale of 1 to 4 is used to reflect the rate of break of emulsions. Fast break is 1 and slow break is 4. A K1 material is a fast-breaking cationic emulsion. Break will be seen as a change in the colour of the emulsion. Break is only complete, however, when coalescence occurs with the films. This will be controlled by the evaporation of the residual water within the system. The cohesive strength of a broken emulsion film will be weak until coalescence takes place. The stiffness of a bituminous mixture will consequently be low until binder films coalesce.

The percentage bitumen content is also used to specify an emulsion. A K1-70 is a fast-setting cationic emulsion which contains 70% by mass of bitumen. Cationic emulsions have a higher bitumen content than anionic emulsions.

The viscosity of emulsions depends on the content of bitumen, the particle size distribution of the bitumen droplets and the viscosity of the bitumen. Many emulsions use cut-back bitumen. With the move to controlling emissions to the atmosphere, there is an interest in developing emulsions containing no cutter and producing the equivalent of conventional bituminous mixtures.

Emulsions are used more commonly in the recycling of bituminous materials, such as with the re-tread process, tack coats, spray coats with surface dressing operations and surface treatments, such as slurry seal. In surface dressing, K1-70 emulsion with a 200 pen or 300 pen base bitumen and 1% kerosene is the most common product because of its viscosity and fast break.

Chemically modified bitumens
Pavement structures have a finite service life. Roadbases will crack from repeated loading, layers will be subject to (very limited) additional traffic compaction leading to a loss of surface evenness, and wearing course layers are subject to abrasion, loss of texture and displacement. Wearing course layers nowadays can be expected to have a design life of ten years. The extension of the service life of a wearing course or the certainty of a ten year service life is becoming more critical. In wearing course binders, modifiers are used to enhance the stiffness modulus of the material making it less prone to deformation and making it less temperature susceptible.

The addition of sulphur increases the workability of the binder and enhances the stiffness modulus of a bituminous mixture. The sulphur content is high, 15−18%. The temperature of the bitumen must be controlled to avoid a chemical reaction which produces asphaltenes and gives off hydrogen sulphide. At lower temperatures, the sulphur reacts with the resin compounds within the bitumen, marginally changing the rheological properties of the material. The action of the sulphur is to reduce the viscosity of the bitumen as a result of its fluidity. On cooling, the sulphur which has not chemically reacted with the bitumen acts as a filler within the mixture of aggregate and binder. It occupies the residual void spaces and consequently enhances the stiffness modulus of the bituminous mixture. Higher penetration grade binders can be employed because of the modifying effect of the sulphur, further enhancing the workability properties of a mixture. Thermopave is one proprietary material used which is machine-laid but does not require roller compaction. Thermopatch is used to repair surface damage; it is poured, levelled and, when it has cooled, can be trafficked.

Ethylene Vinyl Acetate (EVA) is the most common thermoplastic co-polymer used to produce a modified bitumen. The co-polymer, produced by the co-polymerization of ethylene and vinyl acetate, is generally compatible with the chemistry of most petroleum bitumens. EVA has a random physical structure which fits within the chemical structure of the bitumen. The effect of the co-

Table 1.8. Effect of the addition of EVA co-polymers on binder and hot-rolled asphalt wearing course properties

Binder	Binder properties		Marshall properties			Wheel tracking rate at 45°C mm/h
	Pen at 25°C dmm	Softening point (IP) °C	Stability kN	Flow mm	Quotient kN/mm	
70 pen	68	49.0	6.3	3.3	1.9	4.4
70 pen + 5% EVA 150/19	50	65.5	7.6	3.2	2.4	0.8
70 pen + 5% EVA 45/33	57	58.0	8.0	2.7	3.0	1.0

polymer is to increase the viscosity of the bitumen, but there is little change to the elastic properties of the bitumen. The most common blend is 5% EVA in a 70 pen bitumen. The higher penetration produces increased workability in a hot-rolled asphalt wearing course during laying, and the inclusion of the EVA increases the viscosity of the binder at ambient temperatures. Table 1.8 shows the effect of the inclusion of EVA with a 70 pen bitumen on the physical and mechanical properties of the binder and a wearing course mixture.

Different types of EVA exhibit a variety of properties. The properties of an EVA are controlled by molecular weight and vinyl acetate content. Molecular weights are defined by the Melt Flow Index (MFI). The higher the value of the MFI, the lower the molecular weight and the viscosity of a material will be. The vinyl acetate content controls the structure of the co-polymer. The vinyl acetate groups break up the crystalline structure of what are closely-packed ethylene chains. The ethylene chains pack in a manner similar to that of asphaltene structures. The more crystalline a co-polymer remains, the stiffer the structure will be. The vinyl acetate structures are more 'rubbery' in behaviour. The more the crystalline structures are broken up, the more rubbery the response of the co-polymer. The vinyl acetate content is specified with the co-polymer. A specification of EVA 150/19 means an EVA with an MFI of 150 and a vinyl acetate content of 19. This is the most common grade of EVA.

Care must be taken when using EVA-enhanced bituminous mixtures because of their increased workability, which is principally the result of the increase in the base penetration grade of the binder. EVA is often specified as a modifier with bitumen for cold weather working because of the increased workability properties of the mixture. However the rate of change of material stiffness from reducing material temperature can cause difficulties. The crystallization of the co-polymer results in a marked stiffening of the material and increased resistance to compaction.

Styrenic block co-polymers are the more commonly used thermoplastic rubbers with pavement grade bitumens. These thermoplastic rubbers create three dimensional internal cross-linked networks. The cross linking results in strong and elastic structures. When these are compatible with a bitumen, they disperse within the bitumen structures, increasing the viscosity and stiffness

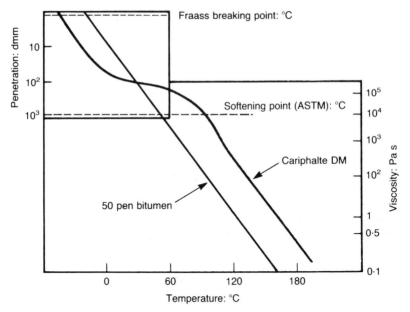

Fig. 1.18. Comparison of viscosity-temperature profiles with a 50 pen bitumen and Cariphalte DM[43]

of the binder at ambient temperatures. Fig. 1.18 shows the nature of the change in the BTDC profile between a 50 pen bitumen and one styrene block co-polymer, styrene-butadiene-styrene (SBS) which is produced by Shell Bitumen and marketed as Cariphalte DM. The improved elasticity of the binder makes it suitable for use with hot-rolled asphalt wearing courses overlying cementitious bases, lean concrete roadbases or old concrete carriageways. Table 1.9 shows the stiffness improvement with the inclusion of the SBS type binder.

Shell Bitumen produces an alternative binder, Cariphalte DA, which is used for friction course mixtures. The higher stiffness and elastic properties ensure the stability of such open-graded mixtures when they are subject to high stressing.

The compatibility of the blend with a bitumen and a styrene block co-polymer is critical. Where compatibility is low, the effect may be to reduce the stiffness of a mixture rather than increase it.

Polymer-modified binders are used for surface dressing application at difficult locations, such as roundabouts and bends. Emulsions are made of co-polymer modified bitumens which are also used for surface dressing applications, and, in terms of research, for the binding of aggregate mixtures.

Thermosetting binders are formed of two-component epoxy resins blended with bitumen. Thermosetting means that the flow properties of a material are not reversible with a change in temperature after the material has cooled for the first time. The two liquid components are a resin and a hardener. The key features of a thermosetting binder are that it is an elastic material with no viscous flow component, its properties are substantially unaffected by an increase in

Table 1.9. Effect of Cariphalte DM on binder and wearing course properties

Binder	Binder properties		Wheel tracking rate at 45°C mm/h
	Penetration at 25°C dmm	Softening point °C	
50 pen	56	52.0	3.2
Cariphalte DM	84	90.0	0.7

ambient temperature in service and it is highly resistant to chemical attack such as fuel spillage.

As a product, its strength increases with time after laying, but the binder, by its nature, has a limited usable life after blending. Epoxy binders are used with hot-rolled asphalt wearing courses where the stiffness modulus can be increased by a factor of ten and the deformation can be measured as zero. When bonded to steel plate, such as in bridge decking, fatigue cracking will take substantially longer to happen than would be the case with conventional asphalts. The binder is also used with surface dressing at particularly difficult sites such as approaches to pedestrian crossings or traffic signals.

Other bituminous binders
Petroleum bitumens are the most common binders used with aggregate to produce a road pavement mixture. Around 2.2 million tonnes of bitumen were produced in 1990. There are also other binders available such as tar and lake asphalt.

There is a deposit of between 10 and 15 million tonnes of 'pitch lake' asphalt in Trinidad. The lake material, called Trinidad Epuré, is dug and refined and consists of 10% organic matter, 36% mineral matter and 54% binder. For a wearing course hot-rolled asphalt the epuré is blended with 200 pen bitumen, 50% by mass. The blend has a penetration value of about 50 pen. The advantages of the blend is its 'natural' weathering qualities and its ductility. Lake asphalt-bitumen blends should comply with BS 3690.[44]

Tar is derived from the manufacture of coke or smokeless solid fuel. Tar produced from coke is described as coke oven tar and is produced at temperatures of around 1200°C. Tar produced from smokeless fuel is described as low temperature tar as it is produced at temperatures between 600°C and 700°C. The residue from either process is called pitch. The pitch is blended with lighter oils to provide refined tar. BS 76[45] defines eight grades of road tar. Tar is graded by viscosity, as with bitumen. The measure of viscosity in tars is its equi-viscous temperature (EVT). The eight grades are from 30°C EVT to 58°C EVT in 4C° increments. Higher values of EVT are more viscous tars, as opposed to lower penetration values which are more viscous petroleum bitumens. Refined tar is used as an alternative to bitumen for macadam products. More viscous tars are used in the summer and with mixes which must have a higher stiffness modulus such as dense roadbase macadams for heavily-trafficked routes. Less viscous tars are used in the winter and with mixes where a high stiffness modulus is not necessary such as open-graded macadams. In asphalts, pitch-bitumen blends are alternatives to bitumen or lake-asphalt bitumens. The specification of tar (pitch)/bitumen blends are covered in BS 3690.[44]

There are two reasons why tar is used in bituminous mixtures. First, tar-bitumen blends have better adhesion properties with roadstone than bitumen. Chemical adhesion agents are alternative products to blend with bitumen and enhance the wetting and therefore the adhesion qualities of a binder. Second, refined tar is more resistant to attack by fuel spillage. It is an alternative binder specification with dense surfacings for vehicle parking areas.

Refined tar is more temperature-susceptible than penetration grade bitumen. At high ambient temperatures, a mixture using refined tar as a binder will be less resistant to plastic flow. Its high temperature susceptibility also means that refined tar will be more brittle at low ambient temperatures.

1.4.8.   Summary and selection of bituminous binders
A wide range of bituminous binders is available. Straight-run penetration grade binders comprise the largest proportion of products. Their properties and specification of straight run bitumens can be summarized as follows:

(a)   the product is specified by penetration grade
(b)   common grades of bitumen are 35 pen, 50 pen, 70 pen, 100 pen and 200 pen
(c)   the product is thermo-plastic, viscosity changes with temperature being reversible
(d)   the viscosity-temperature profile is a fingerprint of the product
(e)   the viscosity of a product is critical to its mixing performance and its placement performance in a bituminous mixture; the viscosity of different products will be achieved at different temperatures
(f)   the product is visco-elastic, meaning it will exhibit an elastic and viscous response to an applied stress
(g)   the stiffness modulus of a product defines its service performance and is a measure of its visco-elasticity at service temperatures and the stiffness modulus of the bitumen directly influences the stiffness modulus of a bituminous mixture
(h)   bitumen is unstable, it will oxidize in contact with oxygen in the atmosphere, and the rate of oxidation is directly related to temperature
(i)   the effect of oxidation on the properties of a volume of material will depend on the surface area exposed to volume of material
(j)   in the period between mixing and placement, the penetration value of a bitumen will reduce by one grade e.g. a 70 pen material will become a 50 pen material.

Straight-run bitumens are used to coat aggregate in a wide range of bituminous products for pavement construction. The structure of bitumens are modified for particular applications and use. Historically, products have been developed which reduce the viscosity of the base binder for ease of application. On application the base binder is recovered. Two products have this specific objective: cut-back bitumen and bitumen emulsion.

Other bitumen products have been developed for use with wearing courses and surface dressing. These products have modified mechanical and rheological properties to sustain the high stressing imposed by direct tyre action and aggression of exposure to chemicals and the atmosphere. Alternative bituminous binders, which may be used independently or as a blend with bitumen, are tars and TLA.

Selection of a bituminous binder is shown in Tables 1.10 and 1.11, by material type for DBMs and by layer type for hot-rolled asphalt. The specification of penetration grade bitumens, cut-back bitumens and TLA are defined in BS 3690[44] and the technical data sheets of the Refined Bitumen Association. Bitumen emulsions are specified in BS 434[46,47] and the technical data sheets prepared by the Road Emulsion Association. The specification of tar is defined in BS 76[45] and the technical literature of the Refined Tar

Table 1.10. Use of binders in bituminous macadams

Type of bituminous macadam	Type of binder	Grade of binder
Dense, close-graded and single course	Bitumen	100/200s 300 pen 200 pen 100 pen
Open-graded and pervious	Bitumen	50s 100/200s 300 pen 200 pen 100 pen
All types	Tar	C30/34 C38/42 C46 C50 C54 C38
Pervious macadam	Modified bitumen	6% SBS additive

Table 1.11. Use of binders in hot rolled asphalts

Type of hot-rolled asphalt	Type of binder	Grade of binder
All courses	Bitumen/lake asphalt bitumen	70 pen 50 pen 40 HD 35 pen
Wearing course	Modified bitumen	70 pen + 5% EVA 150/19
Wearing course	Modified bitumen	7% SBS additive

Association. The specification of modified bitumens is defined in the special products literature of the binder producers, e.g. Shell Bitumen and Esso Bitumen.

References

1. BRITISH STANDARDS INSTITUTION. *Coated macadam for roads and other paved areas. Specification for constituent materials and for mixtures*. BSI, London, 1993, BS 4987: Part 1.
2. BRITISH STANDARDS INSTITUTION. *Hot-rolled asphalt for roads and other paved areas. Specification for constituent materials and asphalt mixtures*. BSI, London, 1992, BS 594: Part 1.
3. DEPARTMENT OF TRANSPORT. *Specification for highway works*. HMSO, London, 1992, **1**.
4. THE SCOTTISH OFFICE ROAD DIRECTORATE. *Aggregates in flexible road pavements*. Technical Memorandum SH8/87, The Scottish Office, Edinburgh, 1987.
5. BRITISH STANDARDS INSTITUTION. *Specification for air-cooled blast furnace slag, aggregates for use in construction*. BSI, London, 1983, BS 1047.
6. O'FLAHERTY C. A. *Highway Engineering*. Arnold, London, 1976, **2**, 2nd edn.
7. BRITISH STANDARDS INSTITUTION. *Methods for determination of particle size and shape*. BSI, London, 1975, BS 812: Part 1.
8. BRITISH STANDARDS INSTITUTION. *Methods for determination of physical properties*. BSI, London, 1975, BS 812: Part 2.
9. BRITISH STANDARDS INSTITUTION. *General requirements for apparatus and calibration*. BSI, London, 1990, BS 812: Part 100.
10. BRITISH STANDARDS INSTITUTION. *Guide to sampling and testing aggregates*. BSI, London, 1984, BS 812: Part 101.
11. BRITISH STANDARDS INSTITUTION. *Methods for sampling*. BSI, London, 1989, BS 812: Part 102.
12. BRITISH STANDARDS INSTITUTION. *Method for determination of particle size distribution. Sieve tests*. BSI, London, 1985, BS 812: Section 103.1.
13. BRITISH STANDARDS INSTITUTION. *Method for determination of particle shape. Flakiness index*. BSI, London, 1989, BS 812: Section 105.1.
14. BRITISH STANDARDS INSTITUTION. *Method for determination of moisture content*. BSI, London, 1990, BS 812: Part 109.
15. BRITISH STANDARDS INSTITUTION. *Method for determination of aggregate crushing value (ACV)*. BSI, London, 1990, BS 812: Part 110.
16. BRITISH STANDARDS INSTITUTION. *Methods for determination of ten per cent fines value (TFV)*. BSI, London, 1990, BS 812: Part 111.
17. BRITISH STANDARDS INSTITUTION. *Method for determination of aggregate impact value (AIV)*. BSI, London, 1990, BS 812: Part 112.
18. BRITISH STANDARDS INSTITUTION. *Method for determination of aggregate abrasion value (AAV)*. BSI, London, 1990, BS 812: Part 113.
19. BRITISH STANDARDS INSTITUTION. *Method for determination of the polished stone value*. BSI, London, 1989, BS 812: Part 114.
20. BRITISH STANDARDS INSTITUTION. *Method for determination of water-soluble chloride salts*. BSI, London, 1988, BS 812: Part 117.
21. BRITISH STANDARDS INSTITUTION. *Methods for determination of sulphate content*. BSI, London, 1988, BS 812: Part 118.
22. BRITISH STANDARDS INSTITUTION. *Method for determination of acid-soluble material in fine aggregate*. BSI, London, 1985, BS 812: Part 119.

23. BRITISH STANDARDS INSTITUTION. *Method for testing and classifying drying shinkage of aggregates in concrete.* BSI, London, 1989, BS 812 Part 121.
24. BRITISH STANDARDS INSTITUTION. *Method for determination of frost-heave.* BSI, London, 1989, BS 812: Part 124.
25. BROWN S. F. *et al.* Development of a new procedure for bituminous mix design. *Proc. Eurobitume symposium*, Madrid, 1989, 500−504.
26. WHITEOAK D. *The Shell Bitumen Handbook.* Shell Bitumen UK, 1990, 123.
27. WHITEOAK D. *The Shell Bitumen Handbook.* Shell Bitumen UK, 1990, 126.
28. BRITISH STANDARDS INSTITUTION. *Bitumen for building and civil engineering. Specification for bitumens for road purposes.* BSI, London, 1989, BS 3690: Part 1.
29. BRITISH STANDARDS INSTITUTION. *Penetration of bituminous materials.* BSI, London, 1983, BS 2000: Part 49.
30. HEUKELOM W. An improved method of characterising asphaltic bitumen with the aid of their mechanical properties. *Proc. Ass. Asphalt Paving Tech.*, 1973, **42**, 62−98.
31. WHITEOAK D. *The Shell Bitumen Handbook.* Shell Bitumen UK, 1990, 70.
32. *Stiffness of bitumen and bituminous mixes — nomographs.* Shell International Petroleum Company Ltd, 1977.
33. LUBBERS H. E. *Bitumen in de weg-en waterbouw.* Nederlands Advies Bureau Voor Bitumen Toepassingen, April 1985.
34. JACOBS F. A. *Hot-rolled asphalt: effect of binder properties on resistance to deformation.* Transport and Road Research Laboratory, Crowthorne, 1981, LR 1003.
35. WHITEOAK C. D. and FORDYCE D. *Asphalt workability. Its measurement and how it can be modified.* Shell Bitumen UK, Sept. 1989, Shell Bitumen Review **64**, 14−17.
36. FORDYCE D. and WHITEOAK C. D. *Asphalt workability. A fundamental technique for measuring asphalt workability.* Shell Bitumen UK, Nov. 1990, Shell Bitumen Review **65**, 16−19.
37. BRITISH STANDARDS INSTITUTION. *Hot-rolled asphalt for roads and other paved areas. Specification for the transport, laying and compaction of rolled asphalt.* BSI, London, 1992, BS 594: Part 2.
38. BRITISH STANDARDS INSTITUTION. *Coated macadam for roads and other paved areas. Part 2. Specification for transport, laying and compaction.* BSI, London, 1993, BS 4987: Part 2.
39. BRAUNSCHWEIG P. R. The effect of the compaction temperature on the compactibility of rolled asphalt mixtures. *Die Asphaltstrasse*, 218−225.
40. O'DONNELL E. *et al.* Guidance on the design of fine dense bituminous material. *Proc. 7th Int. conf. Asphalt Pavements*, Nottingham, 1992, 186−200.
41. VAN DER POEL C. A general system describing the visco-elastic properties of bitumen and its relation to routine test data. *J. Appl. Chem.*, 1954, **4**, 221−236.
42. BONNAURE F. *et al.* A new method of predicting the stiffness of asphalt paving mixtures. *Proc. Ass. Asphalt Paving Tech.*, 1977, **46**, 64−104.
43. WHITEOAK C. D. *Shell Cariphalte DM: An SBS modified bitumen.* Shell Bitumen UK, 1989, Shell Bitumen Review **64**, 2−5.
44. BRITISH STANDARDS INSTITUTION. *Bitumens for building and civil engineering. Specifications for bitumens for industrial purposes.* BSI, London, 1983, BS 3690: Part 3.
45. BRITISH STANDARDS INSTITUTION. *Specification for tars for road purposes.* BSI, London, 1974, BS 76.
46. BRITISH STANDARDS INSTITUTION. *Bitumen road emulsions (anionic and*

cationic). Specification for bitumen road emulsions. BSI, London, 1984, BS 434: Part 1.

47. BRITISH STANDARDS INSTITUTION. *Bitumen road emulsions (anionic and cationic). Code of practice for use of bitumen road emulsions.* BSI, London, 1984, BS 434: Part 2.

2. Properties of road layers

2.1. Introduction

Road construction can involve various combinations of layers between the wearing surface, in contact with vehicle tyres, and the ground over which the road is built. Fig. 2.1[1] shows a selection of these combinations from simple compacted aggregate unsurfaced road layers for low volume roads in developing countries to the thick multi-layer bituminous or concrete roads used for heavily-trafficked routes in Europe.

This chapter concentrates on the typical British construction shown in Fig. 1.1, which consists, essentially, of a wearing course supported by a combination of basecourse and roadbase, all of which are bituminous. These are placed over a foundation usually built from unbound aggregates placed on the sub-grade soil which may be natural ground in a cutting or compacted fill in an embankment.

The design of pavements for trunk roads and motorways in the UK is described in the Department of Transport's *Design manual for roads and bridges*[2] while materials are specified in a companion document.[3]

Each layer in the pavement has a specific role and its properties have to relate to that role. These properties must be examined in the context of pavement design, since layer thicknesses and the mechanical properties of the layers are intimately linked. Consequently, although chapter 3 deals with design, it is necessary in this chapter first to provide a philosophical design framework within which to discuss the properties of road layers.

This chapter describes the significant properties of materials used to construct asphalt pavements, from the soil to the wearing course, in the context of the role each layer of material plays within the pavement. The basic mechanics of pavements, therefore, provide a backdrop to the discussion of material properties, and emphasizes the need to understand soil mechanics' principles for subgrades, capping and sub-bases and asphalt mechanics for the bituminous layers above.

The concepts of an ideal pavement and an ideal pavement foundation are discussed to highlight key features desirable for the materials in the various layers. These are mainly mechanical properties, which will be gradually introduced into end-product specifications as European normalization is implemented.

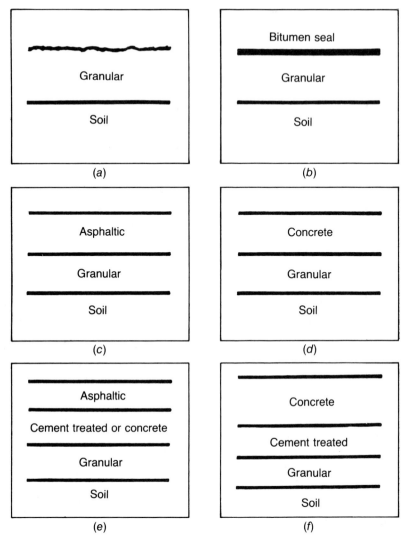

Fig. 2.1. Different types of road construction: (a) gravel road (b) sealed granular road (c) asphalt pavement (d) concrete pavement (e) composite pavement (f) heavy duty concrete[1]

2.2.   The ideal pavement

Brown and Barksdale's 'ideal' pavement[4] (Fig. 2.2) highlights the key roles of the main component layers.

The wearing surface is essentially a cosmetic treatment in that it is not required to have a structural role. However, the surface must provide the appropriate conditions for safe and comfortable contact with vehicle tyres. It must also fulfil a waterproofing function since moisture in pavements is

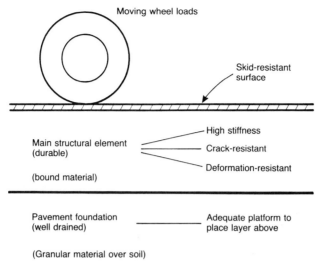

Moving wheel loads

Skid-resistant
surface

High stiffness

Main structural element
(durable)

Crack-resistant

Deformation-resistant

(bound material)

Pavement foundation
(well drained)

Adequate platform to
place layer above

(Granular material over soil)

Fig. 2.2. The ideal pavement[4]

generally undesirable as it can cause various types of damage to the lower layers. Porous asphalt surfacing is an important exception to this requirement and this material is discussed later.

The main structural layer of the pavement is the roadbase (Fig. 1.1). This is usually covered by a basecourse which is a practical expedient to achieve more accurate finished levels than are possible with the thick-lift construction used for base layers. The combination of basecourse and roadbase must have the ability to spread the wheel load so that underlying layers are not over-stressed. Damage through cracking or permanent deformation within the layer itself must also be prevented.

The bituminous layers are supported by a pavement foundation which must be adequate to carry construction traffic and the paving operation. The foundation is most highly stressed during the construction phase and as it usually consists of granular materials and soils, the principles of soil mechanics apply which reveal the need for good drainage to depress the water table and prevent a build up of water in the granular layers.

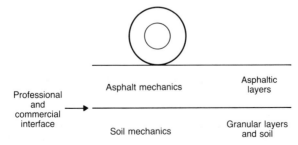

Professional
and
commercial
interface

Asphalt mechanics

Asphaltic
layers

Soil mechanics

Granular layers
and soil

Fig. 2.3. Bituminous base/pavement foundation interface[4]

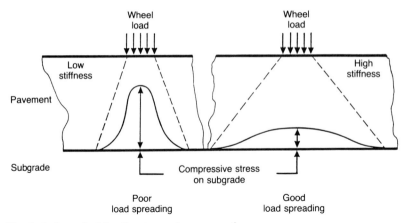

Fig. 2.4. Spread of loading in road pavements[4]

The interface between the pavement foundation and the bottom of the bituminous roadbase is an important one. (Fig. 2.3).[4]

Below this level, work is generally carried out by the main contractor and the principles of soil mechanics apply. Above this level, specialist bituminous material sub-contractors usually operate and the principles of asphalt mechanics are relevant. This interface is, therefore, both a professional and a constructional/commercial one. The complete pavement engineer needs the technical knowledge to operate on both sides of this divide. In practice there are, unfortunately, few individuals who qualify.

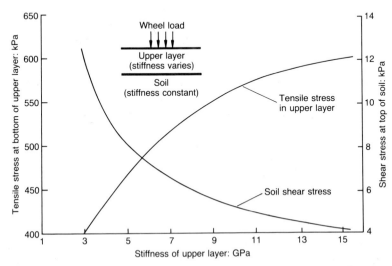

Fig. 2.5. Relationship between shear stress and elastic stiffness in a road pavement[4]

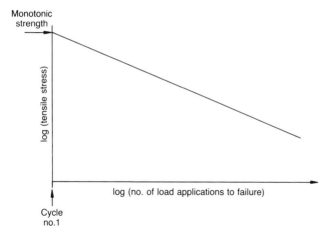

Fig. 2.6. Typical relationship between tensile stress and the number of load applications to failure in a road pavement[1]

2.3. Design concepts

Figure 2.4 illustrates the load-spreading principle in pavement engineering.[4]

The level of stress induced in the supporting layers depends on the elastic stiffness of the roadbase for a given layer thickness. More quantitatively, Fig. 2.5 shows how the shear stress in the foundation decreases as the elastic stiffness of the roadbase increases.[4] The decrease in shear stress is, however, accompanied by an increase in the tensile stress induced at the bottom of the roadbase layer, so there is a possibility that in providing good protection to the foundation, the roadbase layer itself may crack. Repeated loading by traffic can cause fatigue failure. This arises at a stress level much lower than that which the material can sustain under a single load application. Fig. 2.6 shows the characteristic relationship between the tensile stress or strain in the bituminous material and the number of load applications to failure by cracking. Sustainable stress decreases as the number of load applications increases.

The interaction between the elastic stiffness of the base, the shear stress in the foundation and the tensile stress in the roadbase provides the essential key to pavement design. The other related parameters are the roadbase thickness and its composition, the latter dictating the stiffness and fatigue cracking characteristics which are exhibited in practice.

The pavement foundation has a dual function. It must provide short-term service as a haul road and construction platform and long-term service as a support to the bituminous construction above. For this reason, the concept of a flexible pavement designed in two stages has been developed in the UK. Fig. 2.7 shows the design considerations. Stage one involves the foundation, which has to carry a limited number of heavy wheel loads.

It is essential that over-stressing and plastic deformation in the soil are prevented. This is done by ensuring that the combination of thickness and elastic stiffness of the granular material placed above it is adequate. In present UK practice, this granular layer usually consists of two parts. A lower quality

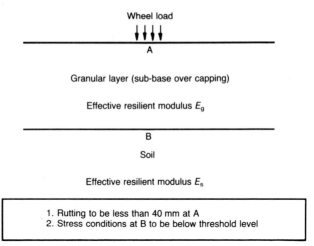

Fig. 2.7. Road pavement design considerations

(capping) layer is placed over the soil and forms part of the earthworks. This is covered by a better quality crushed rock sub-base of standard thickness (225 mm). The capping layer is not required when the assessed strength (measured by a parameter called the California bearing ratio (CBR)) for the soil exceeds 5%. The CBR of a soil is an empirical parameter based on a test which measures the resistance to penetration of a standard plunger into a compacted specimen of the soil contained in a standard mould.[5] The design principles remain the same, whether the granular layer is in two parts or only one. In addition to protecting the soil from over-stressing, the layer itself must not develop serious ruts under the action of construction traffic. For practical purposes, the rut depth should not exceed 40 mm.

Stage two of the design process involves selection of the correct combination of roadbase material and thickness to ensure that the foundation is not over-stressed and that the layer itself does not crack. In addition, the roadbase, basecourse and wearing course combination must not develop permanent deformation which leads to wheel track rutting. This is most effectively dealt with by correct mix design and attention to quality control on site, particularly with regard to compaction and the temperature of materials at the time of compaction.

2.4. Subgrade

The subgrade is the material upon which the road is constructed. It can be virgin soil or imported material. Although the sub-grade is not formally a pavement layer, its properties must be fully understood to be able to design and construct a satisfactory pavement over it. The soil will either be in cut or fill and its stress history will influence the mechanical properties exhibited once the pavement is constructed.

For clays, the consolidation stress history is important. Fig. 2.8[4] shows the stress path which a typical over-consolidated soil will have experienced in

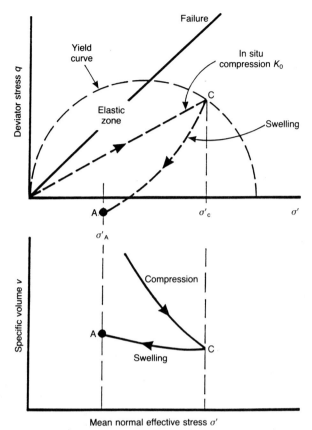

Fig. 2.8. Stress path of a typical over-consolidated soil[1]

geological history and as a result of the construction process, where removal of overburden is involved.

The pre-consolidation stress σ_c' is the highest stress to which the soil has been subjected in the past. The current, relatively lower stress σ_A' is related to it by the over-consolidation ratio (OCR) which is σ_c'/σ_A'. The principle of effective stress must be appreciated to properly understand soil characteristics (Fig. 2.9). The simple equation is

$$\sigma' = \sigma - u \tag{1}$$

where σ' = effective stress
σ = total stress
u = pore water pressure.

In simple terms, effective stress is the normal stress carried by the particle to particle contacts. Since soil is a frictional material, its strength depends on this normal stress. Fig. 2.10 allows the establishment of an equation connecting these parameters.

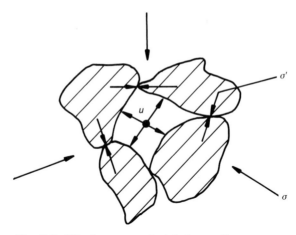

Fig. 2.9. Effective stress principle in a soil

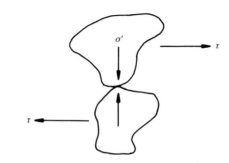

Fig. 2.10. Normal stress and shear stress between soil particles

The relationship between the normal stress σ', the shear stress τ and the coefficient of friction is given by

$$\tau = \sigma' \mu \qquad (2)$$

Since soil particles interlock and interact in a complex three dimensional manner, the effective coefficient of friction is expressed as

$$\mu = \tan \phi' \qquad (3)$$

where ϕ' is known as the angle of shearing resistance with respect to effective stresses. The shear strength is, therefore, expressed as

$$\tau = \sigma' \tan \phi' \qquad (4)$$

Effective stress always has to be determined by equation (1). The total stress is caused by the externally-applied load, while the pore water pressure depends on the hydraulic conditions in the soil at the point concerned. Fig. 2.11 illustrates the principle.

For an element of soil at depth z below formation level when the water table (the natural level of water in the soil) is at depth h, the total stress σ is given by

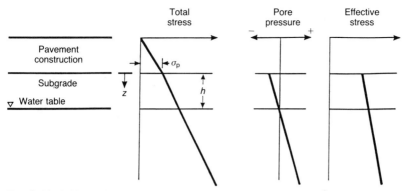

Fig. 2.11. Self-weight stresses in a road pavement and subgrade[1]

$$\sigma = \sigma_p + \gamma z$$

where σ_p = stress caused by the weight of the pavement layers on the
           subgrade

γ = unit weight of the soil.

The pore water pressure u at depth z is given by

$$u = \gamma_w(z-h)$$

where γ_w = unit weight of water.

Hence, the effective stress σ', by equation (1) is

$$\sigma' = \sigma_p + \gamma z - \gamma_w(z-h) \qquad (5)$$

For pavement construction, interest is generally centred on the soil near
formation level (top of the subgrade) which will be above the water table.
For clays, water can be held in the soil pores under capillary action above
the water table and saturated conditions can apply for some height above the
water table. In these circumstances, the pore water pressure will be a suction
and equation (5) becomes

$$\sigma' = \sigma_p + \gamma_z + \gamma_w(h-z) \qquad (6)$$

Soil strength, by equation (4), increases with effective stress. It follows then,
from equation (6), that a low water table is desirable i.e. a large value of h.

When constructing roads in cuttings it may be necessary to depress the natural
water table level and control it, if satisfactory strength is to be maintained in
the soil. Fig. 2.12 shows a typical cross-section where side drains are used
to this end.

The side drains fulfil the additional roles of intercepting surface water, which
may run down the sides of the cutting, and providing a drainage path from
the sub-base in the event of water levels building up there. This could occur
during construction or after many years use when water will leak into the
pavement even if it is well maintained.

The effective stress at formation level is often equated to the soil suction
S. This parameter has been used for many years in the UK to determine design

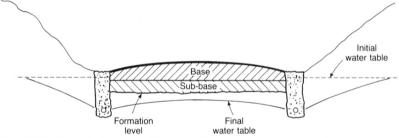

Fig. 2.12. Control of water table in road construction in a cutting[1]

CBR values, on the basis of extensive research by the Transport Research Laboratory (TRL).[6]

Failure in the subgrade is defined, for pavement design purposes, in terms of the permanent (plastic) deformation which can develop if the shear stress, repeatedly applied by wheel loading, is too high. The concept of a threshold level for this shear stress has been developed. Below this critical value, negligible plastic strains will accumulate. Recent studies[7] have indicated that, for design purposes, the threshold shear stress may be regarded as directly proportional to the soil suction. This, therefore, provides a useful basis for design, since the granular layer thickness/stiffness combination must be sufficient to prevent the threshold stress in the subgrade being exceeded under the action of construction traffic.

For a subgrade consisting of compacted fill, the same principles apply. However, there are uncertainties about stress history and water table position which will only be resolved by conducting field observations. Compacted clay fill is, however, thought to be in an effectively over-consolidated state.

The elastic stiffness of the soil is a mechanical property which is important for pavement design, since the stress induced in the upper layers, as well as in the soil itself, will depend on it. It is significant for both stages in the design process and for interpretation of deflection testing results when pavements are subsequently being evaluated in-service.[8]

The term elastic stiffness has been used in a generic way for pavement materials and soils. It may be regarded as broadly comparable with Young's Modulus for materials such as steel or concrete. However, since the stress−strain relationships for pavement materials are affected by several variables, alternative terminology is appropriate. The term resilient modulus is commonly used as equivalent to elastic stiffness. The general definition is the applied stress pulse (due to wheel loading) divided by the recoverable (resilient) strain. More specifically, for soils, the resilient modulus E_r is given by

$$E_r = \frac{q_r}{A} \left(\frac{\sigma'}{q_r} \right)^B \tag{7}$$

where q_r = shear stress pulse
σ' = effective stress
A, B = constants which depend on the soil type.

Equation (7) demonstrates that a particular soil does not have a unique value

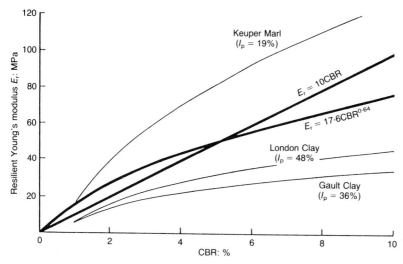

Fig. 2.13. Relationships between CBR and resilient Young's Modulus for three types of soil[4]

of resilient modulus but that the value is influenced by the stress conditions. These include both the static effective stress, due to overburden and water table position σ' and the shear stress caused by traffic loading q_r. The implication of equation (7) is that soil stiffness increases with depth. For pavement design purposes, a value appropriate to formation level is normally used. However, for pavement evaluation purposes, the profile with depth is crucial.[8]

The CBR of a soil is not directly comparable with the resilient modulus.[4,9] A unique relationship between the two parameters does not exist. However, for typical stress conditions, Fig. 2.13 shows relationships for three different soil types, together with empirical equations often used for design. The parameter I_p is the plasticity index of the soil.

For sand subgrades, reference should be made to the following section dealing with granular materials.

2.5. Granular materials

The main difference between granular materials (crushed rock or gravel) and clays lies in particle sizes. The basic principles of soil mechanics still apply. Granular materials will, generally, exist in a partially-saturated state within the pavement, unless drainage provision is completely ineffective. The concepts of stress history are not the same. However, the effects of compaction, usually involving high-quality vibrating rollers in modern construction, are thought to induce some pre-compression in the horizontal direction which is broadly comparable with the over-consolidation of clays.

Equation (4) for shear strength is equally applicable for granular materials and the concept of a threshold stress also appears to hold. In this case, it is defined as 70% of the shear strength.[1]

The stress dependence of resilient modulus for granular materials is considerable. The influence of effective stress σ' is more important than that of the shear stress. Since overburden stresses are very low in the granular layer, the effective stress induced by wheel loading becomes significant. A simple expression for resilient modulus in granular materials is

$$E_r = C.\sigma'^D \tag{8}$$

where C and D are material constants.

For pavements with bituminous roadbases, the non-linear resilient properties of the sub-base do not present a major problem in design. Experimental and theoretical studies[10] have shown that, for a good quality crushed rock sub-base, the effective in situ resilient modulus is about 100 MegaPascals (MPa). This figure will decrease if saturated conditions are approached. Higher figures can arise if the layer has self-cementing properties, such as those exhibited by the softer limestones. Materials used in capping will generally have resilient moduli of about 80 MPa. Laboratory testing[11] has clearly demonstrated that there is no real relationship between CBR and resilient modulus for granular materials.

Compaction is an important factor in the construction of granular layers. High density will increase strength and, hence, resistance to rutting, as these are related. If the material has a high fines content, particularly the proportions less than 75 μm, then it will be susceptible to weakening by water ingress. In a highly compacted state, such materials have a greater affinity for water as the small particle sizes and void spaces increase suction forces. Conversely, drainage becomes difficult as permeability will be low.

There is a danger in over-compacting dense granular materials if they are too wet relative to the compactive effort applied. Fig. 2.14 shows how the compacted state can move towards saturation under high compactive effort i.e. from A to B. In general, it is better to place granular materials dry of optimum water content (to the left of the peak in the curve) since this allows the full potential for high quality compaction to be realised giving high density and, hence, high strength.

Because of the problems which water can cause in soils and granular materials, a drainage blanket of high permeability material can be useful. The ideal pavement foundation incorporating such a layer is illustrated in Fig. 2.15.[1]

The high permeability can be achieved through control of the aggregate grading. It is very important, however, that the layer is properly compacted. Geotextiles can be usefully employed to separate the soil from the granular layer to minimize contamination, which would reduce permeability and, if severe, shear strength too. In the USA, several states use open-textured bituminous materials to form stable drainage blankets below compacted dense granular sub-bases and 'no fines' concrete has been tried in the UK.

Reinforcing grids of high tensile polymeric material can be effective in granular layers by increasing resistance to rutting from construction traffic. It is important that such grids are placed relatively near the surface of the layer where the shear stresses causing rutting are highest. A rule of thumb suggests a depth equal to half the width of the tyre contact patch e.g. 150 mm. These

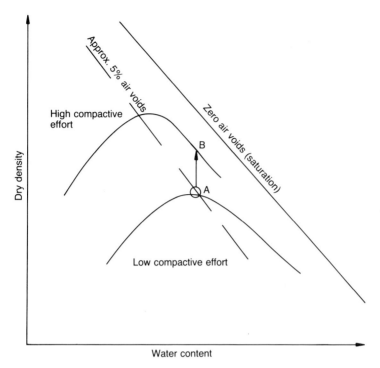

Fig. 2.14. Effect of compaction on degree of saturation

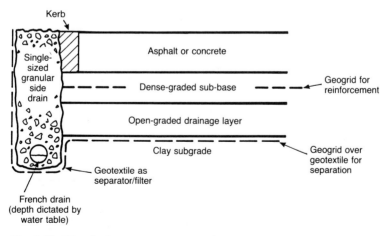

Fig. 2.15. The ideal pavement foundation[1]

grids are effective when they interlock with the aggregate particles. The correct aperture size must, therefore, be selected to match the coarse aggregate dimensions or vice versa.

2.6. Bituminous roadbases and basecourses

The combination of roadbase and basecourse provides the main structural layer in the road. The essential requirements are that the material should offer the following mechanical properties

- (a)   High elastic stiffness ensures good load spreading ability
- (b)   High fatigue strength prevents the initiation and propagation of cracks due to repeated loading by traffic
- (c)   High resistance to permanent deformation ensures that these layers do not contribute significantly to surface rutting.

In addition to these mechanical properties, the materials must be durable, which implies that they are resistant to the effects of air and water.

Under current British practice, roadbase and basecourse materials are specified in British Standards[12,13] using a recipe approach. This provides an aggregate grading envelope, a grade of bitumen and its quantity in the mix. No mechanical properties are measured. The success of the materials in practice is based on past experience and good workmanship in construction. A new method of mix design is in the process of being developed for roadbases and basecourses and has been described in various papers.[14-16] The essential features of the method are outlined below.

Although bituminous mixtures for roads are all essentially combinations of aggregate and bitumen, with some air in the voids following compaction, two generic types of mix have emerged from historical developments in the UK. Dense bitumen macadam (DBM) is a continuously-graded material with a relatively low binder content and a relatively soft binder. Hot-rolled asphalt is a gap-graded material with higher binder content and harder binder. The grading curves for typical examples of these materials are compared in Fig. 2.16.

The elastic stiffness of a mix is influenced by its volumetric proportions and, in particular, the quantity and stiffness of the bitumen which is used. DBM and hot-rolled asphalt roadbases tend to have similar elastic stiffnesses when properly compacted, since the small amount of soft binder in the DBM has the same effect as the larger amount of hard binder in the hot-rolled asphalt. Resistance to fatigue cracking is provided by the volume and hardness of the binder. As hot-rolled asphalt has more binder of a harder grade, it offers much better resistance to cracking than DBM.

Resistance to permanent deformation depends mainly on the aggregate grading and the particle characteristics. Minerals which are rough and angular when crushed offer better resistance than smooth, rounded materials. A low but adequate binder content is needed. Consequently, DBM resists permanent deformation better than hot-rolled asphalt. Fig. 2.17 summarizes the relative permanent deformation characteristics of continuously-graded DBM and gap-graded hot-rolled asphalt expressed in terms of the permanent shear strain that they would accumulate with increasing numbers of load applications. Particular

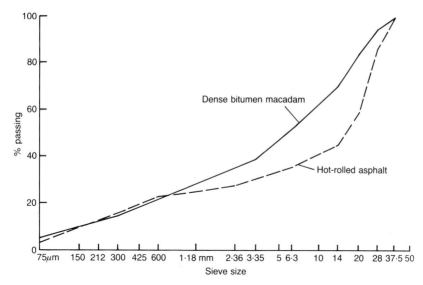

Fig. 2.16. Typical grading curves for hot-rolled asphalt and DBM

mixes can be expected to exhibit a variety of results from the appropriate tests, so Fig. 2.17 should only be taken as a general indication of relative values for typical examples of the two mix types.[17]

It is possible to improve the mechanical properties of the two traditional British Standard materials by simple adjustments to the mix formulations. DBM properties are improved by using a higher binder content and a harder grade of binder. In addition, it is essential that the material is properly compacted. Table 2.1 demonstrates the potential changes to pavement thickness for a given life when adjustments of various kinds are made to a basic DBM with 3.5% of 100 penetration grade bitumen (Design no. 1).

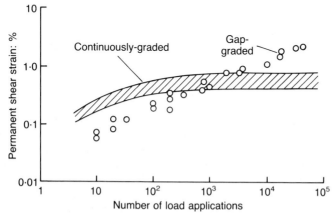

Fig. 2.17. Permanent deformation characteristics of typical roadbase mixtures[17]

Table 2.1. Structural and economic comparisons of alternative designs for 50 million standard axles (msa) on a sub-grade CBR of 5%[17]

Design no.	1	2	3	4	5
Binder grade, pen	100	100	100	100	50
Binder content, %	3.5	3.5	4.5	4.5	4.5
Void content, %	10.0	8.0	8.0	5.5	6.0
Elastic stiffness, MPa	5600	7500	5100	7300	10600
Thickness, mm	360	300	280	220	190
Relative costs of surfacings plus roadbase	100%	87%	84%	71%	63%

Surfacing: 50 mm hot-rolled asphalt, stiffness 4900 MPa, average annual air temperature 9.5°C
Sub-base: 300 mm type 1, vehicle speed 80 km/h

The mix offering best performance is similar to that recently made available in the relevant British Standard as DBM 50 but with a higher binder content (Design no. 5). A similar material, Heavy duty macadam (HDM), contains a higher filler content, which is the percentage of the aggregate passing the 75 μm sieve in the grading analysis. For hot-rolled asphalt, desirable adjustments include a reduction in binder content and an increase in filler content.

These pragmatic changes to recipe specifications have resulted from extensive research in the laboratory and through field trials.[17-19] The results have demonstrated that DBMs with higher elastic stiffness and improved resistance to cracking can be achieved with little or no reduction in resistance to permanent deformation. For hot-rolled asphalt, a formulation with enhanced stiffness and resistance to permanent deformation is produced with an acceptable reduction in resistance to cracking.

Durability and compactibility are important properties of a bituminous mixture. Durability is mainly influenced by the binder film thickness on the aggregate relative to the air void content in the mixture. For the dense mixes used in roadbases and basecourses with air void contents of 2–8%, durability should be adequate. Compactibility of DBMs is improved by the higher binder contents, while that for hot-rolled asphalt remains high, even when the binder content is decreased.

The mechanical properties of a roadbase or basecourse mixture consequently depend on several variables. The objectives of good mix design are to proportion the constituent materials and control the site operation so that satisfactory mechanical properties are realised. This requires a two-stage process involving compaction of trial mixtures in the laboratory while varying the key parameters of aggregate grading, degree of compaction and binder content. The subsidiary variables of aggregate type (whatever is available) and binder grade can also be studied.[15,16]

A chart such as Fig. 2.18 can be used to establish which formulations are likely to produce satisfactory mixtures. This is a plot of void content V_v against VMA showing lines of equal binder content by volume V_B. (VMA $= V_v + V_B$). The target specification box is shown for $V_v = 3-7\%$ and

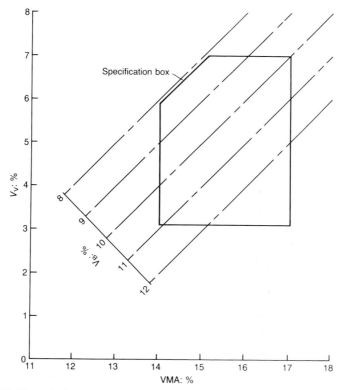

Fig. 2.18. Typical relationship between void content and voids in the mineral aggregate for different binder contents[15]

VMA = 14−17%, while V_B should be greater than 8%. A full discussion of these criteria is available.[15,16] Those mix formations which fall in the target area are taken forward for tests of their mechanical properties. These should include elastic stiffness and resistance to permanent deformation. A device known as the Nottingham Asphalt Tester can conveniently be used for carrying out the tests. This equipment and the associated mix design method have been described in detail elsewhere.[14−16] A procedure of this type is needed to provide the end-product specifications needed in the future and which European normalization is demanding. Results from the US Strategic Highway Research Program (SHRP) will assist with these developments.

2.7. Wearing courses

Bituminous wearing courses have to provide resistance to the effects of repeated loading by tyres and to the effects of the environment. They must also offer adequate skid resistance in wet weather as well as a comfortable vehicle ride. They must also be resistant to rutting and cracking. It is also desirable that they be waterproof, except in the case of porous asphalt. This demanding list of requirements results in practical compromises in terms of the material types which are used.

2.7.1. Hot-rolled asphalt

This is a gap-graded material with less coarse aggregate than the basecourse version shown in Fig. 2.16. In fact it is essentially a bitumen/fine aggregate/filler mortar into which some coarse aggregate is placed. The mechanical properties are dominated by those of the mortar. This material is extensively used for surfacing of major roads in the UK. It provides a durable layer with good resistance to cracking and one which is relatively easy to compact. It is specified in BS 594: Part 1.[12] Since the coarse aggregate content is low (typically 30%) when compacted, this material has a smooth surface, the skid resistance of which is inadequate. Consequently, pre-coated chippings are rolled in to the surface at the time of construction.

Rolled asphalt wearing course mixtures may be designed using a modified version of the Marshall method, developed in the US for continuously-graded aggregate mixtures.[20] The details are fully described in BS 598: Part 107.[21] Alternatively, BS 594: Part 1[12] provides various recipe specifications for this material based on experience. An improved method of mix design based on mechanical properties and resistance to water and ageing will be required in the future and research is underway.

2.7.2. Porous asphalt

This is a uniformly-graded material which is designed to provide large air voids so that water can drain to the verges within the layer thickness. The principle is illustrated in Fig. 2.19.

For this surfacing to be effective, the basecourse below must be waterproof and the porous asphalt must have the ability to retain its open textured properties with time. Thick binder films are required to resist water damage and ageing of the binder. In use, this material minimizes vehicle spray, provides a quiet ride and lower rolling resistance to traffic than dense mixes. In the UK it has been used for many years on airfield pavements to prevent aquaplaning and following field trials[22] it is being introduced for highway surfacing. Many other European countries use this material for roads, but little research has been done on its mechanical properties.

2.7.3. Asphaltic concrete and DBM

These are continously-graded mixtures similar in principle to the DBMs used

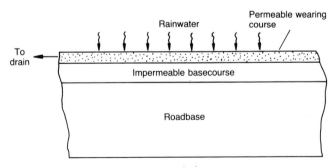

Fig. 2.19. Principle of porous asphalts

in roadbases and basecourses but with smaller maximum particle sizes. Asphaltic concrete tends to have a slightly denser grading and is used for road surfaces throughout the world, with the exception of the UK. DBM surfacing is specified in BS 4987[13,23] but is generally only used on minor roads.

It is more difficult to meet UK skid resistance standards with DBMs than with hot-rolled asphalts or porous asphalts. However, developments in separate surface treatments have offered new possibilities. The durability of DBM surfacings in the UK has been less satisfactory in practice than hot-rolled asphalt. Continuously-graded materials may become more widely used in the UK with the introduction of end-product specifications based on mechanical properties and the use of rational mix design methods.

2.7.4. Stone mastic asphalt

Stone mastic asphalt (SMA) was pioneered in Scandinavia and Germany and is attracting serious attention in the USA as a replacement for asphaltic concrete. SMA has a coarse, aggregate skeleton, like porous asphalt, but the voids are filled with a fine aggregate/filler/bitumen mortar. The result is a gap-graded material with good resistance to rutting and high durability. It differs from hot-rolled asphalt in that the mortar is designed to just fill the voids in the coarse aggregate, whereas in hot-rolled asphalt, coarse aggregate is introduced into the mortar and does not provide a continuous stone matrix. The higher stone content rolled asphalts, however, are rather similar to SMA but are not widely used as wearing courses in the UK, being preferred for roadbase and basecourse construction.

2.7.5. Comparison of wearing course characteristics

The mechanical properties and durability characteristics of the various generic mix types can be compared with the aid of Fig. 2.20.

This shows how low stone content hot-rolled asphalt, asphaltic concrete, SMA and porous asphalt mixes mobilize resistance to loading by traffic. Asphaltic concrete (Fig. 2.20(a)) presents something of a compromise when well designed, since the dense aggregate grading can offer good resistance to the shear stresses which cause rutting, while an adequate binder content will provide reasonable resistance to the tensile stresses which cause cracking. In general, the role of the aggregate dominates. DBMs tend to have less dense gradings and properties which, therefore, tend towards good rutting resistance and away from good crack resistance.

Hot-rolled asphalt (Fig. 2.20(b)) offers particularly good resistance to cracking through the binder-rich mortar between the coarse aggregate particles. This also provides good durability but the lack of coarse aggregate content inhibits resistance to rutting.

SMA and porous asphalt are shown in the same diagram (Fig. 2.20(c)) to emphasize the dominant role of the coarse aggregate. In both cases, well-coated stone is used. For porous asphalt the void space remains available for drainage of water, while, for SMA, the space is occupied by fine aggregate : filler : bitumen mortar. Both offer good rutting resistance through the coarse aggregate content. The tensile strength of porous asphalt is low, while that of SMA is probably adequate though little mechanical testing data have been reported.

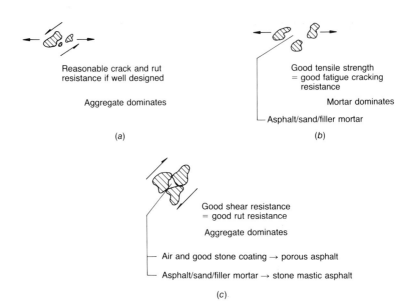

Fig. 2.20. Mechanical properties and durability characteristics of various bituminous mixtures: (a) asphaltic concrete (b) low stone content hot-rolled asphalt (c) SMA and porous asphalt

2.8.   Skid resistance

The principles involved in providing good resistance to vehicle skidding in wet weather are well established and reference may be made to the detailed literature[2] for a description of current practice and requirements.

Low speed skid resistance relies on adequate surface texture for the aggregate particles exposed on the pavement surface. This is referred to as microtexture. Suitable mineral types can provide resistance to the polishing action of traffic in the long term, so that adequate microtexture is preserved. This is quantified by the Polished Stone Value (PSV) test.[24]

At higher speeds, the removal of bulk water from the tyre/stone interface is the major requirement so that good contact is maintained. This is achieved by adequate macrotexture, measured in terms of the texture depth of the surfacing combined with deep treads in tyres. Recent work at TRL[25] has demonstrated that adequate texture depth does reduce accident risk at speeds in excess of 64 km/h on bituminous surfaces. When porous asphalt is used, the bulk water is removed by drainage within the surfacing layer thickness so the concept of texture depth is not relevant.

References

1.   BROWN S. F. and SELIG E. T. (O'REILLY M. P. and BROWN S. F. (eds).) Cyclic loading of soils: from theory to design. *The design of pavement and rail track foundations.* Blackies, London, 1991, 6, 249−305.
2.   DEPARTMENT OF TRANSPORT. *Design manual for roads and bridges, 7: Pavement design and maintenance.* HMSO, London, 1994.

3. DEPARTMENT OF TRANSPORT. *Specification for highway works*. HMSO, London, 1992, **1**.

4. BROWN S. F. and BARKSDALE R. D. Pavement design and materials. *Proc. 6th Int. Conf. structural design of asphalt pavements*, Ann Arbor, Michigan, 1987, **2**, 118–148.

5. BRITISH STANDARDS INSTITUTION. *Methods of test for soils for civil engineering purposes*. BSI, London, 1990, BS 1377: Part 4.

6. POWELL W. D. *et al. The structural design of bituminous roads*. Transport Research Laboratory, Crowthorne, 1984, LR 1132.

7. BROWN S. F. and DAWSON A. R. Two-stage approach to asphalt pavement design. *Proc. 7th Int. Conf. Asphalt Pavements*. Nottingham, 1992, **1**, 16–34.

8. BROWN S. F. *et al.* Development of an analytical method for the structural evaluation of pavements. *Proc. 2nd Int. Conf. bearing capacity of roads and airfields*, Plymouth, 1986, **1**, 267–276.

9. BROWN S. F. *et al.* The relationship between California Bearing Ratio and elastic stiffness for compacted clays. *Ground Engineering*, 12990, **23**, No. 8, 27–31.

10. BROWN S. F. and PAPPIN J. W. *Modelling of granular materials in pavements*. Transp. Res. Record 1022. 1985, 45–51.

11. SWEERE G. T. H. *Unbound granular bases for roads*, Delft Univ of Tech, 1990.

12. BRITISH STANDARDS INSTITUTION. *Hot-rolled asphalt for roads and other paved areas. Specification for constituent materials and asphalt mixtures*. BSI, London, 1992, BS 594: Part 1.

13. BRITISH STANDARDS INSTITUTION. *Coated macadam for roads and other paved areas. Specification for constituent materials and for mixtures*. BSI, London, 1993, BS 4987: Part 1.

14. COOPER K. E. and BROWN S. F. Development of a simple apparatus for the measurement of the mechanical properties of asphalt mixes. *Proc. Eurobitume symposium*, Madrid, 1989, 494–498.

15. COOPER K. E. *et al. Development of a practical method for the design of hot mix asphalt*. Transp. Res. Record 1317, Trans. Res. Board, Washington DC, 1991, 42–51.

16. BROWN S. F. *et al.* Application of new concepts in asphalt mix design. *J. Ass. Asphalt Paving Tech.*, 1991, **60**, 264–286.

17. BROWN S. F. and COOPER K. E. Improved asphalt mixes for heavily trafficked roads. *Proc. 4th Conf. asphalt pavements for South Africa*, Cape Town, 1984, 545–560.

18. BROWN S. F. and COOPER K. E. The mechanical properties of bituminous materials for roadbases and basecourses. *Proc. Ass. Asphalt Paving Tech.*, Minnesota, 1984, **53**, 415–437.

19. BROWN S. F. *et al.* Improved roadbases for longer pavement life. *Proc. 3rd Eurobitume symposium*, The Hague, 1985, **1**, 217–222.

20. ASPHALT INSTITUTE. *Mix design methods for asphalt concrete and other hot-mix types*. Asphalt Institute, Maryland, 1991, MS-2.

21. BRITISH STANDARDS INSTITUTION. *Sampling and examination of bituminous mixtures for roads and other paved areas*. BSI, London, 1990, BS 598: Part 107.

22. DAINES M. E. *Pervious macadam: trials on trunk road A38 Burton by pass, 1984*. Transport Research Laboratory, Crowthorne, 1986, RR 57.

23. BRITISH STANDARDS INSTITUTION. *Specification for coated macadam for roads and other paved areas*. BSI, London, 1993, BS 4987: Part 2.

24. BRITISH STANDARDS INSTITUTION. *Methods for sampling and testing of mineral aggregates, sands and fillers*. BSI, London, 1975, BS 812: Part 3.

25. ROE P. G. *et al. The relation between the surface texture of roads and accidents*. Transport Research Laboratory, Crowthorne, 1991, RR 296.

3. Design of roads

3.1. Introduction

The several layers which make up a road are called the road pavement. A road pavement is a complex engineering structure which has to perform a number of functions which are not always compatible. Its primary objective is to support the applied traffic loads and distribute them to the underlying soil (subgrade). The ultimate aim is to ensure that the transmitted stresses are sufficiently reduced so that they will not exceed the supporting capacity of the subgrade. In addition, the pavement structure should be sufficiently impermeable to avoid the draining ability of the subgrade or sub-base being exceeded. If this happened the pavement would become waterlogged and the structural integrity greatly diminished. The upper surface of the pavement should also be abrasion resistant, so that it does not wear away unduly during the life of the road, and skid resistant to provide a safe running surface.

The pavement structure consists of a number of superimposed layers of selected, processed, laid and compacted materials which together form a pavement structure with the required characteristics. Pavement design is essentially, therefore, a structural evaluation process which ensures that the traffic loads are distributed in such a way that the stresses and strains developed at all levels in the pavement and the subgrade are within the capabilities of the materials at those levels. Since the load-carrying capacity of a layer is a function of both the thickness of the material and its stiffness, pavement design should involve the selection of materials for the different layers based on their known engineering properties and the calculation of the individual thicknesses of each of these materials. The mechanical properties of the materials used in pavement layers, which are fundamental to the design of those layers, are covered in chapter 2.

Climatic conditions also influence the performance of the whole pavement. While moisture can affect the subgrade, sub-base, unbound base materials (those which do not contain bitumen or cementitious binder) and open-textured bituminous mixtures, temperature and air have a significant effect on all bituminous mixtures.

3.2. Pavement layers

Bituminous pavements have essentially three components (Fig. 1.1), which act together to protect the subgrade from climatic and traffic stresses both during construction and the working life of the road.

Each of these layers may, in turn, be sub-divided into two further layers. The surfacing layer consists of a wearing course and a basecourse. The base may be divided into upper and lower roadbase, each of which may be the same or different materials. The sub-base may be divided into two layers called the sub-base and a capping layer. Capping layer material may be a broadly specified type of material similar to sub-base or it may be the upper part of the soil improved by some form of stabilization e.g. with lime or cement. The interface between the sub-base and the natural soil is called the formation.

Generally, the lower the layer is within the pavement, the less processing it has received and the less closely it is specified i.e. larger aggregate sizes may be used and unbound rather than bound materials are more commonplace. This is primarily because such materials are cheaper and because the demands on the layer are less. Towards the top of the pavement, achievement of a good riding quality and an impermeable surface leads the designer towards thinner layers of binder-rich materials which, of necessity, contain a smaller maximum aggregate size and a closer specification to satisfy all the conflicting design requirements.

3.2.1. Sub-base
The sub-base performs four main functions.

(a)   It provides a structural layer which distributes loads to the subgrade and is particularly important where the bound layers are thin.

(b)   It provides a working platform for construction traffic and a compaction platform for the subsequent laying of bituminous mixtures.

(c)   It acts as an insulating layer in conjunction with the bituminous mixtures to protect the subgrade from frost.

(d)   It may provide a drainage layer to remove water from the pavement. The extent to which a sub-base material should be free draining depends on the amount of water likely to enter that layer. In thick layers of bituminous mixtures used for heavily-trafficked pavements, water volumes are likely to be small. It is therefore crucial to have a layer which is dense and which will withstand and transmit construction and traffic stresses. However, for thin pavement layers and the less dense surfacing materials commonly used in car parks, a more open-graded sub-base may be advantageous since water penetration is more likely, especially as a material deteriorates with age and the passage of occasional heavy vehicles may lead to the layer cracking. On fine-grained subgrade soils (clays, silts etc.) the use of a geotextile filter membrane is beneficial to prevent migration of soil fines into the sub-base material, destroying its drainage ability and structural characteristics.

3.2.2. Roadbase
From a structural point of view the roadbase is the most important layer in a flexible pavement. It is expected to bear the burden of distributing the applied surface load so that the bearing capacity of the subgrade is not exceeded. Since it provides the pavement with its stiffness and resistance to fatigue, the material

used must always be of a high quality. Flexible construction is the term used for a pavement where the roadbase is constructed of bituminous-bound materials. A pavement where the roadbase is constructed of a material bound with Portland cement and the surfacing layers are bituminous mixtures is known as a composite construction. Rigid construction is the term used for a pavement where both the roadbase and surfacing layers are constructed of materials bound with Portland cement. On very heavily-trafficked roads, including a number of motorways in the UK, continuously-reinforced concrete roadbase has been used to avoid the undesirable effect of joints.

The main requirements of a roadbase are that the material should be stiff, i.e. provide good load-spreading capabilities and, particularly in the lower part, be resistant to fatigue. The lower part should also have resistance to water degradation as it is likely that, particularly in later life, there will be significant quantities of water at the base of the bituminous layer.

3.2.3. Surfacing

The surfacing has the largest number of criteria to satisfy. It is frequently laid in two layers. The purpose of the lower basecourse or binder course is to provide a well-shaped surface upon which the upper wearing course is constructed and also, to an extent, to distribute the traffic loads over the roadbase. The wearing course must satisfy the following requirements

- (a)   resist deformation by traffic
- (b)   resist cracking as a result of thermal movement or traffic stresses
- (c)   be impervious, thus protecting the lower pavement layers (unless the pavement has been designed to have a permeable wearing course such as porous asphalt)
- (d)   be resistant to the effects of weather, abrasion and fatigue
- (e)   provide a skid resistant surface, either in itself, by the application of embedded pre-coated chippings or by surface dressing
- (f)   provide a surface of acceptable riding quality
- (g)   provide a surface with acceptably low levels of noise generation
- (h)   contribute to the strength of the pavement structure
- (i)   be capable of being laid and compacted throughout the whole year in the UK.

The methods by which these can be achieved are discussed in section 3.7 and chapter 2.

3.3. Road pavement design considerations

A pavement design flow chart is shown in Fig. 3.1.

3.3.1. Foundation design

It is crucial that the design of a road pavement structure is based on a sound evaluation of the degree of support offered to the pavement by the sub-grade. Failure of the road foundation is expensive to remedy as complete removal and reconstruction of the pavement is necessary.

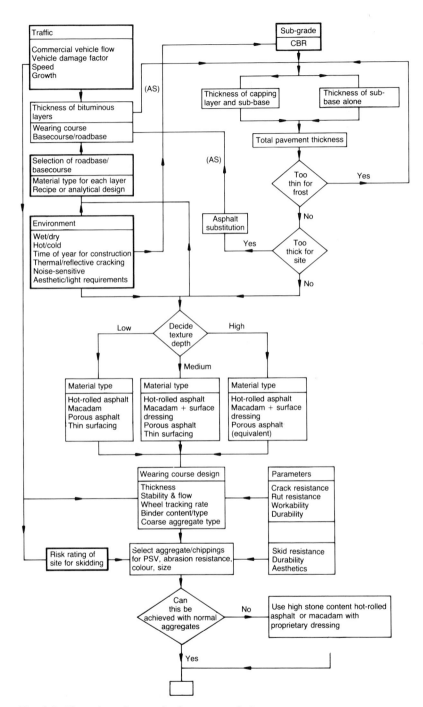

Fig. 3.1. Flow chart for standard pavement design

California Bearing Ratio

There are a number of test methods for evaluating the strength of a subgrade. Some of these are in situ measurements, some are strength tests carried out in the laboratory and others compare granulometry (particle shape, size and/or grading envelope) with bearing capacity. Some analytical pavement design programs use the elastic stiffness of a soil i.e. its strength, which is obtained using triaxial equipment. The majority, however, use empirical data based on known thicknesses to give acceptable performance over a period of time.

The most common sub-grade strength measurement is the California Bearing Ratio (CBR). While it may be criticized for lack of rigour, it has the great advantage of simplicity, cheapness and a vast array of engineering experience to support it. The equipment and test procedure for CBR tests are described in detail in BS 1377.[1]

In summary, the soil under test is compacted into a standard mould at the moisture content and dry density which, it is estimated, will be achieved in the prepared subgrade. Using a plunger 49.6 mm in diameter, the soil is penetrated at a standard rate of entry. A load ring measures the load required to cause penetrations of 2.5 mm and 5 mm which are expressed as ratios of the loads to cause the same penetration in a standard crushed rock material. Surcharge rings are added on the top of the mould containing the specimen to simulate the confining effect of the road pavement. Since the test is carried out with a relatively small plunger, the method demands the removal of aggregate particles greater than 20 mm. This can create a significant distortion in CBR results where such particles form a significant percentage of the soil. The CBR test is, therefore, most appropriate for fine-grained soils.

Some inaccuracies can be generated as a result of the re-orientation of the soil particles by the re-compaction in the CBR mould. This is more likely where CBR tests have to be carried out on bulk samples taken during investigation in a cutting, or in excavated soils where they are to be transferred to form the subgrade on an embankment. Laboratory CBR values tend to be higher than those in situ because of mould restraint factors. This is particularly true for less cohesive soils with low air voids.

It may be possible to estimate the CBR of a soil in situ either by carrying out a field test using the laboratory CBR testing equipment transferred to site and using a vehicle to provide the reaction or, more simply but less accurately, using a MEXE cone penetrometer. This is a sharp, pointed cone mounted on a spring-loaded device which is pushed into the ground by hand. The simplicity of the latter device enables a large number of tests to be carried out, and the use of extension spindles enables CBR values to be determined down to a depth of around half a metre. Considerable expertise is required where mixed granular/fine-grained soils are present.

A plate-bearing test may also be used to overcome the problems associated with larger aggregate size. The load/deformation relationship using a 760 mm dia. plate was used by Westergaard[2] to define the modulus of subgrade reaction K expressed in $MN/m^2/m$. The measurement of the K value presents practical difficulties on site because the required reaction demands a fully loaded lorry. The size of this plate creates a pressure bulb which goes down about half a metre below the surface, depending on the soil type. Smaller plates are sometimes used, particularly when evaluating the K value for sub-base and capping layer materials. The load/deformation curve is not, in general, linear, and on removal of the load at completion of the test, part of the deflection is found to be non-recoverable. Therefore, the test is not elastic and it cannot be closely related to the elastic modulus of the soil. There is an approximate relationship between the K value and CBR of the form

$$K = 13.2 \ \mathrm{CBR}^{0.63}$$

although this should be treated with caution.

Engineers in receipt of site investigation data must look very closely at the moisture content and location where the sample was taken. Site investigation data taken from trial pits in summer in the UK will probably over-estimate the CBR of the soil because the top metre of a site is likely to have been affected by climatic conditions. Also, it is probable that the equilibrium soil conditions beneath the road pavement will not be the same as that pertaining on site. Whenever site CBR values are below 7%, it is necessary to measure the moisture content/CBR relationship to ascertain the sensitivity of the soil to moisture content change.[3]

The need for a thorough, but costly, site investigation to determine the CBR should be assessed following an evaluation of the likely variation of soil within a contract. Small schemes on compact sites are less likely to be affected than major new works. For the purposes of the UK design method, sub-base and capping layer do not increase in thickness linearly with CBR. The soil has to be placed within ranges of CBR, which are currently: under 2%; 2−3%; 3−5%; 5−7%; 7−14%; over 14%.

Alternatively, the equilibrium suction index CBR can be determined from a knowledge of the plasticity index, as shown in Table 3.1. This is based on the soil suction method described by Black and Lister[4] for the fine-grained soils that are most common in the UK. Table 3.1 indicates reasonable estimates of equilibrium values of CBR for combinations of poor and average construction conditions, high water table and thick and thin pavements.

Table 3.1. *CBR for cohesive soils with a high water table*

Type of soil	Plasticity index: %	Construction			
		Poor		Average	
		Thin	Thick	Thin	Thick
Heavy clay	70 60 50	<2%	2%−3%		
Silty clay Sandy clay	40 30 20 10	<2%		>3%−<5%	5%−7%
Silt	—	<2%			

'Average' conditions exist where the subgrade is protected promptly and the site is well-drained with adequate falls. 'Poor' conditions exist where there is little or no subgrade protection and rainfall is likely to occur on a poorly drained site. A high water table is one 300 mm or less below formation level and is consistent with effective sub-soil drainage laid at a depth of 0.6 m. A thick pavement is 1.2 m in depth including any subgrade improvement layer, whereas a thin pavement is 300 mm.

To estimate the CBR during the life of the road, often known as the equilibrium CBR, the effect of any proposed drainage systems may also have to be considered. However, evidence indicates that such systems become less effective just when they are most needed i.e. when the road surface is beginning to fail and water ingress is increasing. A pessimistic view of the likely degree of saturation of the subgrade is, therefore, prudent. In practice, the moisture conditions during construction in the UK are likely to cause greater changes than the in-service condition. This may not always be the case because of the UK practice of carrying out roadworks throughout the year.

Black and Lister[4] showed that subgrades which become saturated during construction subsequently reach lower equilibrium strengths than those which are kept dry and well-drained and thus shorten the life of the road. It has been estimated that the critical life of roads built in extremely wet weather has been only about half of that for roads built under normal conditions, while roads built in dry weather can reach a life of double the average road. Even when a protective capping layer and sub-base have been laid, an increase in the moisture content of the soil can result after a period of wet weather. Therefore, there can be no reasonable assurance that the subgrade is adequately protected until the first layer of bound material has been laid. The aim should be to carry out carriageway construction as quickly as possible.

On sites where the CBR varies from place to place, the lowest recorded value should be used or, alternatively, appropriate designs should be provided for different parts of the site using the lowest CBR recorded in each part. It may be possible to remove soft spots and therefore ignore those low CBR values if such areas are not extensive.

While the contractor has responsibility under the contract for protecting and adequately draining the formation during construction, the designer should recognize the practicalities of site working with the soils encountered and make suitable provision in the design.

The degree and method of compaction can have a significant effect on the strength of a soil. Appropriate tests should be carried out in accordance with BS 1377[1] to determine the density/moisture relationship. These are BS light test 3.6 & 3.7, known colloquially as the Proctor test after the originator, BS heavy test 3.4 & 3.5, or vibrating hammer test 3.8. These enable the engineer to specify the moisture content range suitable for compacting the material.

The Proctor test was developed to simulate the plant of the period but the advent of heavier plant led to the development of the BS heavy test using the 4.5 kg hammer. The increased energy leads to a lower optimum moisture content and higher maximum dry density. For granular soils with a small silt/clay content, including granular sub-bases, the vibrating hammer test is particularly useful in defining the moisture content and density targets as these materials are commonly compacted by vibrating rollers.

Frost

Some UK soils are susceptible to frost action. This can have a differential effect along and across a pavement leading to cracking and, by loosening the subgrade and sub-base after thawing, significant foundation weakening. The extent of damage is related to the presence of moisture and the type of soil in the subgrade. Fine-grained soil and colliery shales may be particularly frost susceptible. A frost-susceptibility test[5] is available for checking the propensity of a soil to heave. In severe winters, frost has been known to penetrate 600 mm below the road surface, although this is only likely to happen once every ten years and current design guidance for most of the UK recommends that materials incorporated within 450 mm of the finished road surface should be non-frost susceptible. This requirement is not normally insisted on in footways where the cost of damage is unlikely to justify the higher initial cost of construction.

3.3.2. Sub-base and capping layer design

The capping layer is intended to provide a working platform upon which to lay and compact the sub-base without risk of contamination from a weak subgrade, i.e. one having a CBR value of less than 5%. Capping layer material is intended to be a low-cost option to make the best use of locally available materials or excavated materials from the site. The strength of a subgrade may also be improved by mixing in lime or cement in situ using stabilization techniques. The capping layer plus sub-base option is particularly suitable for new construction. The sub-base only option may, however, be more attractive for small areas of reconstruction or where public utility services would encroach within the pavement if a capping layer was used. The thickness of sub-base and capping layer in a combined layer are given in Table 3.2.

The initial function of the sub-base and capping layers is to provide a road for construction traffic but it also regulates and insulates the subgrade from frost damage. The required thickness will depend on the amount of construction traffic which, in itself, depends on the length of the sub-base which will be trafficked. It is therefore possible to reduce sub-base thicknesses where there is little or no construction traffic likely to use the layer e.g. haunching or reconstruction works. Table 3.3 outlines the suggested thickness of sub-base for haunching, local reconstruction, new works, or other major areas of reconstruction.

Table 3.2.   Thickness of combined sub-base and capping layer

CBR of subgrade (%)	Thickness		
	Sub-base	Capping layer	Total
14–30	150	—	150
5–14	225	—	225
>2–<5	150	350	500
<2	150	600	750

Table 3.3.   Thickness of sub-base (mm)

	CBR of subgrade, %				
Type of works	2–3	>3–<5	5–<7	7–<14	14+
Haunching of local construction	275	225	175	150	150
New works or major areas of construction	400	325	225	200	150

For CBR values less than 2% a capping layer from Table 3.2 should always be used. The ultimate aim of the sub-base layers is to achieve a strength which is at least equivalent to that of a material having a CBR of 30% at the top of the layer uniformly throughout the site.

In cases where adequate foundation cannot be provided using unbound material e.g. if public utilities' services are present, substitution of some of the unbound material with bitumen bound material is possible using the equivalence factors as found in Table 3.8.

Current UK practice does not measure the strength of sub-base layers in situ, although this practice is commonplace on the continent. Contractors following the recipe specifications for material composition, moisture content if any, compaction method and thickness, are deemed to have provided an adequate pavement foundation.

Engineers should evaluate and estimate CBR values of pavements prudently to avoid problems associated with the actual values being lower than anticipated. This is particularly wise in contracts which are let on a site rental basis where large claims payments may be generated by delay, and on some small projects where supervision levels may be low and the pavement left exposed or trafficked. In these circumstances it is wise to avoid materials which are susceptible to loss of strength when wet, such as Department of Transport Type 2 sub-base.

3.3.3.   Design of bound layers
The subgrade and its overlying pavement structure must support the number, weight and speed of load repetitions from the tyres of commercial vehicles during a design life of the road.

Traffic
The overall thickness of the pavement is primarily affected by the number, configuration and loading on the axles of vehicles. The maximum statutory limits for axle loads vary (see Table 3.4), so continental design guide information should be used with care. However, axle load measurements carried out close to UK ports have indicated that UK limits are regularly ignored.

The tyre pressure affects the stresses and strains developed at the surface in the surface layers of the pavement. A typical commercial highway vehicle tyre pressure is 0.5 MPa. Variations can have a significant influence on the performance of the materials used for road surfacing and airfield pavements

Table 3.4. Variation in permissible axle loading

Country	Permissible axle loading, t
UK	10.17
Germany, Japan, Netherlands	10.00
European Community (proposed)	11.05
Belgium, France	13.00

as it affects the contact area. This is particularly important in the design of low speed road and parking areas. The design of airfields and ports demands a more accurate knowledge of the axle loads and tyre contact areas of the aircraft or container handling machines.

Clearly, as the wheel load increases, the depth of pavement for a given composition must also be greater if the stress transmitted to the subgrade is not to be increased.

The effect of total axle loading was one outcome of the American Association of State Highway and Transportation Officials (AASHTO) road pavement tests carried out between 1959 and 1960 in the USA.[6] This developed an equation of the form

pavement damage = axle load to the power n.

The data concluded that for heavy wheel loads on pavements of medium and high strength, the value of n lay in the range $3.2-5.6$. In practice, however, a value of n=4 is very often used. The effect of a fourth power law is demonstrated in Table 3.5.[7]

In determining the structural life of a road pavement, the effect of cars and similar vehicles is negligible. Given a standard axle of 8 t, a 6 t axle causes damage equivalent to less than 0.3 of this standard axle, whereas a 10 t axle is equivalent in damaging power to 2.3 standard axles. For airfield construction, one Boeing 747 jumbo jet is equivalent to 137 heavy commercial vehicles.

It is therefore the heaviest axles in the stream of commercial vehicles which cause a disproportionately large amount of structural damage to a flexible pavement. Axle measurements have shown that while the overall weight of a truck may be within the legal limit, poor distribution of the load within the vehicle may make individual axle loads very high. Vehicles with air suspension are now permitted to have a higher axle load as the effect is counteracted. Conversely, vehicles with wide single tyres rather than dual-wheel assemblies have the maximum permitted load on each axle reduced from 10 t to 9 t in the UK because of the reduced contact area.

On conventional roads, normal traffic is mixed in composition and axle loads. The UK road design information has been developed from empirical studies using this spectrum of vehicles.[8,9] However, design engineers carrying out work on schemes where they believe an unusual spectrum of light/heavy vehicles is likely e.g. in bus lanes or industrial hardstandings, are recommended to carry out a detailed analysis of traffic. Thus, detailed calculations of standard axles per commercial vehicle can be undertaken.

Table 3.5. Equivalence factors and damaging power of different axle loads

Axle load		Equivalence factor
kg	(lbs)	
910	(2000)	0.0002
1810	(4000)	0.0025
2720	(6000)	0.01
8160	(18000)	1.0
9980	(22000)	2.3
10890	(24000)	3.2
11790	(26000)	4.4
12700	(28000)	5.8
13610	(30000)	7.6
16320	(36000)	15.0
18140	(40000)	22.8

The concept of millions of standard axles (msa) during the design life of a pavement has been developed as the key guideline for thickness design in standard design techniques. This has four factors. The first factor is the number of commercial vehicles per lane per day. On a maintenance scheme this may be measured by traffic sensing techniques or by carrying out a simple count of commercial vehicles over a short period and using expansion figures based on experience. For new schemes, figures must be developed using traffic modelling or experience of the traffic generated by a particular development e.g. Table 3.6. Where traffic is channelled e.g. at toll booths, bus stops and where pedestrian refuges or other obstructions are placed in the highway, this has the effect of increasing the number of axles passing over one part of the pavement since normal steering wander is constrained. In such circumstances, it may be necessary to double the traffic flow in these areas for the purposes of structural and wearing course design. On dual carriageways and motorways, as the volume of traffic increases, the proportion of the total flow in the slow lane falls from 80% to 50% as some heavy vehicles move into the overtaking lanes. Further information is provided in TRL Report LR 910.[7]

Table 3.6. Housing development traffic

Type of road	Traffic, cv/day	Standard axles/cv
Cul-de-sac	10	0.75
Estate and rural roads	50	0.75

Table 3.7.   Vehicle damaging factors

Commercial vehicle	Standard axles/vehicle
>2000	3.6
1000–2000	3.2
250–1000	2.5
<250	1.3

Second is the vehicle damaging factor i.e. the number of standard axles per vehicle. Typical information is available in Table 3.7 for a mid-life of year 2005. Mid-life is ten years after the road is opened to traffic for a bitumen-bound pavement or composite construction or 20 years for a rigid pavement. As traffic volumes increase, the proportion of heaviest vehicles in the flow also increases and therefore the vehicle damaging factors increase.[10]

The third factor is the anticipated growth in commercial vehicles over the design life. This may be a national figure e.g. 2%, or knowledge of the potential for commercial vehicle growth in the area served by the pavement.

The final factor is the design life of the pavement. Work reported in LR 1132[11] suggests that for pavements with a bituminous roadbase, the optimum design life that minimizes the cost, discounted over 40 years, is close to 20 years, with a 0.85 probability that roads will survive that period without requiring a strengthening overlay to extend their lives. In the case of wearing courses, life is based on the structural deterioration associated with a 10 mm rut in the wheel track. The total cost of a road over 40 years is not very sensitive to a comparatively large increase in design life because of the relatively low initial cost of providing extra life and the discounting of future costs of reconstruction. However, the consequences of specifying too short a life can be considerable, especially on heavily-trafficked roads where the traffic management involved in maintenance intervention can be a large percentage of the whole. Nevertheless, in an urban area a design life of less than 20 years may be justifiable where there are frequent openings by the public utilities.

A doubling of life from 500 cv/day to 1000 cv/day (in each direction) i.e. 10–20 msa, can be achieved with the additional thickness of 40 mm of material on a 280 mm thick pavement of bitumen-bound materials such as hot-rolled asphalt or DBM (asphaltic concrete). Therefore, clients must consider carefully whether such economies are justifiable given the scatter of results which the design guide is based on, and variable construction tolerances at the site. It is possible, for example, that a pavement designed for a 20 year life could need structural maintenance any time between 15 and 25 years from the time of construction.

Fig. 3.2 gives a simple nomograph to relate commercial vehicles per day to millions of standard axles outlining the following.

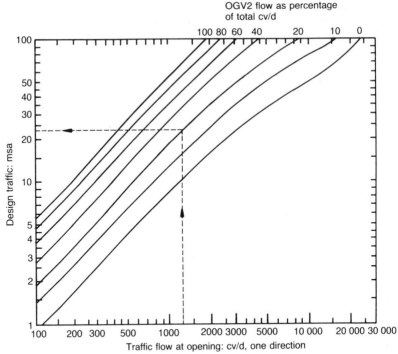

Fig. 3.2. Design traffic for flexible and composite pavements: dual carriageway

(a)   OGV2 are commercial vehicles exceeding 15 kN unladen weight with more than 4 axles (rigid or articulated). Total commercial vehicles include buses, coaches and 2 and 3 axle vehicles exceeding 15 kN unladen weight.

(b)   The high growth prediction from the National Road Traffic Forecast (NRTF 1989) is used, with average growth rates for commercial vehicles.

(c)   Year of opening is assumed to be 1995.

(d)   A 40 year design life is assumed with major maintenance after 20 years.

(e)   Curved lines allow for lane transfer at high vehicle flows.

Vehicle speed

The effect of vehicle speed may be important, particularly when speeds are below about 30 km/h e.g. in urban areas, on roundabouts, close to traffic lights and on steep uphill gradients. Where lateral forces are also present because of braking and turning, the design of the wearing course layer in particular requires more consideration.

The resilient modulus of bituminous mixtures depends on the rate at which they are loaded, with the modulus increasing with increased rates of loading i.e. reduced pavement deflection takes place with increasing vehicle speed. Typical speed values chosen for commercial vehicles and used in empirical design methods are in the range 60–80 km/h.

In designing wearing course materials, the formulae used for stability and flow or wheel tracking rate were based on an analysis of the performance of a high speed road[12] and a vehicle speed correction factor is required as discussed in section 3.7.2.

The analytical approach to the structural design of highway pavements requires a determination of the dynamic stiffness values for the various layers of the pavement and takes into account traffic speed. However, temperature is far more significant than vehicle speed e.g. increasing the vehicle speed by a factor of two produces the same change in effective modulus as a reduction in temperature of $2.5C°$. Consequently for structural layers, the speed value selected is not critical provided the loading frequency, i.e. the traffic spectrum discussed above is realistic.

3.4. Road pavement design methods

There are two basic approaches to pavement design, empirical and analytical. The empirical method is most commonly used worldwide. In the USA, design methods[6] were based on the AASHTO road trials which used normal forms of construction and normal road vehicles but at only one site in Illinois. These are shortly to be revised by SHRP which has installed numerous test sites throughout the USA and some in Europe to calibrate new design methods and materials tests in accelerated and long-term trials. In the UK, pavement design thicknesses have been determined from a long-term study of the actual performance of roads in service,[11] dating back to 1957.

Empirical design methods do not demand a detailed knowledge of engineering principles and are satisfactory as long as there is no significant change in the traffic composition or in the performance of materials. The method cannot be reliably used for extrapolation purposes. The rapid growth in commercial vehicles since both the AASHTO and UK trials made methods to extrapolate this field work necessary.

These methods had to be based on sound engineering principles using a theoretical analysis of the stresses in the various layers of a road pavement, the mechanical properties of the bituminous mixtures and the relevant climatic conditions. Such an analytical pavement design method should, in principle, be capable of dealing with any design solution. It would also provide the opportunity for incorporating new materials into road pavements e.g. those incorporating polymer modified bitumens with known mechanical properties different from those of traditional materials.

Current UK design practice follows the publication in 1984 of LR 1132[11] which forms the basis of the Department of Transport Design Standard HD14/87.[13] LR 1132 contains design recommendations, derived from the observation of the long-term performance of 34 sections of experimental road containing well-compacted dense roadbase macadam manufactured to BS 4987 (1973) and containing 100 pen bitumen. The deflection under a rolling wheel load moving at creep speed was measured using a deflection beam. The rut depth was also measured. While the scatter of results given in Fig. 3.3 is very wide, a design curve (Fig. 3.4) was drawn so that 85% of the roads will achieve the design life.

The performance of 29 sections of experimental road with rolled asphalt

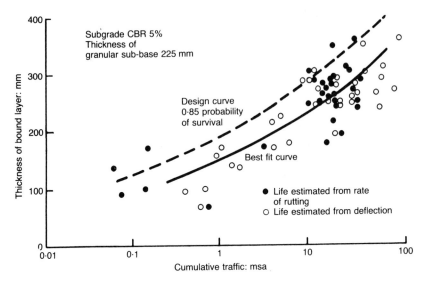

Fig. 3.3. Relationship between thickness and experimental roads with DBM roadbases[11]

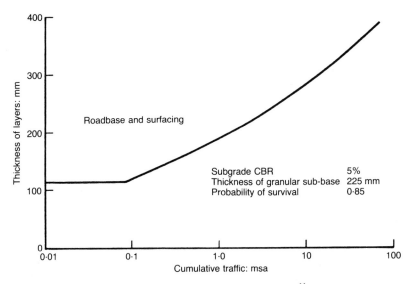

Fig. 3.4. Design curve for roads with bituminous roadbases[11]

roadbases was investigated in the same manner. The composition of the rolled asphalt covered a wide range of aggregates and gradings but all complied with BS 594 (1973). The performance of the roads with rolled asphalt roadbase was indistinguishable from those with dense bitumen macadam roadbase and therefore the same design rules can be used for both. However, this does not necessarily mean that both types of pavement deteriorate in the same manner.

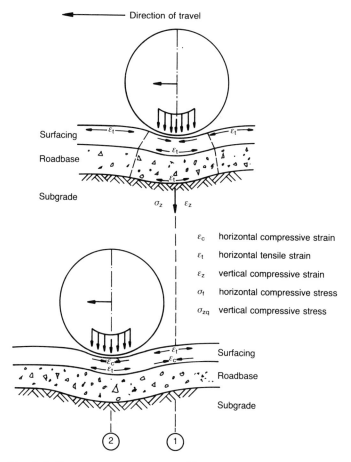

Fig. 3.5. Forces in a road pavement[15]

Macadam is more likely to fail through fatigue, failure of aggregate interlock or bitumen debonding, whereas asphalt is more likely to fail in deformation or embrittlement of the binder.

Work was also carried out by Lister, Powell & Goddard[14] using repeated loading tests on a full scale pavement in a circular road machine (Fig. 3.5), and the results correlated with laboratory fatigue testing. A fatigue-resistant material at the top and bottom of the bituminous layer is necessary, particularly on thin pavements where deflections beneath individual heavy vehicle wheels may be larger than on thicker pavements used for numerous heavy vehicles. As the wheel passes by, the top and bottom of the pavement moves rapidly from compression to tension as it returns to its original profile.

Information from this work was used to generate conservative design criteria for the maximum permissible vertical strain in the sub-grade and the maximum permissible horizontal strain at the bottom of the bitumen-bound material. This data provided the necessary input into an analytical pavement design computer

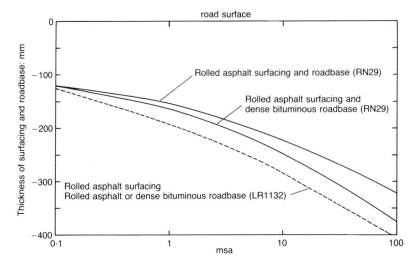

Fig. 3.6. Comparison of Road Note 29 with TRL report LR 1132 (1984): asphalt surfacing and bituminous base

program developed at the TRL so that the necessary extrapolation of the data could be carried out.

Current UK design guide information is, therefore, a combination of empirical data and analytical pavement design techniques. It also provides the opportunity for new materials and alternative designs to be used. The concept of probability of survival recognizes the variability in the road structure and the uncertainties of traffic forecasting.

The charts and graphs used in the completely empirical Road Note 29[9] method were based upon a rut depth of 20 mm or severe cracking at the end of the design life. However, strengthening such a pavement by overlay did not always result in an acceptable subsequent performance. LR 1132[11] recommends a more stringent critical condition of 10 mm rut depth or when cracks are first observed. The slightly different interpretation of the original field work carried out on the A1 at Alconbury Hill in 1957[16] and the Wheatley Bypass in 1964,[17] and the analytical design approach led to an increase in pavement thickness (Fig. 3.6). Overlaying at the onset of critical conditions maximizes the investment in the existing road pavement.

3.4.1. Analytical pavement design
Shell method
The Shell pavement design method is possibly the most widely known analytical approach. This design process, originally published in 1963[18] regards the pavement as a three-layer system in which the lowest layer represents the subgrade, the middle layer represents the combined unbound roadbase and subbase, and the uppermost layer includes all bitumen-bound materials above the roadbase.

BISAR

The Shell design guide approach, now published as a computer program BISAR[19] (bitumen stress analysis in roads), involves estimating the bitumen-bound and unbound layer thicknesses required to satisfy the following strain criteria

(*a*) the compressive strain in the surface of the subgrade i.e. if this is excessive, permanent deformation will occur at the top of the subgrade and this will cause deformation of the pavement surface

(*b*) the horizontal tensile strain in the bituminous bound layer, generally at the bottom of the layer which, if excessive, will cause cracking.

Permissible values for compressive subgrade strain were derived from an analysis of the AASHTO road test pavements. The permissible strain in the bituminous layer was determined from extensive laboratory measurements for various bituminous mix types. Other criteria taken into account include the permissible tensile stress or strain in any cementitious material in the middle layer and the integrated permanent deformations at the pavement surface due to deformations in the individual layers. The engineer is required to enter the following input data into BISAR: the subgrade modulus; the bituminous mix design; traffic volume, and mean annual air temperature. The program derives various combinations of pavement structures which satisfy the criteria given above.

Since bituminous layers will, in themselves, deform under traffic when some alternative structural designs have been selected, the next stage in the design process is to estimate the extent of the permanent deformation i.e. the rut depth, anticipated in the surface layer for each potential pavement structure during its design life. The permanent deformation directly attributable to the bituminous layer is estimated by the program from the thickness of the layer, the average stress in the layer, the stiffness of the layer, and a correction factor for dynamic effects.

Permanent deformation in the unbound roadbase and sub-base layers is also estimated. The estimated total surface deformation in each potential pavement design is compared with the allowable depth of rut to determine whether it is or is not acceptable. In practice, rutting of a wearing course can occur through plastic deformation in the layer itself and may not be a major factor in the overall structural performance of the pavement.

The BISAR program may also be used to back-analyze existing road pavements given the necessary information, based on the analysis and measurements of cores taken from the pavement layers for the purposes of overlay design.

Nottingham design method

Nottingham University,[15] under a research programme sponsored initially by Mobil which led to the publication of their design guide,[20] has developed an analytical pavement design method based on the philosophy that a pavement structure could be treated in the same way as the design of any other civil engineering structure. The method is to

(a)   specify the loading
(b)   estimate the proportions of the structure
(c)   consider the materials available
(d)   carry out structural analysis using theoretical principles
(e)   compare critical stresses, strains or deflections with allowable values
(f)   carry out an iterative process until a satisfactory design is achieved
(g)   consider the economic feasibility of the result.

The Nottingham design method[21] considers two modes of failure. Cracking is considered to arise from repeated tensile strain with the maximum value occurring at the bottom of the layer as shown in Fig. 3.6. The crack, once initiated, propagates upwards and causes a gradual weakening of the pavement.

Rutting arises from the accumulation of permanent strain throughout the pavement structure. If the vertical strain in the subgrade is kept below a certain level, excessive rutting will not occur, assuming adequate bituminous-mix design and compaction during laying.

To carry out the structural analysis, the pavement is divided into a bituminous layer with a value of stiffness modulus (E) and Poisson's ratio (v), both of which are variable, but v is generally close to 0.4. This layer overlays a granular sub-base where $E = 100$ MPa and $v = 0.3$. Since bituminous materials are not linearly elastic, the parameter of stiffness modulus is used in place of Young's Modulus.

The actual design process is summarized in the flow diagram in Fig. 3.7

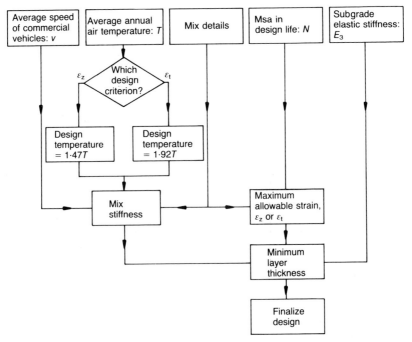

Fig. 3.7. Flow diagram for the Nottingham design method[15]

and a computer program has been devised to carry out the computations.[21] Details of the relevant properties of the materials are given in chapter 2.

Standard UK method
The standard UK pavement design method is based on LR 1132.[11] This is similar to the Nottingham method in that the following structural criteria need to be satisfied

(*a*) the sub-grade must be able to sustain traffic loading without excessive deformation, and this is controlled by the vertical compressive stress or strain at formation level

(*b*) bituminous materials used in roadbase design must not crack under the influence of traffic, and this is controlled by the horizontal tensile stress or strain at the bottom of the roadbase

(*c*) in pavements containing considerable thickness of bituminous materials the internal deformation of these materials must be limited, therefore their deformation is a function of their creep characteristics

(*d*) the load spreading ability of granular sub-bases in capping layers must be adequate to provide a satisfactory construction platform.

Both the Nottingham and TRL methods assume that the mode of failure is the generation of a fatigue tensile crack at the base of the bituminous layer, which is common sense and engineering logic, particularly for a brittle material like Portland cement bound material. However, it is a difficult hypothesis to prove in practice since by the time the crack is visible, it has, by definition, reached the surface.

Where the horizontal tensile strain at the bottom of the bituminous layer is not induced by bending, but by thermal movements at a crack in an underlying cement-bound material, crack propagation does not begin at the bitumen-bound/cement-bound interface but at the surface of the road itself.[22,23] This would appear to be as a result of the visco-elastic nature of the bitumen binder and its ageing characteristics. These can vary with the bitumen source and may be modified to an extent by the introduction of polymers into the bitumen. The tensile force introduced by thermal forces will be largest at the surface in material which has embrittled fastest because of environmental factors. Cracks will therefore be initiated at the surface and propagated downwards, exacerbated by water effects. They are generated in the winter months when thermal forces from material contraction are largest and the bitumen binder is at its most brittle because of cold weather. Cracks in high binder content materials such as hot-rolled asphalt may 'heal' during the warmer weather. Current analytical design methods do not take account of ageing characteristics or this healing effect directly, but the models may be calibrated against pavements of known characteristics and loading.

For the purpose of calibrating analytical techniques to in-service performance, specially-designed field trials are being undertaken in the USA by SHRP and in the UK by the TRL using a full-scale pavement test facility. Unfortunately, the latter is not able to model the effect of long-term ageing of bitumen binder which is so important in terms of the cracking and deformation of road pavements.

Analytical pavement design techniques require a knowledge of the fundamental properties of the asphalt mix which are

(*a*)   elastic stiffness (stiffness modulus)
(*b*)   fatigue resistance
(*c*)   tensile stress/strain relationship
(*d*)   deformation resistance (creep resistance)
(*e*)   the variation of these with loading frequency and temperature.

While information on these properties for a number of mixtures has been evaluated by researchers, until recently, equipment has not been available for this to be carried out on a routine basis. The measurement of fatigue resistance is still proving difficult. Nonetheless, a major step forward has been the introduction of the Nottingham Asphalt Tester (NAT) method[24] and the ARRB (Australian Road Research Board) method.[25] These have made the measurement of deformation and the stiffness modulus possible on a routine basis on both laboratory samples and on cores taken from a road.

The work being carried out under the auspices of the Comité Européen de Normalization (CEN), as discussed in chapter 11, means that in the future, the properties of bituminous mixtures will be specified in terms of end-product performance characteristics which will be suitable for providing the input for analytical pavement design procedures.

The administrative and financial structure for highway construction in the UK, which involves both central and local government, makes the installation and monitoring of field trials difficult. Thus the calibration of new design techniques and end-product tests, which is fundamental to the accurate prediction of in-service performance, is not executed as efficiently as it could be.

3.4.2.   Structural design for maintenance
Principles for maintenance intervention
While these pavement design techniques are satisfactory for new construction, when maintenance is required it is also necessary to evaluate the structural worth of the existing pavement so that appropriate maintenance can be designed. Currently 30% of all money spent on roads is spent on road maintenance. While some parts of the highway network have reached their anticipated design, nearly all roads have been subjected to a dramatic increase in damage due to increases in both the gross weight of vehicles and the heavier axle loadings. This means that many carriageways have carried substantially more loading than was allowed for in their design.

It is common practice to strengthen a road once significant rutting or cracking becomes evident at the road surface. However, when this occurs, the road construction has probably deteriorated to the extent that total reconstruction or replacement of all of the upper layers is necessary. Where damage to the existing pavement is large, its structural strength will have become so low that large overlays are required. This may demand significant work at the road edges and may not be possible over or under existing bridges.

As required by current UK design practice for new roads,[13] the cheapest solution is to carry out a thin overlay treatment before major surface deterioration is apparent and before the structural integrity of the pavement

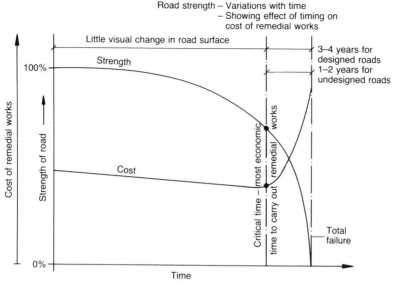

Fig. 3.8. Evaluation of the critical time for major road maintenance

is seriously compromised. The identification of the optimum time at which an overlay should be undertaken is known as the critical condition (Fig. 3.8). This 'doomsday curve' takes cracking and rutting in wheel tracks into account to define the critical condition.

Strengthening by overlay has the additional advantage that it is quicker, reduces traffic delays and is less sensitive to weather conditions as the road formation remains protected. Furthermore, the possibility of damage to existing drainage and kerbing during reconstruction is avoided.

Where a road, containing a lean-mix concrete roadbase, is suffering from reflective cracking which is visible at the road surface, a considerable thickness of unmodified bituminous mixture may be required to delay recurrence of this type of failure during the design life. The thickness is likely to be 200 mm or more. Alternative strategies, at increased cost, include a stress-absorbing membrane inter-layer (SAMI) e.g. a bitumen impregnated non-woven geotextile, or the use of a polymer modified binder in the bituminous mixture. The sequential tasks required to carry out a structural examination of a bituminous pavement are shown in Fig. 3.9.

The structural design of an overlay may be based on knowledge of the existing materials in the road pavement alone, or in conjunction with measurements carried out using a deflectograph or Benkelman beam (named after its inventor). Both of these methods are largely empirical. Information about an existing pavement can be obtained by the use of the falling weight deflectograph (FWD) and/or core removal from the existing road pavement and analysis using the wheel tracking test[26] or the NAT.[24] Armed with this information, an experienced engineer can employ analytical pavement design techniques to evaluate an existing road. This method is particularly suitable for sites with special problems, however it is not used as often as it could be.

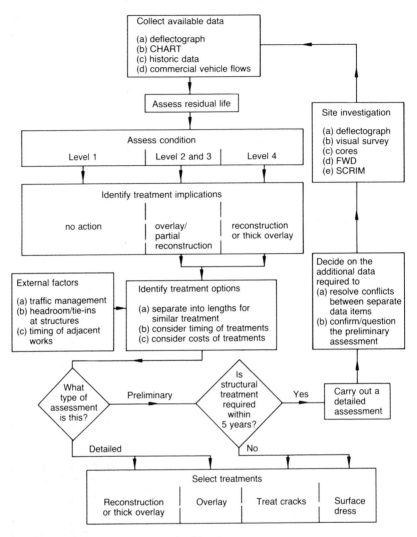

Fig. 3.9. Structural examination of a bituminous pavement

Site investigations which measure layer thicknesses, the condition of the existing pavement layers and the CBR of the sub-grade are likely to be cost effective for even the smallest scheme.

Simplified method for pavement analysis
This method, which is particularly suitable for lightly trafficked roads carrying less than 200 commercial vehicles per lane per day, is carried out by assigning to each layer of the existing pavement an equivalence factor which is used to convert the existing thickness of the layers to an equivalent thickness of DBM. The equivalence factors, which are based on those developed by the Asphalt Institute,[27,28] are given in Table 3.8.

Table 3.8.   Conversion factors for evaluating highway pavement materials

Category of material	Material conversion factor	
	Suggested value*	Range
Cement-bound material 1 (CBM1)	0.4	0.2–0.6
Cement-bound material 2 (CBM2)	0.5	0.3–0.7
Cement-bound material 3 (CBM3)	0.7	0.5–0.9
Cement-bound material 4 (CBM4)	0.7	0.5–0.9
Pavement quality concrete	1.7	1.5–1.9
Heavy duty macadam	1.1	1.0–2.0
Dense bituminous macadam	1.0	0.9–1.1
Hot-rolled asphalt base	1.0	0.9–1.1
Hot-rolled asphalt wearing course	0.8	0.8–1.0
Cold bitumen stabilization	0.8	0.8–1.0
Open-textured macadam	0.7	0.5–0.9
Porous asphalt	0.65	0.6–0.8
Wet-mix or dry-bound macadam	0.45	0.3–0.6
Type 1 granular sub-base on a subgrade with CBR >5%	0.3	0.15–0.4
Type 1 granular sub-base on a subgrade with CBR < or = 5%	0.2	0.1–0.25
Type 2 granular sub-base on a subgrade with CBR > 5%	0.2	0.1–0.25
Type 2 granular sub-base on a subgrade with CBR < or = 5%	0.1	0.05–0.15
Subgrade improvement material	0.1	0.05–0.15

*The suggested values relate to the stated materials in their common situations. In unusual situations, alternative values may be substituted but will not, generally, depart from the range. Departures from the suggested values will occur only when there is a specific reason for change. The factors in this Table have been derived by adapting those figures initially developed by the Asphalt Institute to ensure that they are comparable with relative thickness given in TRL Report 1132 and are based on elastic stiffness.

Additionally, two condition factors, CF1 and CF2, may be used to make due allowance for the condition of the material and the degree of any localized rutting or settlement.[29] The equivalence factor is multiplied by either or both of the condition factors for each of the relevant pavement layers. In the case of CF1, a degree of judgement is required in establishing the condition of the material, especially for those courses beneath the surface. In the absence of any evidence to the contrary, it may be assumed that the conditions at the surface continue through all the pavement layers. It is unlikely that they will be in a poorer condition, unless flint gravel bitumen-bound material has been used (which tends to strip if water gets into the layer), and this is, therefore, a conservative assumption. The values of CF1 and CF2 are given in Tables 3.9(a) and (b).

Benkelman beam/deflectograph investigation
The Benkelman beam and the deflectograph work on the same principle. The deflection beam measures the elastic response to the surface of a pavement as it is subjected to a known load. In the Benkelman beam, the load is 14000 lb (6350 kg).[30] It is then assumed that this elastic response can be used to

Table 3.9(a). Value of Condition Factor CF1

Condition of material	CF1
As new	1.0
Slight cracking	0.8
Substantial cracking	0.5
Fully cracked or crazed and spalled	0.2

Table 3.9(b). Value of Condition Factor CF2

Degree of localized rutting or localized settlement (mm)	CF2
0−10	1.0
11−20	0.9
21−40	0.6
> 41	0.3

determine the overall condition and the remaining life of the pavement. It measures the combined elastic response of all the various paving layers including the sub-base and subgrade.

However, the stiffness of a bituminous layer changes with the temperature of the binder. Consequently, the magnitude of the deflection measured will vary according to the temperature of the constituent bituminous layers. For simplicity, the temperature is taken at 40 mm below the surface and corrected to a temperature of 20°C. However, deflection measurements are best made when the pavement temperature is close to 20°C and measurements outside the range 10−30°C should be avoided because of the great temperature correction which would be required.

Cores from the pavement have to be taken to measure the thickness of the pavement layers and the material type although in the future Ground Radar may provide sufficient information. Unbound and Portland cement-bound materials have a significantly different elastic response i.e. they have a very different stress/strain relationship to bitumen-bound materials. Cement-bound materials have significantly less deflection for the same load than bitumen-bound materials. The different layers and their thicknesses must be identified so that the appropriate design charts can be used.

The deflection beam is slow and expensive for collecting large masses of data. Depending on the condition of the surface, deflections are measured at intervals of 12−25 m, permitting, at most, 1 km of pavement to be surveyed in a working day. The deflectograph works at a speed of 2 km/h and permits 10−12 km of pavement to be surveyed on a working day. This was developed in the mid 1960s by the Laboratoire Central des Ponts et Chausses (LCPC)

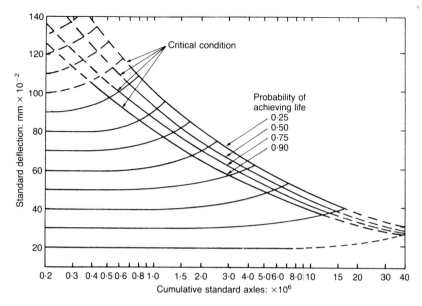

Fig. 3.10. Relationship between standard deflection and life of pavements with bituminous roadbases[32]

in France, and modified for use in Britain by the TRL and operated in accordance with a Department of Transport Advice Note.[31]

The machine consists of a special lorry and a deflection beam assembly located beneath it, and an automatic recording system. When measurements are taken, the beam rests stationary on the carriageway aligned between the front and rear axles of the vehicle. Deflections are measured as a pair of rear wheels, each loaded to 3.175 t, passes by with the beam between the wheels. After measurements are taken, the beam is lifted and moved forward and laid down again approximately 3.8 m further forward. It is then ready to take further readings as the vehicle continually moves forward slowly. The measurements taken are converted to equivalent deflection beam results. A 100 mm $\times$ 10^{-2} Deflectograph result is approximately equal to 115 mm $\times$ 10^{-2} deflection beam. All relevant data is automatically corrected for temperature and recorded by an on-board computer.

Studies, made largely on the A1 at Alconbury Hill in 1957,[16] have enabled deflection histories to be built up for a variety of construction types and these have been consolidated into a suite of TRL design charts[32] which relate standard deflection to cumulative standard axles. The chart for flexible pavements is reproduced in Fig. 3.10. This is valid for flexible pavements surfaced with bituminous materials in common use in the UK and is a means of evaluating the probability that a road will achieve its designed life span. Normally a probability of 0.85 is used to agree with the criterion for new roads. Further charts in LR 833[32] permit the design of a suitable overlay to allow the road to reach its design life. A typical chart is given in Fig. 3.11.

The thickness derived from the design chart applies to overlays using hot-

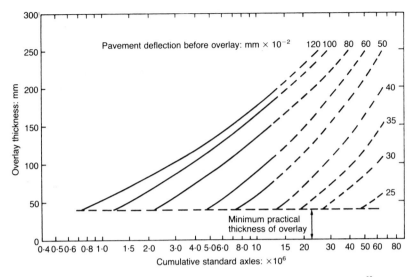

Fig. 3.11. Overlay design chart for pavements with bituminous roadbases[32]

rolled asphalt wearing course and dense coated macadam or hot-rolled asphalt roadbase (if necessary). For other materials the equivalence factors in Table 3.8 may be used. The design charts recommend a minimum thickness of 50 mm for asphalt overlays because of the difficulty in predicting the performance of thinner layers and because of the uncertainty in the observed structural performance of thin overlays.

The value of CF2 is substantially reduced if significant rutting has taken place in the wearing course. The presence of materials containing delayed set (storage grade) or soft binders, such as those used by public utilities for temporary patching, in an existing road, may provide inadequate support to a proposed overlay. This could cause excessive strains therein and premature fatigue failure. It may, therefore, be necessary to remove this material and introduce a basecourse or upper roadbase material beneath the overlay.

Around the time deflection measurements are carried out, the structural condition of the pavement should be assessed visually using the Computerized Highways Assessment Ratings and Treatment Survey method (CHART) and trial pits/cores taken as described later in this section. If there is a gross mismatch between the measured deflections and the visual condition of the pavement, a check should be made to rule out mistakes in measurement and ensure that the analysis of the deflection results has been properly carried out using the appropriate table from LR 833[32] for the roadbase type. The tables are only intended for sub-grade CBR values in the range 2.5%−15% and values outside this range should be applied with caution.

Possible errors in measurement
It is possible that inaccurate estimates of deflection can occur when pavements are sited on very strong or very weak sub-grades, or are tested at temperatures

close to the extremes of the working range, or the types and properties of the bitumen-bound material and sub-base vary. It is assumed that the deflectograph has been properly maintained and calibrated and the software verified.

The deflection of a pavement at a given point does not remain constant with time but increases slowly as the bituminous binder hardens with age. Bitumen hardening occurs quite rapidly during the first months after laying and thereafter more slowly. Deflection measurements are therefore not recommended when a pavement is less than a year old.

The reasons for this hardening and the rate at which it occurs are primarily related to the degree of access of oxygen to the bitumen i.e. it will occur less in dense binder-rich materials like hot-rolled asphalt. It is also affected by bitumen/aggregate/filler interaction probably by absorption, volatile evaporation, slow crystallization of waxes, steric hardening or photo-oxidation by ultraviolet light. The penetration test carried out on a normal 50 pen bitumen recovered from road pavements can fall from 35 pen on laying to below 20 pen in five years. Specimens of bitumen, initially at 50 pen, from hot-rolled asphalt wearing course with voids of 2−4% were found to have fallen to between 14 and 21 pen after storage at 60°C for two months. The propensity of a bitumen to hardening by oxidation and volatile evaporation depends on the source of the bitumen and the method used to refine it.

Different bitumens and bitumen-bound mixtures also respond to speed of loading in different ways. Although the Benkelman beam and deflectograph both have a slow loading speed this has been calibrated at the high speed A1 contract[16] for the mixtures incorporated therein.

Despite bitumen hardening, as Fig. 3.10 shows, the deflection does not diminish during the life of the pavement. This is concluded from the fact that the phenomenon was not observed with the materials used at Alconbury Hill, which test forms the basis of the chart. The mechanism which governs the relationship between deflection and traffic is a complex interaction of strengthening due to binder hardening and/or compaction of the layers, weakening of the pavement due to cracking (both thermally and fatigue induced) and, to a lesser extent, shearing of unbound layers. The shape of the deflection/traffic curve will depend on the relative influence of 'positive' and 'negative' effects which may be slightly different on the site under consideration from the standard chart.

Traffic considerations
The use of the standard design charts (Fig. 3.11) requires knowledge of the cumulative number of standard axles carried by the existing highway, since its construction or since its last structural maintenance, and also the traffic expected during the anticipated life of the overlay. The methods for predicting future traffic levels are the same for maintenance as they are for the construction of new roads.

Overlays predicted from deflectograph investigations often show significant variations along the length of the road. These diversities have to be resolved by the design engineer so that a reasonably consistent and cost-effective overlay treatment can be prescribed. Extremely weak locations requiring very large overlays are normally the subject of full reconstruction locally, while very strong areas are still overlaid to maintain levels and provide a consistent wearing surface.

FWD investigation

The FWD is based on the principle of subjecting the pavement to an impact load of duration, intensity and area of impact similar to that of a travelling vehicle. The deflection of the road is measured at the point of impact and also at exact distances away from that point so that the deflected shape of the road can be measured in addition to the total deflection. Using computer software, an experienced operative armed with knowledge of the existing pavement construction i.e. the thickness of the bitumen-bound or cement-bound layers, the sub-base and the sub-grade CBR, can deduce the elastic stiffness of the pavement and, sometimes, the individual layer. This information can then be fed into one of the existing analytical pavement design programs to determine the overlay required to produce the same characteristics as a new pavement of known traffic-carrying capacity. Work is still in progress to validate this process.

The FWD is more accurate in measuring the effects of cracks in a road pavement than the deflectograph, particularly those exhibited in a cement-bound roadbase. However, the deflectograph can be modified to measure deflections at each side of a crack in order to investigate load transfer.

Trial pit and core investigation

Both deflection beam and FWD require information on the existing pavement layer thicknesses, in the case of the former, so that the appropriate design chart can be used and in the case of the latter, to ensure that essential information is fed into the computer program. The number of holes/cores to be taken depends on the perceived variability of the site assessed from data collected and/or visual inspection.

In addition, cores provide an opportunity for carrying out dry wheel tracking rate tests[26] which measures the deformation resistance of the wearing course. This is essential if the main reason for maintenance intervention is to restore skid resistance on the grounds of safety. Cores also enable measurement of the deformation resistance and elastic stiffness of the bound-base materials. If these are inadequate, it may affect the choice of overlay treatment. The wheel tracking rate test uses 200 mm dia. cores. Deformation and elastic stiffness determinations using the NAT[24] uses cores of diameter 100 mm, 150 mm or 200 mm.

A 200 mm dia. core also provides enough material for granulometric analysis and measurement of the quantity and penetration of the bitumen recovered on analysis. If the penetration of the bitumen is below 15 pen it is possible that the long-term fatigue properties of the layer will be inaccurate, leading to premature fatigue failure of the pavement. Information on binder properties is also essential for recycling of bituminous materials either in situ in the case of wearing course using the remix/repave processes,[33] or by planing and re-using bitumen-bound pavement materials, generally at a central hot-mix plant.[34]

A visual inspection of cores identifies any unusual combination of materials which may affect the performance of the pavement or may have affected the deflection beam result. These could be that

(a)   a layer of soft surface dressing may be providing a slip plane within the pavement so that the layers do not act homogeneously

(b)   a layer of binder-rich material may have been introduced as a regulating layer which is providing inadequate support to the upper layers leading to alligator cracking

(c)   materials may be poorly coated, stripped, honeycombed or poorly compacted.

The investigating engineer must decide whether or not it is prudent to leave such materials in place. Larger core holes and trial pits permit the thicknesses of the sub-base and capping layer to be investigated and granulometric analysis carried out to ensure that the layers can carry out a drainage function (if necessary) and have not been over-contaminated with fines to the extent that the materials have become non-load bearing or frost susceptible.

Finally, using either a MEXE probe or other in situ CBR device and/or by taking a sample for reconstruction in the laboratory, the CBR of the subgrade must be assessed together with its moisture content. As the sub-base/subgrade is being investigated, the presence of water may be noted in the hole, the source of which should be investigated. In any event, drainage which is more than 20 years old will probably need replacement, particularly where combined surface water (over-the-edge) and ground-water (french drains) systems have been installed, as the latter tends to introduce fines into the filter material.

Drainage investigation

An evaluation of the moisture content/CBR relationship of the subgrade will indicate the potential benefits of reducing the possibility of water entering and remaining under the carriageway. Such measures include: sealing the central reservation (if any) with blacktop or surface-dressed sub-base material; installing kerbing and gullies to feed surface water straight into a piped system; restoring or improving the effectiveness of the french drainage (sub-surface water drains) system by replacing the drainage material; introducing additional pipes and filter material or installing fin drains in embankment slopes. Fin drains may also be installed where pavement widening or hard shoulder construction has caused a discontinuity in the pavement layers which is damming up the lateral flow of water to the drains.

A closed circuit television survey will check for pipe damage and root infiltration into drains to verify their effectiveness. Where an increase in impermeable area is envisaged, pipe sizing may have to be checked.

In view of the large increase in pavement thickness required when CBR values drop below 7%, it is nearly always cost effective to carry out subgrade drainage improvement measures. It may even be the case that with these in place, and given some time to take effect, a subsequent survey will indicate no structural overlay is required.

SCRIM investigation

The sideways coefficient routine inspection machine (SCRIM)[35-37] is now used on a regular basis to monitor the skid resistance of a pavement. In the UK, regulations have been introduced[38] which dictate the steps to be taken if the SCRIM value falls below that permitted for that site. This does not automatically mean that surface treatment has to be carried out. The desirability of treatment also depends on the number and types of accidents at the site. However, slippery road warning signs should be erected at such locations.

Normally, the cheapest maintenance treatment for inadequate skid resistance is surface dressing using a binder type, chip size(s) and polished stone value (PSV) appropriate for the site.[39] However, before such treatment is carried out, it is prudent to investigate the other data collected about the state of the pavement since it may be that advancement of planned maintenance involving resurfacing, with or without structural strength improvement, would be a more cost-effective solution.

The principle of SCRIM is that a smooth tyre is held at 20° to the road surface and, while the road surface is kept wet by a bowser (mobile water tank), the loaded wheel is towed at 50 km/h. The Sideways Friction Coefficient (SFC) is the force at right angles to the plane of the inclined wheel (the sideways force) expressed as a fraction of the vertical force acting on the wheel. The SFC varies with the climate (summer and winter) and, to a lesser extent, with the weather conditions prevailing immediately preceding the SCRIM run. The mean summer value is normally used, this being the mean of three readings taken between June and September. The most important factors affecting the skid resistance of the road surface are the aggregate PSV, a measure of microtexture, the volume, speed and mass of traffic, and the extent to which this traffic is flowing freely or is required to brake or turn sharply. Typically a change of one unit in PSV corresponds to a change of .01 unit of SFC.

Recent studies indicate that the wear resistance, (abrasion resistance), of an aggregate affects its ability to maintain its PSV. Softer aggregates wear away constantly exposing new surfaces to the tyre. Harder aggregates, particularly those with a homogeneous molecular composition, may, given time, polish to give a very low SFC. The PSV test simulates, consistently, a polishing regime typical of a medium-trafficked road. The actual performance in service will depend on the complex interactions of traffic, polishing and wearing away of the surface and the restorative effect of frost, the nature and volumes of wet and dry detritus, and rainfall. Guidance given in the UK[38] is, therefore, only valid for UK climate and typical UK aggregates and is constantly reviewed. Studded tyres are not used in the UK.

In the absence of SCRIM information the skid resistance of a surfacing on a straight site may be predicted from the following

$$SFC = 0.24 - \frac{0.663C_v}{10^4} - \frac{1}{PSV10^2}$$

Where C_v = number of commercial vehicles per lane per day
     PSV = polished stone value of the chippings

It has been confirmed that skidding resistance as measured by SCRIM is normally independent of texture depth but may be reduced when the macrotexture is unusually low e.g. on a worn surface dressing. Both macrotexture, as discussed in section 3.9.3 of this chapter, and skidding resistance, as measured by SCRIM, are essential for safe roads.

A device known as the UK Griptester has been shown to correlate well with SCRIM. A small wheel is towed behind a van and is continually braked generating a horizontal force at the road surface permitting the coefficient of friction to be measured. The compact nature of the device, the fact that it can operate at a range of speeds and in wet or dry conditions makes it particularly useful for the measurement of friction on restricted sites, roundabouts or even road markings.

CHART

The examination of surface defects is usually carried out using CHART[40] involving procedures which are fully documented for use by survey technicians.[41] Hand-held computers which permit large quantities of data to be captured are used on site. For the purposes of determining a suitable maintenance treatment, the number and spacing of surface transverse, longitudinal and multiple cracks and fretting, and the wheel track structural and non-structural rutting are observed and recorded.

The deflectograph and CHART information may be used to make a preliminary condition assessment, although the degree of confidence will reflect the limited information available on the pavement layers below the wearing course. Following the site investigation using cores or trial pits, a more competent assessment can be made when all the information is available. Four levels of courses of action can be taken to strengthen the pavement

(*a*)   Level 1: no action
(*b*)   Level 2: overlay with crack sealing
(*c*)   Level 3: overlay with crack sealing and local reconstruction
(*d*)   Level 4: total reconstruction.

For levels 2 and 3, there may be circumstances where the wearing course would provide inadequate structural support for the overlay, or, particularly where cement-bound material is present, it may be necessary to remove some of the wearing course and basecourse before overlaying in order to remove excessively cracked material. Remix/Repave may be a suitable alternative.

To speed up the survey process, crack surveys may now be carried out at high speed, typically 50 km/h, using vertically-mounted video cameras, with rut measurements taken simultaneously using a bank of lasers mounted transversely across the front of a road monitoring vehicle.

An important characteristic for skid resistance is the texture depth (macrotexture) of the road surface and this can be measured by means of a laser. All the data may now be recorded together with location information using a high speed road monitoring vehicle[42] so that, when linked with skid resistance measurements from the SCRIM machine, a complete picture of the state of the existing network can be maintained within a computerized pavement management system. A comparison of year-on-year information enables the correct moment for maintenance intervention using the doomsday curve (Fig. 3.8) to provide the most cost-effective method of maintaining a high quality road network. It also permits generalized data (Fig. 3.12) to be presented and changes compared over a period of time.

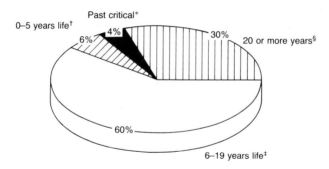

* Past critical — advanced state of structural deterioration, beyond economic repair
† 0–5 years — limited structural strength, best time for economic maintenance
‡ 6–19 years — some structural strength, could require surface treatment
§ 20 or more years — adequate structural strength, could require surface treatment

Fig. 3.12. Remaining life of category A roads — 1989

Digitized map-based presentations permit engineers to prioritize maintenance treatments and rapidly identify sections of the network which require surface dressing or crack treatment, or major or minor structural improvement. Combined with historical cost data, the maintenance of the highway network can be accurately planned and adjusted to suit political and financial constraints. The flow chart shown in Fig. 3.9 gives the sequential actions for maintenance management.

3.5. Selection of materials for pavement layers

The selection of materials is included in the pavement design flow chart (Fig. 3.1. Investigations have shown that DBM and hot-rolled asphalt, typically specified in the UK using BS 4987: Part 1[43] and BS 594: Part 1,[44] can have very similar stiffnesses.[11] Research work and field observation have also shown that hot-rolled asphalt has a higher resistance to fatigue[45] than DBM. These properties are fully covered in chapter 2.

A road pavement flexes under HGV traffic loading (Fig. 3.5). Therefore, the bottom and top layer of the pavement are more subject to cyclic tension and compression. This may lead to fatigue failure. As in a structural beam, the central section is primarily responsible for load spreading and this is where stiffness modulus is of prime importance.

For heavily-trafficked roads (in excess of 3000 cv/lane/day (80 msa)),[13] the Department of Transport has recognized the benefits of sandwich construction i.e. placing hot-rolled asphalt roadbase as the lower roadbase and one or more layers of DBM as the upper roadbase. Each course is compacted to an end-result density specification. Sandwich construction has the additional advantage that an impermeable and water-resistant material, hot-rolled asphalt, is placed on the sub-base, which may otherwise become waterlogged in later life as the sub-base becomes less effective and/or pavement cracking permits excess quantities of water to enter the sub-base. Engineers have recognized that the use of hot-rolled asphalt lower roadbase is possible down to traffic levels of 100 cv/day. Indeed, since the traffic axle load diminishes only slightly on thinner pavements, the deflection introduced into the pavement may actually increase as the pavement thickness reduces. At present, sandwich construction has not been monitored long enough to determine whether its theoretical advantage provides a genuine increase in pavement life. However, in parts of the country where hot-rolled asphalt is not significantly more expensive than DBM, it is an option to be considered.

The introduction of heavy duty macadam (HDM) or DBM with 50 pen rather than 100 pen bitumen has provided a material with increased stiffness modulus.[46] The benefits of this are a reduction of the pavement thickness as indicated by the equivalence factors in Table 3.8 and, hence, a slight reduction in overall cost by extending the life of the structural layer. The material is slightly more expensive than DBM because of increased temperatures and mixing time required, and there are some difficulties with laying. Given the problems of estimating traffic intensity, it may be prudent to take advantage of HDM for increased life rather than savings in the form of reduced layer thicknesses.

3.6. Asphalt substitution

When carrying out reconstruction, it may be necessary to reduce the thickness of the pavement to avoid disturbance of statutory undertakers' apparatus. To make sure that the subgrade is not over-stressed, this must be achieved by increasing the effective stiffness of the pavement layers. The weakest layers in a pavement are the sub-base and capping layer materials (with equivalences of 0.2 and 0.3 in Table 3.8). Therefore, if these layers can have their stiffness increased, a reduction in thickness should be possible. This can be achieved by substituting the sub-base with bitumen-bound roadbase[47] using the equivalence factors in Table 3.8 i.e. 10 mm DBM is equivalent to about 30 mm type 1 sub-base over material with a CBR exceeding 5%.

Leech and Nunn[47] showed that, provided thick-lifts of hot-rolled asphalt roadbase were used, adequate compaction could be achieved on material having a CBR as low as 3%. Where the subgrade has a CBR value lower than 5%, a granular sub-base material 150 mm thick should be introduced to increase this value and provide a compaction platform with the underside protected from fines migration by means of a non-woven geotextile, if necessary.

3.7. Wearing course design

Choice of material

Wearing courses can be any one of a number of materials which satisfy the criteria listed in section 3.2.3. The design of wearing courses is included in design flow chart (Fig. 3.1). Each has certain advantages as shown in Table 3.10,[48,49] which is based on the performance after six years in service, and are only approximate.

The characteristics of the material can be significantly affected by the properties of the constituents, which can be adjusted by mix-design and recipe changes. The selection of an appropriate wearing course is governed by the relative importance that the design engineer places on each of the various factors. A high emphasis on noise or spray reduction, for example, will lead to the selection of porous asphalt despite its shorter life. A high emphasis on

Table 3.10. Relative properties of bituminous mixtures

Characteristic	Effectiveness ratio (1 = poor, 6 = very good)				
	Open-textured macadam	Porous asphalt	Dense macadam	Hot-rolled asphalt	Surface dressing
Structural strength	2	3	3	6	1
Rut resistance	2	3	4	4	1
Crack resistance	3	3	3	4	1
Skid resistance, low speed	5	5	5	4	5
Skid resistance, high speed	4	5	4	4	5
Texture depth	2	2	3	4	6
Spray suppression	3	6	1	2	3
Low noise	4	4	3	2	1
Ease of application	3	3	3	2	5
Area treated/unit cost	2.5	1.5	2	1.5	6

a long life, high skid-resistant surface will lead to the selection of hot-rolled asphalt with pre-coated chippings to provide the high texture depth (macrotexture) currently required for higher speed roads (in excess of 80 km/h) or dense bitumen macadam or stone mastic asphalt for low speed roads (less than 50 km/h) where a lower texture depth may be tolerated.

Skid resistance
Since the Department of Transport is primarily responsible for the design and maintenance of the UK high speed, high traffic volume network, its primary concern is with the specification and use of hot-rolled asphalt. In South-East England, aggregate with high PSV for skid resistance has to be imported by rail or sea and is, therefore, expensive. Hot-rolled asphalt with pre-coated chippings is relatively less expensive than macadam because of the smaller proportion of aggregate in the former. It is, therefore, also used on sites with slower speeds where a lesser texture depth is permissible on the grounds of capital cost. However, as traffic speeds fall and loading time increases, the deformation caused by an individual vehicle pass increases, and problems can be caused by premature rutting of hot-rolled asphalt in the vicinity of traffic lights, in the channelled flow in narrow urban streets, in bus lanes, and even on the uphill stretches of inter-urban roads carrying high numbers of commercial vehicles. As traffic turns sharply, the sideways force from three-axle bodies or heavy commercial vehicles exerts extreme sideways force on all surfacings. Chippings in hot-rolled asphalt must be well embedded and dense macadam well compacted to withstand such forces.

Table 3.11. Design criteria for hot-rolled asphalt wearing course from BS 594: Part 1[44]

Traffic flow, cv/day	Stability, kN	Flow, mm
< 1500	2–8	5
1500–6000	4–8	5
Over 6000	6–10	5 or 7

Marshall stability

Hot-rolled asphalt wearing course primarily depends on the binder for its properties. The Marshall design method, incorporated in BS 594: Part 1,[44] is used in the UK to establish the required percentage of bitumen (normally 50 pen) in the mixture and to measure the stability and flow of the proposed mix. Work by Jacobs[12] on the A30 Winchester bypass related Marshall stability and flow to the UK criteria of a 10 mm maximum rut in 20 years. Table 3.11 is taken from BS 594: Part 1[44] and is based on this work. In addition, work by Lees[50] recommends a minimum flow of 2 mm to avoid brittle mixtures.

In the north and west of the UK, manufactured sand (crushed rock fines) is used in the mixture. This ingredient makes the achievement of these design requirements readily possible, using 50 pen unmodified binder, although there is a substantially increased risk of the material failing in fatigue. During winter months in Scotland and the north of England, 70 pen bitumen is substituted to provide adequate workability and still achieve the requirements. The surface temperature in summer is lower than in South East England, so that warm weather deformation is less likely as a consequence.

In South East England, manufactured sand is not readily available, so natural sand is selected on economic grounds. The effect of this change on stability is shown in Fig. 3.13. Achieving higher stability can, therefore, only be achieved by use of a harder (e.g. 35 pen) bitumen. Work is in progress to evaluate a TLA/bitumen blend of 35 pen, initially to ascertain if the better ageing properties reported by Nunn[23] make this a viable proposition. Using 35 pen material in a layer 50 mm thick makes it particularly difficult in winter with increased risk of high winds to get chip adhesion and embedment.

The other alternative is to alter the rheology (composition) of the bitumen by the use of polymers such as Ethyl Vinyl Acetate (EVA). However, their visco-elastic properties make measurements by the Marshall test a less reliable prediction of field performance.[51] In general, a material using polymer-modified bitumen needs about two-thirds of the Marshall stability value of an unmodified bitumen to achieve the same rut resistance measured by the wheel tracking test.[52]

Wheel tracking rate

A test was developed by TRL in 1977 to simulate the in-service rutting of hot-rolled asphalt. This has been published as a British Standard draft for development[26] and the results correlate with field performance in practice.

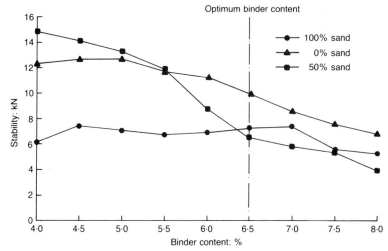

Fig. 3.13. Variation in stability of hot-rolled asphalt with change in proportion of sand/crushed rock fines

The test involves subjecting a slab of material mixed in the laboratory, either using a re-heated sample from the paver (not recommended), or a 200 mm dia. core taken from the wearing course after laying, to a rolling wheel load which traverses the specimen at a constant temperature (normally 45°C) under a standard load. Rutting under the wheel is measured over a period of time. Maintaining a constant test temperature is very important for the reproducibility and the accuracy of the result.

The test can also be used for design purposes using plant-processed dry aggregates, mixed with binder and compacted in the laboratory to known density using a vibrating hammer.

TRL developed a relationship based on work on the A30 Winchester[12] which related the results from the test to the number of commercial vehicles travelling at 60 mph required to form a 10 mm rut at the end of a 20 year life.

$$\text{WTR} > \frac{14\ 000}{C_v + 100} \cdot \frac{x}{60}$$

Where WTR = wheel tracking rate
C_v = the number of commercial vehicles per lane per day
x = speed of commercial vehicles in mph

The formula has been found to correlate reasonably well with in-service performance for sand-based hot-rolled asphalts especially when its linear speed correction factor is included to allow for the difference between measured traffic speeds and 60 mph.

Skatkowski,[53] however, showed that the wheel tracking rate measured on hot-rolled asphalt i.e. its deformation resistance, is very sensitive to sand type, binder and filler contents. For example, a 1% increase in both binder and filler increased the wheel tracking rate of a laboratory mix from 2.2 mm/h to

Table 3.12. Comparison of wheel tracking rates of hot-rolled asphalt wearing course with 50 pen bitumen, EVA modified binders, SBS modified binder and 50/50 TLA/50 pen bitumen blend

Binder	Penetration value	WTR at 45°C (mm/h)
50 pen	56	3.2
70 pen + 5% EVA (45/33)	57	1.0
200 pen + 7% SBS	84	0.7
50/50 TLA + 50 pen bitumen	24	1.6

4.5 mm/h. Work by Choyce and Woolley in 1988 with EVA polymer[54] and work by Denning and Carswell in 1981 using SBS[51] polymer has shown the extremely beneficial effects on rut resistance of modifying bitumen with polymers when used in hot-rolled asphalt.

In designing wearing course mixtures for large numbers of heavy vehicles, slow, traffic, or highly stressed sites, the engineer may specify a particular polymer-modified mixture or may decide to give the contractor a choice of other binders such as those given in Table 3.12. Hot-rolled asphalt mixtures with the same wheel tracking rate should give approximately the same rut-resistance performance and allow the economics of the various options to be determined by the contractor. Polymer modification reduces the temperature susceptibility of the bitumen and, therefore, probably also improves the low temperature characteristics thus reducing the propensity of hot-rolled asphalt wearing course made with such binders to crack in winter.

Hot-rolled asphalt mixtures containing TLA have been found to have improved deformation resistance at low speeds compared with unmodified bitumen.[55]

The formula for wheel tracking rate has been correlated with the performance in-service, of a 40 mm thick hot-rolled asphalt surfacing containing 30% coarse aggregate. There is insufficient evidence to assess the suitability of this formula for DBM-type mixtures. In DBMs, the method of deformation is not one of plastic flow of the binder/filler/sand matrix during hot weather, but rather the re-orientation of the particles as the softer bitumen ceases to be an adhesive and becomes a lubricant in hot weather. A much larger wheel tracking test has been developed in France using a road tyre and 1000 kg mass to evaluate such mixtures.

Surface texture (macrotexture)
Roe et al.[56] have compared the surface texture (macrotexture) on large sections of highways of all kinds in the UK, using the high speed texture meter,[57] with accidents at those sites. This has shown that texture is an important criterion in reducing skid resistance accidents, not only on wet road surfaces but also when they are dry. It is postulated that this is because the rugous (wrinkled) texture provides grip for the tyre as the rubber moulds around it. Information from three road networks was used to produce the relationship shown in Fig. 3.14. This shows that roads with texture depths (sensor measured)

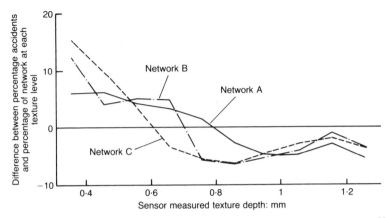

Fig. 3.14. Difference between accident sites and network texture distributions[56]

of 0.6–0.8 mm have a smaller proportion of accidents than those carriageways with texture depths smaller than this range. A sensor-measured depth of 0.6–0.8 mm relates to a reading of around 1 mm obtained by the sand patch test.

All three networks consisted of approximately 15% of class B or unclassified roads with traffic speeds in the 60 km/h range so that the effect of texture depth applies at a wide range of traffic speeds.

While excessive texture significantly increases noise and rolling resistance, the presence of a rugous surface from relatively large chippings also provides a significant reservoir in cases where fatting-up of the binder or embedment of the chippings occurs in hot weather. There can be considerable difficulties in laying a good mat of chipped hot-rolled asphalt wearing course. This is particularly true where the proportion of pre-coated chippings is high or where laying is undertaken in restricted areas or in windy conditions or where roads cannot be closed for the work to be carried out. Despite the difficulties, chipped hot-rolled asphalt wearing course, when laid properly, provides an extremely safe surface throughout the life of the pavement. However, the lack of sufficient pre-coated chippings means that significant areas of binder/filler/sand matrix are visible on the road surface. This will significantly lower the texture and measured SCRIM values.

In parts of the UK where crushed rock aggregate is available in significant quantities and at reasonable cost, much of it with a PSV adequate for the traffic level and risk rating of the site, the cost of macadam mixtures can be less than that of hot-rolled asphalt which is normally higher because of its high bitumen content. In such cases engineers must evaluate whether the possibility of increased accidents outweighs the benefits of using hot-rolled asphalt at higher laying costs.

Macadam mixtures can satisfy all the requirements set out earlier in this chapter except macrotexture, provided they are made with adequate bitumen content and are laid and compacted using vibratory compaction to ensure sufficiently low air voids as to be impermeable. This is normally achieved where air voids are within the range 2−7%. Air voids of less than 2% in the field may lead to fatting-up of the bitumen binder in hot weather. Contractual and climatic arrangements permitting, a DBM, surface dressed after laying with a suitably-designed bitumen spray and chippings in accordance with Road Note 39,[39] can satisfy all these requirements. It may be necessary, however, to erect slippery-road warning signs if the road is opened to traffic between the completion of the DBM wearing course and the application of the surface dressing chippings as the binder film on the aggregate will prevent full development of skid resistance from exposure of the aggregate microtexture.

Microtexture
The microtexture of the surface in contact with the tyre is the principle characteristic at the surface which determines the friction available at this interface. Macrotexture primarily provides drainage paths so that tyre rubber has access to this microtexture. There are currently no acceptable dry methods of measuring this property in-service. The PSV of the aggregate at the surface is deemed to provide a surrogate measure of friction that will be achieved by a new surface, the actual value will only be available after some time using SCRIM or the Griptester (see section 3.4.2). The selection of PSV for a particular site is given by the Department of Transport.[38,58]

Noise
Both hot-rolled asphalt and DBM surfaces are designed to be impermeable and grip has to be provided at the tyre/road interface by the rugocity of the surface. This automatically leads to a noisy road surface in both dry and wet weather. If the water is removed from the surface of the road pavement by other means, only a texture depth suitable for dry surface grip need be provided. Porous asphalt (pervious macadam/drainage asphalt) is, as its name implies, capable of absorbing the rainfall and draining it to the road edge through the thickness of the layer itself. Trials[59] have shown that this material provides a surfacing approximately 3 dbA quieter, i.e. a 50% reduction, than a standard hot-rolled asphalt surfacing, with even greater reductions in wet weather when water hitting vehicle wheel arches is a major source of noise. This is a significant and perceptible reduction in noise.

Following trials on the A38 in 1984,[59,60] mix design criteria for porous asphalt were developed. These are described in chapter 2 and in published work,[61] and show that such a layer can be designed, mixed, delivered and laid on site to produce a durable wearing surface. Durability is reduced in the long term as a result of binder hardening, leading to the possibility of fretting failure and by the clogging up of the pores within the mixture, initially designed to be about 18−25%, by tyre rubber and other detritus. As the pores clog, water increasingly flows over the face of the material and the resulting texture depth is then inadequate in wet weather. Engineers in Austria have developed a modified suction sweeper to clean clogged material.

Porous asphalt also requires careful edge detail design to dispose of the water flowing through the material, and a different salting regime in winter to that which is used on conventional asphalt.

Current design guides recommend the same PSV of aggregate in the material as would be used in a pre-coated chipping in hot-rolled asphalt wearing course. Although this may increase the price of material, this may be offset by the faster rate at which porous asphalt can be laid by a smaller surfacing gang. It is also less sensitive to wind chill as it comes out of the paver almost fully compacted.

Noise reduction measures are already high on the agenda of pavement design engineers on the continent and, as noise becomes an increasingly political issue in the UK, it is probable that the use of porous asphalt will also increase.

3.8. Footway surfacing

The requirements set out in section 3.2.3 may be modified for footway surfacings to

(*a*)   resist penetration by concentrated loads e.g. stiletto heels

(*b*)   be impervious, thus protecting the sub-grade and sub-base which in this case forms the main structural layer

(*c*)   be resistant to the effects of weather and abrasion

(*d*)   provide a slip-resistant surface

(*e*)   provide a surface of acceptable riding quality for babies in buggies with small wheels

(*f*)   be capable of being laid and compacted in small areas and in small quantities and in thin layers.

These requirements can generally be met by a macadam type mixture with dense grading, small aggregate size, with aggregate of a good abrasion resistance, using a binder-rich mixture probably incorporating 200 pen or 300 pen bitumen. A 6 mm nominal size material complying with Clause 5.2 of BS 4987: Part 1[43] has been found satisfactory laid 20 mm thick. The underlying material may also be a dense macadam or hot-rolled asphalt depending on cost, in accordance with BS 4987: Part 1[43] or BS 594: Part 1,[44] laid approximately 35 mm thick.

While the requirements for the sub-base on a footway are modest in structural terms, the pavement still has to provide a resistance to the compacting forces introduced by vibrating rollers to ensure that the resulting surface is impermeable. Crushed rock[62] or demolition waste to Type 1 sub-base gradation or planings from cold planing operations both provide a suitable material. Gravel materials such as Type 2 sub-base materials[62] can be satisfactory provided they are kept reasonably dry during storage and placing.

3.9. Car park surfacing design

Because the structural effect of cars on a road pavement is negligible, the requirements for a car park surface are more akin to a footway than a carriageway. However, as cars turn and brake, sometimes at speed, lateral forces are set up within the upper part of the pavement structure. A DBM or hot-rolled asphalt basecourse 90 mm thick overlaid by 60 mm of bituminous material on a granular sub-base should therefore be adequate to resist such forces. The softer binders used in hand-laid bituminous mixtures for footways are likely to mark and distort in the summer in the UK, so 100 pen bitumen for machine laying work and 200 pen bitumen modified with EVA co-polymer for hand laying work are preferable.

The heaviest load on the surfacing is likely to be during construction using conventional wheeled pavers, so a sub-base layer as described in Table 3.2 to bring the CBR of the formation to a value close to 30% should be provided.

If such a lightweight paving is used it is essential that heavy vehicles are excluded from the pavement by means of a physical barrier. Where this is not possible or where access is required for waste disposal vehicles, school buses, or other delivery vehicles, a proper carriageway design based on a sensible estimate of the number of commercial vehicles is necessary, despite the increased cost this incurs. This, will satisfy requirements for up to 30 commercial vehicles per day.

References

1. BRITISH STANDARDS INSTITUTION. *Methods of test for soils for civil engineering purposes*. BSI, London, 1990, BS 1377: Part 4.
2. WESTERGAARD H. M. *Stresses in concrete pavements computed by theoretical analysis*. Publ. Works Wash., 1926, **7**(2).
3. DAVIS E. H. The California Bearing Ratio method for the design of flexible roads and runways. *Geotechnique*, 1949, **1**(4).
4. BLACK W. P. M. and LISTER N. W. *The strength of clay fill sub-grades; its prediction in relation to road performance*. Transport Research Laboratory, Crowthorne, 1979, LR 889.
5. ROE P. G. and WEBSTER D. C. *Specification for the TRRL frost heave test*. Transport Research Laboratory, Crowthorne, 1979, Suppl. R 829.
6. HIGHWAY RESEARCH BOARD. *The AASHTO Road Test Report 5*. Nat. Acad. Sci., Washington DC, Special Report GIE, 1962, Publication 954.
7. CURRER E. W. H. and O'CONNOR M. G. O. *Commercial traffic; its estimated damaging effect 1945−2005*. Transport Research Laboratory, Crowthorne, 1979, LR 910.
8. LIDDLE W. J. Application of AASHTO road test results to the design of flexible pavement structures. *Proc. Int. Conf. asphalt pavements*, 1962, Ann Arbor, Mich.
9. DEPARTMENT OF TRANSPORT. *A guide to the structural design of pavements for new roads*. HMSO, London, 1970, Road Note 29, 3rd edn.
10. ADDIS R. R. and WHITMARSH L. R. A. *Relative damaging power of wheel loads in mixed traffic*. Transport Research Laboratory, Crowthorne, 1981, LR 979.
11. POWELL W. D. *et al. The structural design of bituminous roads*. Transport Research Laboratory, Crowthorne, 1984, LR 1132.
12. JACOBS F. A. *A30 Winchester Bypass, the performance of rolled asphalts designed by the Marshall test*. Transport Research Laboratory, Crowthorne, 1983, LR 1083.
13. DEPARTMENT OF TRANSPORT. *Structural design of new road pavements*. DTp, London, 1987, HD 14/87.
14. LISTER N. W. *et al.* A design for pavements to carry very heavy traffic. *Proc. 5th Int. Conf. structural design of asphalt pavements*, 1982, Delft (Univ of Mich).
15. BROWN J. F. and BRUNTON J. M. *An introduction to the analytic design of bituminous pavements*. 1986, 3rd edn.
16. THOMPSON P. D. *et al.* The Alconbury Hill experiment and its relation to flexible pavement design. *Proc. 3rd Int. Conf. structural design of asphalt pavements*, 1982, University of Michigan, Ann Arbor.
17. DEPARTMENT OF TRANSPORT. *Road Research (1964)*. HMSO, London, 1964, Annual Report of Transport Research Laboratory.
18. SHELL INTERNATIONAL PETROLEUM COMPANY. *Shell design charts for flexible pavements*. SIPCo., 1963.
19. DE JONG D. *et al. Computer program BISAR, layered systems under normal*

and tangential surface loads. Koninklijke/Shell Laboratorium, Amsterdam, 1973, Report AMSR.0006.73.

20. MOBIL OIL COMPANY. *Asphalt pavement design manual for the UK.* MOCo., 1985.

21. BROWN S. F. and BRUNTON J. M. Computer programs for the analytical design of asphalt pavements. Institution of Highways and Transportation, London, 1984, *Highways and Transportation*, **31**.

22. BURT A. R. M4 Motorway, a composite pavement, surface cracking. Institution of Highways and Transportation, London, 1987, *Highways and Transportation*, **34**.

23. NUNN M. E. An investigation into reflection cracking in composite pavements. *Proc. Int. Conf. reflective cracking*, Liege, 1989.

24. COOPER K. E. and BROWN S. F. Development of a simple apparatus for the measurement of the mechanical properties of asphalt mixes. *Proc. Eurobitume Symposium*, Madrid, 1989, 494–498.

25. SHARP G. and ALDERSON A. *Standard method for the laboratory determination of the elastic modulus of asphalt.* Australian Road Research Board, Report Summary ARR 210.

26. BRITISH STANDARDS INSTITUTION. *Method of determination of the wheel tracking rate of cores of bituminous wearing courses.* BSI, London, 1990, Draft for Development DD 184.

27. AMERICAN ASSOCIATION OF STATE HIGHWAY AND TRANSPORTATION OFFICIALS. *Guide for the design of pavement structures.* AASHTO, Washington DC, 1986.

28. ASPHALT INSTITUTE. *Thickness design — asphalt pavements for highways streets.* Asphalt Institute, Maryland, 1984, MS-1, AI.

29. BRITISH STANDARDS INSTITUTION. *Guide for structural design of pavements constructed with clay or concrete block pavers.* BSI, London, 1991, BS 7533.

30. KENNEDY C. K. *et al. Pavement deflection: equipment for measurement in the UK.* Transport Research Laboratory, Crowthorne, 1978, LR 834.

31. DEPARTMENT OF TRANSPORT. *Deflection measurement of flexible pavements. Operational practice for the deflection beam and the deflectograph.* DTp, London, 1983, Advice Note HA 24/83.

32. KENNEDY C. K. and LISTER N. W. *Prediction of pavement performance and the design of overlays.* Transport Research Laboratory, Crowthorne, 1978, LR 833.

33. DEPARTMENT OF TRANSPORT. *In situ recycling: the repave process.* DTp, London, 1982, Advice Note HA 14/82.

34. DEPARTMENT OF TRANSPORT. *Recycling of road pavement materials.* DTp, London, 1992, Advice Note HA 49/92.

35. HOSKING J. R. and WOODFORD G. C. *Measurement of skidding resistance. Part 1: guide to the use of SCRIM.* Transport Research Laboratory, Crowthorne, 1976, LR 737.

36. HOSKING J. R. and WOODFORD G. C. *Measurement of skidding resistance. Part 3: Factors affecting the slipperiness of a road surface.* Transport Research Laboratory, Crowthorne, 1976, LR 738.

37. HOSKING J. R. and WOODFORD G. C. *Measurement of skidding resistance. Part 3: Factors affecting SCRIM measurements.* Transport Research Laboratory, Crowthorne, 1976, LR 739.

38. DEPARTMENT OF TRANSPORT. *Skidding resistance of in-service roads.* DTp, London, 1987, Advice Note HA 36/87.

39. DEPARTMENT OF TRANSPORT. *Design guide for road surface dressing.* HMSO, London, 1992, Road Note 39.

40. WINGATE P. J. F. and PETERS C. H. *The CHART system of assessing the structural needs of highways.* Transport Research Laboratory, Crowthorne, 1975, SR 153 UC.

41. DEPARTMENT OF TRANSPORT. *CHART 5. An illustrated site manual for inspectors*, DTp, London.

42. COOPER D. R. C. *The TRRL high speed road monitor. Assessing the serviceability of roads, bridges and airfields.* Transport and Road Research Laboratory, Crowthorne, 1985, RR 11.

43. BRITISH STANDARDS INSTITUTION. *Coated macadam for roads and other paved areas. Specification for constituent materials and for mixtures.* BSI, London, 1993, BS 4987: Part 1.

44. BRITISH STANDARDS INSTITUTION. *Hot-rolled asphalt for roads and other paved areas. Specification for constituent materials and asphalt mixtures.* BSI, London, 1992, BS 594: Part 1.

45. GODDARD, R. T. N. *Fatigue resistance of a bituminous road pavement design for very heavy traffic.* Transport Research Laboratory, Crowthorne, 1982, LR 1050.

46. LEECH D. A. *A dense coated roadbase macadam of improved performance.* Transport Research Laboratory, Crowthorne, 1982, LR 1060.

47. LEECH D. and NUNN M. E. *Substitution of bituminous roadbase for granular sub-base.* Transport Research Laboratory, Crowthorne, 1992, RR 58.

48. BROWN J. R. *The performance of surfacing for maintaining bituminous roads: (A1 Buckden (1975—1982)).* Transport Research Laboratory, Crowthorne, 1986, RR 70.

49. SKATKOWSKI W. S. and BROWN J. R. *Design and performance of pervious wearing courses for roads in Britain (1967—76).* Highways and Road Construction Int. Jan/Fen, 1977.

50. LEES G. Asphalt mix design for optimal structural and tyre interaction purposes. *Proc. 6th Int. Conf. Structural Design of Asphalt Pavements* 1987, Ann Arbor Michigan.

51. DENNING J. and CARSWELL J. *Improvements in rolled asphalt surfacings by the addition of organic polymers.* Transport Research Laboratory, Crowthorne, 1981, LR 989.

52. CARSWELL J. *The effects of EVA modified bitumen on rolled asphalts containing different fine aggregates.* Transport Research Laboratory, Crowthorne, 1989, RR 122.

53. SZATKOWSKI W. S. Rolled asphalt wearing courses with high resistance to deformation. The performance of rolled asphalt surfacings. *Proc. Inst Civ. Engrs Conf.*, London, 1980, 107—122.

54. CHOYCE P. W. and WOOLLEY K. G. EVA modified binders. London, 1986, *Highways* **56** (Jan).

55. OTTO H. Wirtschaftlichkeit von asphaltbauweisen. *10th Int. Conf. Trinidad Lake Asphalt*, Berne, 1992.

56. ROE P. G. *et al. The relation between the surface texture of roads and accidents.* Transport Research Laboratory, Crowthorne, 1991, RR 296.

57. ROE P. G. and WEBSTER D. C. *Measurement of the macrotexture of roads Part 3: the development of the high speed texture meter.* Transport Research Laboratory, Crowthorne, 1992, RR 297.

58. DEPARTMENT OF TRANSPORT. *Requirements for aggregate properties for new bituminous surfacings.* DoT, London, 1992, Advice Note HA 45/92.

59. DAINES M. E. *Pervious macadam: trials on trunk road A38 Burton by pass, 1984.* Transport Research Laboratory, Crowthorne, 1986, RR 57.

60. DAINES M. E. *Trials of porous asphalt and rolled asphalt on the A38 at Burton.* Transport Research Laboratory, Crowthorne, 1992, RR 323.

61. DEPARTMENT OF TRANSPORT. *Porous asphalt.* DoT, London, 1992, Advice Note HA 50/92.

62. DEPARTMENT OF TRANSPORT. *Specification for highway works.* HMSO, London, 1992, **1**.

4. Production: processing raw materials to mixed materials

4.1. Introduction
This chapter considers the types of rock used in manufacturing bituminous mixtures in the United Kingdom and examines the methods used to quarry these materials. Different bituminous mixture production methods are then discussed and their advantages and disadvantages summarized.

4.2. Raw materials
The raw materials, or coarse aggregates, used in roadstone production must be hard, clean and durable. The term aggregate is defined in BS 6100: Part 6[1] as 'granular material, either processed from natural materials such as rock, gravel or sand, or manufactured such as slag'. The UK is fortunate since there is an abundant supply of different aggregates. Its availability and relatively low cost contribute to the supply being somewhat taken for granted, although it is obviously not a limitless resource.

The roadstone industry uses crushed rocks, sand and gravel and artifical materials such as blast furnace slag, or steel slag as the main ingredients for the construction of flexible pavements. The decline in the steel industry has limited the availability of slag.

4.2.1. Crushed rocks
Crushed rocks are classified relative to their geological origins and these are termed igneous, sedimentary and metamorphic.

Igneous
Rocks formed from a hot liquid, or magma, cooling either above, or beneath, the earth's surface. These rocks have a crystalline structure. Examples are basalt, diorite, dolerite, gabbro, granite.

Sedimentary
The action of ice, wind or water on a rock's surface, or organic remains, results in the formation of smaller particles. These particles are deposited on lake or river beds, the sea floor or depressions in the earth's surface. This sediment accumulates, becomes buried and compaction together with cementation over the years removes the air and water to form sedimentary rocks. Examples are gritstone, limestone, sandstone.

Metamorphic

Heat, pressure and/or chemical substances deep beneath the earth's surface act on existing igneous and sedimentary rocks to form a new type called metamorphic rock. Examples are gneiss, hornfels, marble, quartzite, slate.

4.2.2. Sand and gravel

Sand and gravel are the products of mechanical and chemical weathering of rocks such as granite and sandstone and are, therefore, a type of sedimentary deposit. Weathering by heat, ice, wind and moving water transports the rock fragments, this movement results in further mechanical working and transforms the sand and gravel into rounded and smooth irregular shapes. The deposits are found along the sea shore, along river valleys and on the sea floor.

Gravels are usually classified by their major rock type e.g. flint gravel. Gravels can also contain a variety of several different rocks.

4.2.3. Blast furnace slag and steel slag

Although limited in availability, BS 594: Part 1[2] and BS 4987: Part 1[3] allow the use of a process manufactured material as the coarse aggregate in a flexible pavement. The British Standards allow the use of '(c) blast furnace slag conforming with BS 1047[4] and (d) steel slag, either electric-arc furnace or basic oxygen slag, which shall be weathered until it is no longer susceptible to falling. The compacted bulk density shall be between $1700 \, kg/m^3$ and $1900 \, kg/m^3$ when tested in accordance with BS 1047: 1983.'

4.3. Resources

The UK has a wide variety of quality aggregates with specific and different characteristics. A simplified geological survey is shown in Fig. 4.1. This shows the distribution of the major rock formations throughout the country. The map indicates that the South East of England is predominantly sand and gravel, clay and chalk with very little hardrock.

The South East is a highly populated region creating considerable development and therefore has a need for quality materials. Although some roadstone production can be met from local resources, the largest percentage is transported from the South West and to a lesser degree from the Midlands and by sea from Ireland and Scotland.

The major suppliers of aggregates have developed sites containing large asphalt mixing plants with capacities from 400 t/h up to 800 t/h in and around the London area. These plants have rail off-loading and storage facilities and they can have aggregates delivered in loads of up to 5000 t capacity by rail as well as deliveries by truck.

The majority of aggregate movement is by road transport since most mixing plants are local to the rock source. With the advent of fewer but larger capacity asphalt mixing plants the use of both rail and sea transport is likely to increase.

4.4. Methods of producing aggregates

There are numerous types of rock making up the igneous, sedimentary and metamorphic classifications. Each rock has its own characteristics e.g. aggregate abrasion value, aggregate crushing value, aggregate impact value,

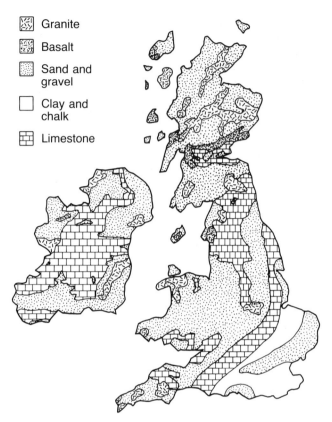

Granite

Basalt

Sand and gravel

Clay and chalk

Limestone

Fig. 4.1. Simplified geological map of the UK showing major distribution of rock types used in the roadstone industry

polished stone value. These properties are discussed in more detail in chapter 1. These inherent qualities together with local geological experience and engineers' expertise within the quarrying industry govern the type and size of equipment selected for an aggregate processing plant.

Plants processing materials for roadmaking have to produce a wide range of sizes with restrictions on shape, the preference being cubical. They should have the ability to operate at varying capacities and be flexible enough to alter the product sizes when the need arises. Therefore, the aim is to produce aggregates to the required sizes, with an optimum shape, at the capacity needed and in an economical manner. Crushed rocks used in roadstone production have a maximum flakiness index of 45,[2] whereas rolled in pre-coated chippings have a maximum allowed flakiness index of 25.[2] The flakiness index is an indication of the aggregates shape, the lower the figure the more cubical the product.

The stone is reduced in size by a crusher. There are four types of crusher: jaw, gyratory, impact and cone. To produce the required size and optimum shape the rock is reduced in stages and plants can have one, two, three or more stages depending on the application. The shape is controlled by the type of crusher employed, the reduction ratio, the crusher setting, the rate of feed and the physical characteristics of the stone itself. The best cubical shape is achieved at, or around, the crusher setting and with a choke feed on the cone, jaw and gyratory machines.

Plants are usually rated at an operational efficiency of 80% to allow for blockages within the crushers, or breakdowns. Therefore, if 240 t/h of material are required the plant would have to be rated at 300 t/h minimum in order to meet this demand.

Aggregate processing plants can be roughly divided into three categories: hardrock, limestone and sand and gravel plants. The industry loosely uses the term limestone crushing and screening plants to describe plants for soft and medium hard limestones where the silica content is less than 5% Limestones can be hard with much higher silica contents.

4.4.1. Hardrock crushing and screening plant

An example of the flow diagram for a hardrock crushing and screening operation is shown in Fig. 4.2.

In preparation to process rock, the overburden, consisting of topsoil, subsoil, clay and small loose rocks, is removed and holes are drilled at predetermined distances for the blast charge. Blasting generally occurs once per day, some of the larger quarries can blast up to 30,000 t or more, at a time. Great care and skill are needed when blasting, not only to ensure that the blast itself is safe and contained but to produce a stable quarry face afterwards. Quarries tend to be deep and therefore, in the interests of safety, the face is worked in levels, or benches. Typically these benches can each be 15 m deep but this depends on the rock characteristics and the depth of the rock seam. Many workings go below the water table and in these instances a pumping system is used.

The maximum size feed material to the plant illustrated is an 800 mm cube. Cost-effective primary blasting means there will be a small percentage of rock which is too large for the primary crusher. Secondary breaking is the term used for reducing these oversize rocks. Examples of this process are further drilling and blasting, the drop ball, the hydraulic hammer, the impact breaker and the rock breaker. The blasted rock is loaded and hauled from the face to the plant by large dump trucks which can have a capacity up to 100 t.

The rock is fed into a lump stone chute equipped with a chain curtain to control the flow. The material discharges onto a reciprocating tray feeder with a built-in grizzly section. A grizzly is a term used by crushing plant manufacturers for a vibrating grid with bars in one direction only. The grizzly bars are set at 50 mm spacing, allowing the natural fines to by-pass the crusher, or go to a dirt stockpile if the quality is poor. The 1200 mm × 1000 mm primary jaw crusher operates at 150 mm closed side setting and therefore produces a maximum size of approximately 230 mm. The crushed rock and the 'good' natural fines are passed over a scalping screen where the minus 20 mm material is removed.

Plant capacities in excess of 600 t/h would use a primary gyratory crusher instead of the jaw. In such applications, the gyratory crusher does not need

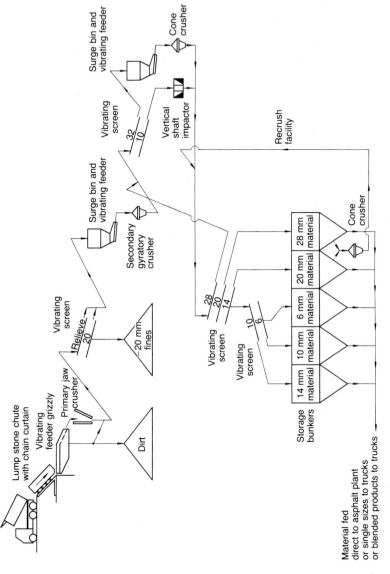

Fig. 4.2. Flow diagram of a hardrock crushing and screening plant[5]

Fig. 4.3. Direct feed of stone into a gyratory crusher. Photograph courtesy of Quarry Management

a feed station since the dump trucks can tip directly into the machine. Although a good cubical shape is produced the primary gyratory is relatively expensive and requires a large headroom for installation. However, at high capacities the cost per tonne of material crushed is lower than with other crushers. Fig. 4.3 shows material being fed into a primary gyratory crusher.

The plus 20 mm material from the scalping screen is transferred to a surge hopper. This facility enables the secondary crusher to have a controlled choke feed. An electric vibrating feeder delivers the material to a secondary gyratory which operates at 40 mm closed side setting. A gyratory crusher is used here as a secondary machine to take advantage of the high reduction ratio available and the good product shape. The crushed material is conveyed to a two-deck vibrating screen.

The plus 32 mm material is fed to a surge bin, electric vibrating feeder and a tertiary cone crusher operating with a 19 mm closed side setting. A cone crusher with this type of non-segregated feed grading has a good reduction ratio and performs effectively with hard abrasive materials. It is economical and crushing results in a cubical product near to the closed side setting providing a choke feed is maintained.

The minus 32 mm plus 10 mm aggregate is directed into a vertical shaft impactor which operates on the crushing principle of throwing stone by centrifugal force against captive stone. Vertical shaft impactors are useful third or fourth stage crushers and produce a good cubical shape. On occasion these machines are used to perform a shaping duty on poor quality materials.

The crushed materials, together with the minus 10 mm from the two-deck screen are conveyed to the final sizing screens and the single sized materials are stored in bunkers. The plus 28 mm material is recirculated to remove the oversize stone.

Most overseas crushing plants store the materials in ground stockpiles instead of bunkers and many transportable and mobile plants in this country also adopt this approach. This layout reduces the cost considerably but can lead to contamination and it exposes the crushed aggregates to adverse weather conditions. A material with a higher moisture content will be more expensive to dry at a later stage in the roadstone production process. Storage of the crushed products also enables a re-crush facility to be more easily incorporated. A cone crusher fed from the 28 mm and 20 mm aggregate bins provides a more versatile arrangement should market requirements change.

It is not unusual for hardrock plants to have extensive storage facilities including larger and smaller sized products. The materials from the bunkers can be fed directly to an asphalt plant, single sizes or blended products to trucks.

4.4.2. Limestone crushing and screening plant

The extraction from the ground of a softer, less abrasive material such as limestone, with a low silica content, is almost identical to the hardrock process. The overburden is removed, the rock is blasted, secondary breaking takes place where necessary and the stones are transported from the quarry face to the plant in large dump trucks. A limestone crushing and screening plant is shown in Fig. 4.4

In the flow diagram shown in Fig. 4.5 the trucks discharge into a lump stone chute. The largest feed size to the plant is a 900 mm cube and a chain curtain is used to control the flow. A vibrating feeder grizzly removes the minus 50 mm material to either bypass the crusher or feed to a dirt stockpile, The plus 50 mm material is discharged into a primary horizontal impact breaker.

The impact breaker works on the principle of reducing the material size by the impact of the rock against hammers which are attached to a high speed rotor. The broken rock then strikes breaker bars located on the inside of the machine resulting in more reduction and the stones fall again onto the revolving hammers for further reduction. The product size is determined by the position of the breaker bars relative to the rotor hammers and the rotor speed can be adjusted to cater for different feed materials.

A high reduction ratio can be achieved with this machine. A maximum feed size of 900 mm results in a product size of minus 100 mm and a good cubical shape. This machine can achieve the equivalent of two stages of crushing in a hardrock plant. The feed material must be relatively soft, not too abrasive and have a silica content of less than 5%. The crushed aggregates are conveyed to a surge bin which provides the remainder of the process with a controlled feed by eliminating the inconsistent dump truck discharges experienced at the primary crushing stage. An electrical vibrating feeder discharges the material from the surge bin to a conveyor and two deck vibrating screen.

Fig. 4.4. Limestone crushing and screening plant. Photograph courtesy of ARC (Southern) Ltd

The plus 20 mm material is fed to a secondary horizontal impact breaker, the minus 20 mm bypasses the crusher. The secondary impactor operates on a similar principle to the primary impactor. However, the product size is better controlled and some machines incorporate grids at the discharge outlet. Although the reduction ratio is high compared with a jaw crusher, it is less than with the primary impactor and again, a cubical shape is obtained throughout the product sizes. The crusher discharge and the bypass material are conveyed to the final sizing screens. Material sizes of 20 mm, 14 mm, 10 mm, 6 mm and 3 mm are stored in bunkers. The plus 20 mm is returned for further reduction.

As with the hardrock crushing and screening plants, a low cost option is for the material to be stored in ground stockpiles but adverse weather can affect the finished products. Contamination can also result and there is the cost of additional drying at the asphalt plant.

A cone crusher is situated between the 20 mm and 14 mm bins to produce a re-crush option should the market needs change. This is not necessarily considered to be a further crushing stage but a design feature to make the plant's production more flexible.

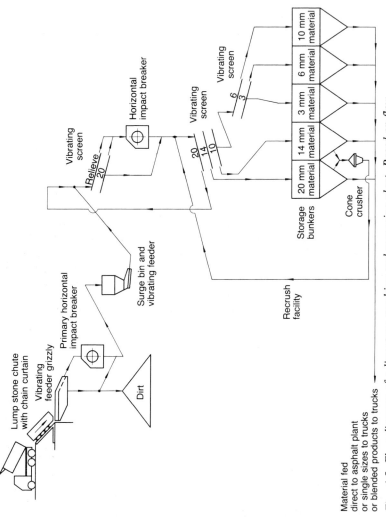

Fig. 4.5. Flow diagram of a limestone crushing and screening plant. Based on flow diagrams courtesy of Goodwin Barsby Division, Leicester[5]

As the setting of an impact breaker and the reduction ratio are not as significant in influencing the product shape with these low silica content materials, the process described only needs two crushing stages. If the feed material could be supplied cost effectively at 100 mm down then only a single stage crushing operation would be required.

4.4.3. Sand and gravel processing plants

Sand and gravel can be dredged from the sea or dug out of the ground. Marine dredged gravels are only used for roadstone production when hardrock and limestone are in short supply. Care must be taken with marine gravels to ensure that the chlorides are removed before the application of bitumen otherwise stripping of the bitumen will occur.

Land-based sand and gravel is found in two forms, a wet pit or a dry pit. A wet pit is where the majority of the deposit is under water, this material can be won by a suction dredger or a drag line excavator. In a dry pit the sand and gravel can be worked using a loading shovel, a drag line excavator or a scraper together with a tractor. The water can be pumped out of a wet pit and the deposit worked as a dry pit. Deposits can be up to, say, 8 m thick but they usually average about 5 m. The overburden is generally shallow, say from 0−2 m thick.

Many boreholes are taken when excavating for sand and gravel and the samples are then tested in the laboratory. The ratio of sand to gravel is of prime interest along with the geological composition of the gravel since these factors govern the design of the processing plant. Dump trucks can be used to transport the sand and gravel from the workings to the plant or a field conveyor system can be installed. Such a plant is shown in Fig. 4.6.

The sand and gravel processing plant flow diagram shown in Fig. 4.7 includes a field conveyor from the deposit to the plant. The as-dug material is conveyed

Fig. 4.6. Sand and gravel plant. Photograph courtesy of Allis Mineral Systems (UK) Ltd

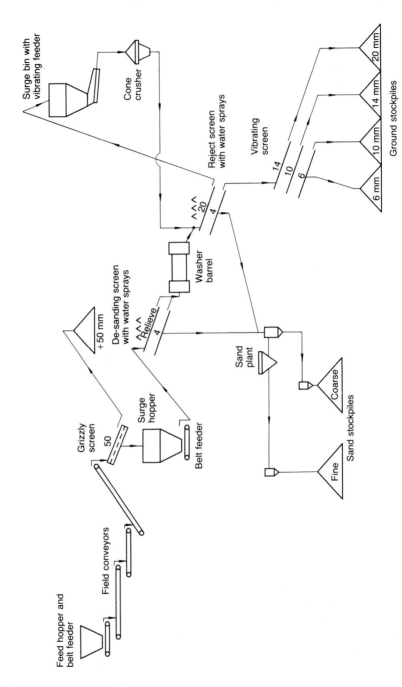

Fig. 4.7. Flow diagram of a sand and gravel plant

to a vibrating grizzly with the bars set at 50 mm spacing. The plus 50 mm is stockpiled and the minus 50 mm material passes to a large surge hopper.

A fixed speed belt feeder extracts the material from the hopper by way of a conveyor to a de-sanding screen equipped with water sprays. The aggregate is discharged into a rotating washer barrel where water is added. The washed aggregate is then passed over a reject vibrating screen, again fitted with water sprays.

The sand and water are flumed to a sump and separated into coarse and fine fractions. The dirty water is pumped to lagoons where the silt settles out prior to the water being recirculated to the plant. The plus 20 mm material from the reject screen is transferred to a surge bin. An electric vibrating feeder discharges the material to a cone crusher and the crushed gravel is returned to the reject screen. The minus 20 mm plus 4 mm material is conveyed to a grading screen separating out 20 mm, 14 mm, 10 mm and 6 mm products to ground stockpiles. These single sizes can be stored in bins and discharged, individually or blended, to trucks. A stockpile is a good means of allowing the gravel to de-water before further processing.

4.5. Methods of producing hot bituminous mixtures[6]

The selection of the type of plant for producing bituminous mixtures is never easy since the circumstances surrounding each purchase are usually different. The plant could be required for a quarry where the aggregates can be controlled and are known to be in plentiful supply, or a depot site where the aggregate source is likely to change. The type of plant chosen is influenced by the markets in which the materials are to be sold since they determine the product requirements. The duty required will not only be directly related to the market potential but also to the number and type of plants in the surrounding area. The budget will play a major role in determining the equipment selected, as will the running costs. Finally, the type of plant will be influenced by local site restrictions such as noise, availability of space, height and even emission regulations should the plant be destined for overseas.

There are two basic categories of production facilities, batch plants and continuous plants. Batch plants are conventional asphalt batch plants and batch heaters while continuous plants can be categorized into drum mixers and counterflow drum mixers.

4.5.1. Conventional asphalt batch plants

A conventional asphalt batch plant is shown in Fig. 4.8.

The aggregates specified in the final mix are loaded into the feed hoppers, a different hopper is used for each ingredient. The feeders control the flow of material from the hoppers and provide a proportional feed. Each feeder is set to give the required percentage in the finished mix. A controlled and accurate feed rate is needed otherwise energy will be wasted on unused materials. The feed materials are conveyed to the dryer, consisting of a rotating cylinder with a burner mounted at the discharge end. The cold and wet aggregates are dried and then heated within the cylinder. Inside the dryer is a series of lifters which are designed to cascade material across the diameter of the drum. The lifters at the feed end produce a dense material curtain to

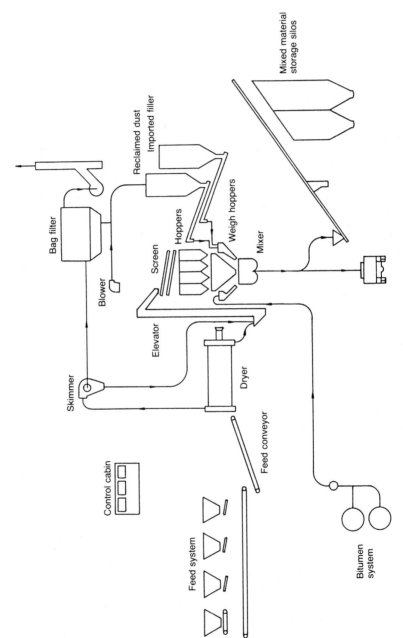

Fig. 4.8. Flow diagram of a conventional asphalt batch plant

Fig. 4.9. Conventional asphalt batch plant. Photograph courtesy of Tracey Enterprises Ltd

expose the aggregate and sand to heat from the burner. At the discharge end, the lifters hold the material around the circumference of the dryer enabling the flame to develop and complete combustion to occur. When combustion takes place within the dryer the gases expand. The products of combustion and the water vapour produced in the drying process are removed by an exhaust system.

Figure 4.9 shows a Titan 2000 conventional asphalt batch plant near Dublin.

A contraflow air system operates on the majority of stone dryers where the hot exhaust gases travel in the opposite direction to the material flow. This arrangement is an efficient process since the exhaust gases pre-heat the incoming material. The exhaust system creates an air speed within the dryer so that fine particles of dust are also picked up. Fig. 4.8 illustrates a dry collection system, the coarse particles, approximately $+75\,\mu$m, are removed by the skimmer and returned to the plant via the hot stone elevator. The fine particles of dust are removed from the exhaust air by a bag filter. Emission levels of $20\,\text{mg/m}^3$ are achieved using this method and all of the dust is reclaimed by the process. The material discharges from the dryer at temperatures ranging from $120-200°$C and is fed into the hot stone elevator. The elevator transfers the material to the screen on the mixing section. Asphalt plant screens used in the British market grade four sizes plus rejects, or alternatively six sizes plus rejects. The material discharges from the screen into the hot stone bins, internal chutes enable the bin compartments to overflow to a single point.

The mix recipe selects the required aggregates to be weighed in the batch hopper and simultaneously the bitumen and filler are weighed off. The ingredients are then emptied into the paddle mixer until the contents are fully coated. Typically, mixing cycle times of 45 seconds and 60 seconds are possible, depending on the specification. The mixer can load directly to trucks or to a skip and mixed material storage.

Ancillary equipment includes heated bitumen storage with a ring main to the bitumen weigh hopper, imported and reclaimed filler systems consisting of a silo, rotary valve and screw conveyor to the filler weigh hopper.

Between 10 and 15% of reclaimed mixed material can be added to a conventional asphalt batch plant by a belt feeder and conveyor directly into the batch weigh hopper. In this arrangement, the reclaimed material is considered to be an additional ingredient in the mix. The reclaimed material can also be added directly into the mixer. The equipment consists of a belt feeder and a conveyor fitted with a belt weigher.

Up to 50% of reclaimed material can be added to the mix using the parallel drum method. The material is heated to between 80 and 120°C, weighed in a separate vessel and then screwed into the paddle mixer, a flow diagram of the process is shown in Fig. 4.10. The exhaust from the reclaimed, or black, dryer is directed into the discharge end box of the virgin drum where the fumes are incinerated. Table 4.1 lists the advantages and disadvantages of conventional asphalt batch plants.

4.5.2. Batch heater plant

The flow diagram for a batch heater plant is shown in Fig. 4.11. This process is different from the conventional asphalt plant since each batch is manufactured individually. Pre-graded aggregates are fed to the plant in a single batch by time controlled feeders, the time is directly related to the proportion set in

Table 4.1. Advantages and disadvantages of conventional asphalt batch plants

Advantages	Disadvantages
Ability to manufacture all materials to BS 594 and BS 4987	Heat wasted on rejected and overflow material
Emission to atmosphere within acceptable limits	Relatively high maintenance costs
Inconsistent feed materials tolerated	High production costs compared with drum mixer process
Small tonnages possible	Relatively high capital cost
Production of materials at all temperatures	Capacity restricted to mixer size and mixing cycle
Mixed material storage is not essential	
High percentage of reclaimed material can be added	

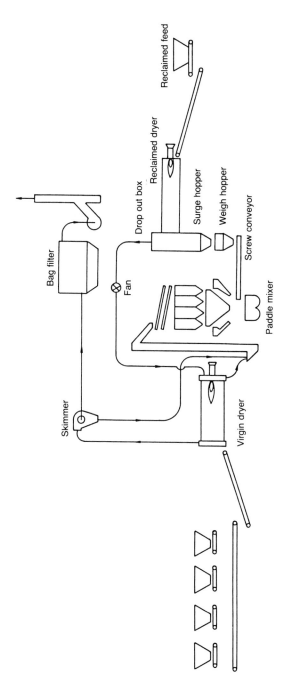

Fig. 4.10. Flow diagram of a batch plant with dryer for recycled material

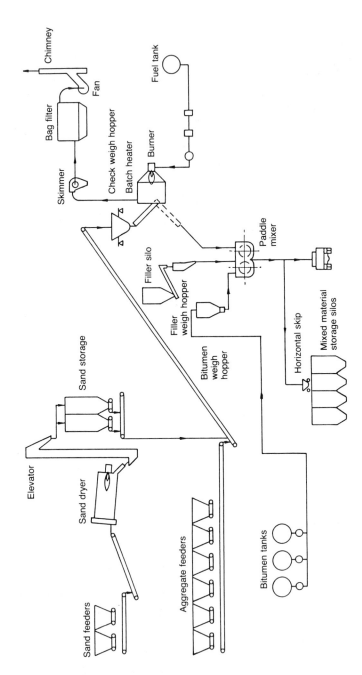

Fig. 4.11. Flow diagram of a batch heater

Fig. 4.12. Twin batch heater plants fed directly from a quarry plant. Photograph courtesy of ARC (S Wales) Ltd

the mix recipe. The materials are conveyed to a check weigh hopper where the wet feed weight is determined. An allowance is made for the moisture content in the feed, the control system then calculates the bitumen and filler contents and proceeds to weigh these ingredients.

Drying and heating of the materials takes place in a short rotary drum, the discharge temperature is controlled by the time the materials are kept in the heater. All materials are discharged into a paddle mixer where a homogeneous mix is produced. A typical mixing cycle time is two minutes but this depends upon the moisture content in the feed and the material specification. The mixer can discharge directly to trucks or, alternatively, to mixed material storage. Some plants have an additional weigh vessel between the batch heater and the paddle mixer to determine the dry aggregate weight. The bitumen content is then calculated from this dry batch weight. The skimmer can be arranged to discharge on a discrete basis into the paddle mixer.

The bitumen storage and ring main, the filler storage and the bag filter exhaust system are generally as described previously. If moisture contents greater than 2% overall are experienced then a pre-dryer is used for the sand, as shown in Fig. 4.11. The sand is then called up from the dry storage. Between 10% and 15% of reclaimed material can be added to the mixer by means of a belt feeder and weigh conveyor.

Fig. 4.12 shows twin batch heater plants fed directly from a quarry plant. Table 4.2 lists the advantages and disadvantages of batch heaters plants.

4.5.3. Drum mixer

A flow diagram of a drum mixer is shown in Fig. 4.13.

The feed section is one of the most important aspects of the drum mixer process. It is essential for the success of the finished product that pre-graded aggregates are used. The feeders control the flow of material to the plant at

Table 4.2. Advantages and disadvantages of batch heater plants

Advantages	Disadvantages
Small batches economically produced	Single-sized feed materials
No material wastage	Maximum 2% moisture in feed materials
Ability to manufacture all materials to BS 594 and BS 4987	Pre-drying required if moisture content greater than 2%
Quick recipe change	Relatively high capital cost
Mixed material storage is not essential	Relatively high maintenance costs
Number of feed materials is only limited by quantity of feeders	
Reclaimed material can be added	

a rate proportional to the quantities called for in the mix. Volumetric feeders are generally employed although an option sometimes preferred is the use of weigh feeders especially for sand and dust. Wet sand and dust can be difficult to handle and they do not always flow evenly or readily. The use of a volumetric feeder cannot therefore be relied on and in these instances a weigh feeder is used.

A scalping screen can be incorporated into the feed system to remove unwanted oversize material when the quality of the aggregates cannot be guaranteed. For example, this would be appropriate if the plant is fed from ground stockpiles.

The materials pass over a belt weigher before entering the drum. The mass passing over the weigher supplies the master signal to the ratio controller which determines the flow rates for the other mix ingredients making allowance for the moisture content in the feed.

The drum mixer consists of a rotating cylinder with the burner mounted at the feed end. Lifters around the circumference of the drum keep the material away from the flame, followed by a lifter pattern which forms a dense cascading material curtain. After drying and heating, the lifters are designed to mix the aggregates with bitumen inside the drum. The binder is usually added from the discharge end to prevent degradation which can occur if it is too close to the burner flame. Imported or reclaimed filler can also be added to the mixing zone.

One of the most accurate methods of providing a continuous flow of filler is to employ a 'loss-in-weight' system. Special attention is given to conditioning the filler and it is continuously reverse weighed against a known calibration rate. A pneumatic conveyor transfers the filler to the drum mixer. An exhaust system removes the products of combustion and a filter prevents the inherent dust particles being emitted into the atmosphere. Due to the nature of the process the dust collected is considerably less than that which is recovered from a conventional dryer. Up to 15% of reclaimed mixed material can be added to the drum mixer and successfully mixed before fuming becomes a problem.

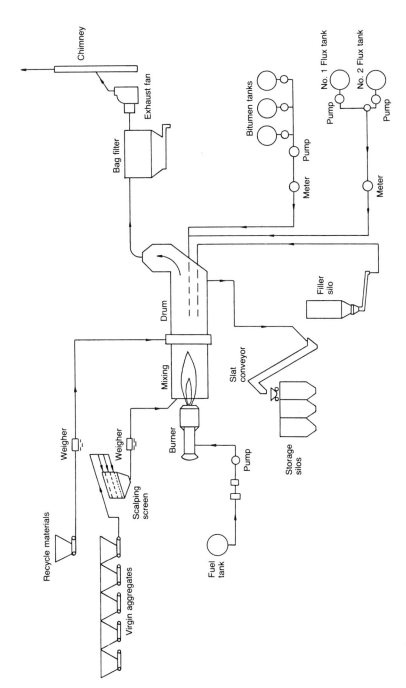

Fig. 4.13. Flow diagram of a drum mixer with recycling facility

Fig. 4.14. Drum mixer plant. Photograph courtesy of Cedarapids USA

Table 4.3. Advantages and disadvantages of drum mixers

Advantages	Disadvantages
Economical plant for long production runs of one material specification	Unable to manufacture all BS 594 and BS 4987 materials
Number of feed materials is only limited by quantity of feeders	Single-sized feed materials
Dust removed from process is minimal	Mixed material system essential
High capacities readily achieved	Small batches are uneconomical
Easily adopted to mobile design	Possibility of fume emission with high temperature mixes
Relatively low maintenance costs	
Reclaimed material can be added	

As the process is continuous, large tonnages are possible in a relatively short timescale, a mixed material system is therefore needed. Belt conveyors, travelling skips and slat conveyors have all been used for these storage schemes.

Figure 4.14 shows a Cedarapids Standard Havens drum mixer plant operating in San Francisco. Table 4.3 lists the advantages and disadvantages of drum mixers.

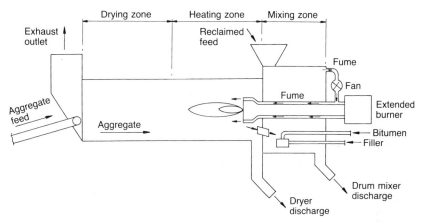

Fig. 4.15. Flow diagram of a counterflow drum mixer

4.5.4. Counterflow drum mixer

Figure 4.15 shows the layout of a counterflow drum mixer.

There are two major differences between the counterflow drum mixer and the drum mixer. Firstly, an extended burner is used and this is mounted at the discharge end of the drum and secondly, the mixing chamber is completely separated from the drying and heating zones.

The drying section of the counterflow drum mixer is identical to a conventional dryer, preheating of the feed materials results giving a more efficient drying method. The burner flame cannot be in direct contact with the bitumen since the mixing zone is isolated. As a result, the possibility of fumes is reduced considerably. A scavenger fan removes the hot gases and any fumes present from the mixing chamber are incinerated by the burner flame.

Bitumen, filler and reclaimed material are added to the mixing zone as called for by the specification. The drying and heating section can be used purely as an aggregate dryer if required. Up to 50% reclaimed material can be added to the counterflow drum mixer.

The principle of operation and associated equipment such as feeders, scalping screen, weigh conveyor, exhaust system, bitumen system, filler storage and mixed material are as described for the drum mixer. This process evolved from the drum mixer to meet the need for a continuous plant with improved emission levels and a reduced fume content when mixing at higher temperatures or using different bitumens.

Table 4.4 lists the advantages and disadvantages of counterflow drum mixers.

4.5.5. Summary

Table 4.5 summarizes the main features of bituminous mixtures production plants.

4.6. Controlling manufacture

4.6.1. Aggregate processing plants

Control of the feed material in an aggregate processing plant is important

Table 4.4. Advantages and disadvantages of counterflow drum mixers

Advantages	Disadvantages
Most economical plant for long production runs of one material specification	Wastage at beginning and end of production
Possibility of fume emission is reduced when producing high temperature mixes	Small batches are uneconomical
Ability to manufacture all materials to BS 594 and BS 4987	Single-sized feed materials
Number of feed materials is only limited by quantity of feeders	Mixed material system essential
High capacities readily achieved	
Relative low maintenance costs	
Reclaimed material can be added	

Table 4.5. Summary of bituminous mixtures production plants

Type of plant	Uses	Remarks
Conventional asphalt batch plant	The most common type of mixing plant in the UK used for small batch production and up to the medium capacity range of 240–300 t/hr	Highly flexible production and capable of manufacturing all British Standard specifications
Batch heater	Ideal for small production requirements with the facility to change specifications quickly	Maximum allowable moisture content of 2%; a pre-dryer is needed if moisture is higher
Drum mixer	Economical plant for the production of large quantities of one material specification	
Counterflow drum mixer	Selected for the continuous production of large quantities of one material specification when the control of fume emission is critical	

for several reasons. Jaw, gyratory and cone crushers should be choke fed to optimize plant performance and obtain good product shape. Maximum feed sizes allowed to enter the crushers should be observed in order to provide continuity of production as a blockage within a crusher, or chute, can cause considerable delays. Unwanted materials like soil, clay and very fine rocks should be prevented from entering the plant since these reduce overall throughputs.

Tramp iron (unwanted iron or steel in the aggregates e.g. an excavator tooth) should be avoided in the feed since this can permanently damage the crusher

Fig. 4.16. Aerial view of a limestone quarry. Photograph courtesy of ARC (Southern) Ltd

wear parts, a metal detector is therefore usually fitted before the secondary crushing stage. An audible warning can be given to detect the metal and/or a magnet used to remove it. The majority of crusher wear parts are made from manganese steel and as this material is non-magnetic, electronic detectors have to be used to identify any of these loose objects within the process.

A limestone quarry is shown in Fig. 4.16.

Larger aggregate plants are controlled by microprocessors and these eliminate the need for costly, extensive electrical relays and their associated wiring. This method of control also enables the status of the plant such as material flow, motors on/off, doors open/closed or water supply on/off to be monitored and alarmed if a fault condition occurs. Individual crushers can be microprocessor controlled to automatically adjust, or maintain, the discharge setting e.g. in a cone crusher. This is a useful feature since a granite with a high silica content can noticeably wear the liners of a cone crusher in a very short time say, four or five hours, and such crushers require regular re-setting to maintain a quality product.

Automatic overload protection is available on jaw, cone and gyratory crushers by means of a hydraulic system. Tramp iron and other uncrushable materials are allowed to pass through the crushers without costly damage to the liners, the original setting is then adopted. This would be in addition to the metal detector discussed earlier. The correct screen mesh sizes need to be fitted to gain full control of the product sizes. The meshes should be inspected regularly for proper tensioning and holes caused by wear. The individual units of new aggregate plants are usually sheeted which makes the plant more aesthetically acceptable as well as reducing noise and dust emissions. The nuisance dust from the conveyor transfer points and housings is collected in a dry bag filter and the 'cleaned air' emitted to atmosphere.

A re-crush facility enables the plant to be more versatile with the sizes being manufactured. This feature incorporated in the design is useful when the market trends change. The larger sized materials 40 mm, 28 mm and 20 mm are reduced to the more popular smaller sizes 14 mm, 10 mm and 6 mm.

The final products are either stored in stockpiles on the ground, or in storage bunkers. In either case, contamination of the aggregates should be prevented. Stockpiles should be on hardstanding such as the quarry floor, or concrete to prevent contamination from below and separated to avoid overspill. Storage bunkers should be fitted with high level indicators or overflow chutes. Having invested substantial amounts of time and money to ensure that the product size and shape is correct, degradation of the material should also be prevented. Rock ladders are used within stockpiles and storage bunkers to control the flow of large aggregates from a height and overcome unwanted breakages.

Laboratory inspection of the materials, both during and after production, is an important aspect of controlling manufacture and quality. The results of the samples from within the process indicate whether the plant is functioning correctly and enable adjustments to be made where necessary. The results from the final material samples determine whether the product is acceptable for sale. Single sized products will be checked against BS 63: Part 1.[7] The sample must be representative of the bulk material and the personnel undertaking the testing must be skilled operatives. Methods of sampling and testing aggregates are given in BS 812: Part 102[8] and section 103.1.[9] More details on this subject are contained in chapter 9.

4.6.2. Asphalt and bituminous macadam plants
Storage of raw materials

Asphalt and bituminous macadam plants can either be located adjacent to the aggregate plant in the quarry, or remote at a depot site. If the plant is remote then the aggregates will be transported by truck or rail for larger operations. The quantity of materials delivered by rail necessitates an automatic off-loading scheme incorporating underground discharge hoppers, tripper conveyor and covered storage bays. The material is recovered from storage by an underground feed system to the asphalt plant. Both truck and rail delivered aggregates should be stored on hardstanding and separated to avoid contamination. This method of storage along with a quality feed material provides a consistent feed to the asphalt plant reducing overflow, wastage and fuel costs.

The ideal situation is for an asphalt plant to be fed directly from the aggregate plant bunkers. The product sizes can be more easily controlled by matching the screen mesh sizes on both plants and, since the material has not been exposed to the weather since entering the quarry plant, the moisture content will be less. This layout can reduce the overall feed moisture content by up to 2% and is therefore more economical.

The filler silos containing imported and reclaimed materials can be subject to the formation of condensation on the inside walls under certain weather conditions. As filler is a hygroscopic material, it has a tendency to absorb moisture and the contents of the silos can become solid if left for prolonged periods. It is good practice to order only the quantity of imported filler needed for planned production runs and leave the silos empty when the plants are not operational for extended periods.

Bitumen needed for production is stored within the temperature ranges laid down by the specification. These temperatures are given in BS 594[2] and BS 4987.[3] The bitumen is usually circulated in a ring main system up to the mixing plant before usage to reduce heat loss in the pipework even though the pipes are heated and insulated. To prevent the bitumen hardening through oxidation the material should not be stored at high temperatures for long periods especially if no new bitumen is added. Prolonged storage with no usage should be at approximately 25C° above the bitumen softening point and recirculation stopped. Bulk reheating of the bitumen should be carried out intermittently to prevent local overheating around the heating member. Bitumen can ignite instantaneously if temperatures above 230°C are reached so such temperatures should be avoided.

Plant capacities

There are two basic constraints on the output of an asphalt and bituminous macadam plant, the drying capacity and the mixing capacity. The drying process is complex. Some of the factors which directly affect the capacity are

(*a*)   altitude
(*b*)   air temperature
(*c*)   type of feed material and grading
(*d*)   feed material temperature
(*e*)   moisture content in feed material
(*f*)   mositure content in discharge material (usually less than 0·5%)
(*g*)   dryer discharge temperature
(*h*)   exhaust temperature
(*i*)   dryer angle
(*j*)   calorific value of the fuel
(*k*)   exhaust air volume available
(*l*)   lifter pattern in dryer
(*m*)   efficiency of burner.

The mixing process is also complicated, and some of the factors affecting the capacity and whether the mix is homogeneous are

(*a*)   mix recipe
(*b*)   mixer live zone
(*c*)   paddle tip speed
(*d*)   size of paddle tips
(*e*)   mixer arm configuration

(f)   method of feeding ingredients
(g)   temperature of materials
(h)   mixing time (some mixes have a minimum mixing time).

One of these processes will determine the overall plant capacity. In a conventional asphalt batch plant and a batch heater the two processes are separate and can therefore be easily identified and considered. In a drum mixer and counterflow drum mixer the drying and mixing are combined and require more detailed consideration. Basically, the feed material must be completely dried and heated before the bitumen is added. The drying constraints are therefore similar to the batch plant dryers. The mixing time in a continuous plant is controlled by adjusting the throughput, altering the drum angle, adjusting the position of the bitumen injection pipe within the drum, altering the design of the mixer lifters and altering the speed of the drum. In practice, once these variables are set during commissioning they are not usually altered.

To determine the equipment sizes for an asphalt and bituminous macadam plant the manufacturer requires the following information

(a)   maximum plant capacity
(b)   type of feed material
(c)   average moisture content of feed material
(d)   type of mixes to be produced
(e)   temperature of mixes
(f)   special constraints, e.g. minimum mixing time.

Plant control

With the exception of the batch heater, which has fixed speed time controlled feeders, most plants have variable speed feeders with a 20:1 turn down ratio (the feeders operate from 5% to 100% of the belt speed or capacity). Volumetric belt feeders generally have a repeatability accuracy of $\pm 0 \cdot 5\%$ and vibrating feeders are around $\pm 2\%$. This repeatability and turn down ratio enables an accurate plant feed to be obtained and, providing the feed materials are within the specified size limits, this also ensures an economic control of the materials, since heat will not be wasted on unused aggregates. Low level indicators can be fitted to the feed hoppers with elevated lights above the hoppers to warn the shovel driver of an impending shortage. The feeders can be equipped with no-flow indicators and after an initial warning and a pre-set time delay, the plant will shut down if prescribed.

The burner mounted on the dryer, or drum mixer can have turn down ratios of up to 10:1. Different plant throughputs and different dryer discharge temperatures can therefore be easily accommodated with an economic use of the fuel. The burner is automatically controlled by a temperature sensing device in the dryer discharge chute. A non-contact infra-red pyrometer is normally used to record the temperature as the response times are a few seconds compared with a thermocouple probe which can take up to 15 seconds.

To sustain combustion an exhaust system is connected to the dryer or drum mixer. As air is drawn through the drum, small particles of dust are picked up. This must be removed before discharge to atmosphere. A dry bag filter dust collector is employed to remove the dust and plants are currently being guaranteed to meet particulate emission levels of 20 mg/m^3. The dust collector also pulls from nuisance points on the plant and controls the overall emission level. The air volume is controlled by a variable damper in the system, usually fitted prior to the fan on the clean side of the filter. The damper is controlled by a pressure reading taken at the burner end of the dryer.

Continuous level indicators can be installed in the hot stone bins of conventional asphalt batch plants. These give an indication of the material level within each compartment and provide a means of trending. From this information the feeders can be set up during commissioning and adjusted when, and if, the feed aggregates become inconsistent. Again this enables the plant to operate more economically with reduced overflow.

Temperatures of the sand in the hot stone bins and the bitumen in the ring main are taken by thermocouple probes and recorded. These readings confirm that the temperatures are correct before mixing commences.

Although the British Standards allow tolerances of $\pm 0 \cdot 6\%$ on the accuracy of the bitumen mass in the mix, plants are expected to perform with an accuracy of $\pm 0 \cdot 1\%$. Bitumen is the most expensive ingredient and therefore economics dictate the importance of this requirement. Batch plants are equipped with load cell weigh systems with repeatable accuracies of $\pm 0 \cdot 5\%$ for the aggregate weigh vessel and $\pm 0 \cdot 3\%$ for the bitumen and filler hoppers. Continuous plants use a load cell in the form of a belt weigher and accuracies of $\pm 1\%$ are guaranteed.

Fully automatic microprocessor control systems are used on all types of asphalt plants. On batch plants the microprocessor provides accuracy by calculating the bitumen content as an actual percentage of the aggregate and filler content achieved. Outputs are maintained at an optimum by weighing off a large percentage of the bitumen at the same time as the aggregate, the required bitumen is then calculated relative to the actual aggregate and filler content. Self-learning in-flights (to provide accurate weighing) for the hot stone bin discharges and alternative bin selection are also features incorporated on conventional asphalt plants which help throughput whilst maintain accuracy.

The use of microprocessor control allows the operator to adopt a passive role during production since the plant is operated from the input data. Providing this data is correct and no faults are recorded, the finished mix will be correct. The operator is relieved to a certain extent of the responsibility and therefore the quality of the product is not operator dependent. The monitoring of process faults like sticking doors, tare (weight) faults, stopped motors, broken chains, or sticking valves reduces down time and improves the quality of the products. In addition, the microprocessor can stop the plant from mixing if acceptable pre-set limits are not met. These faults are logged and recorded on the system again providing quality control.

An infra-red pyrometer can be positioned, at the paddle mixer discharge and the mixed material bin outlets, to read and record the final discharge temperature. Microprocessor control can prevent contamination of different mixes within the mixed material system by remembering the type and quantity of mix in each bin. Storage silos mounted on load cells can accurately weigh out mixed material to trucks, avoiding wastage and saving time.

Full microprocessor control systems offer VDU screens with plant mimic diagrams, the ability to store over two hundred recipes, production data, storage data, a daily job queue, print-out of mix results, date, time, etc. The print-out and storage data are important elements of quality assurance since each batch and truck load can be traced, confirmation of a manual or automatic mix is also recorded. The controls information can be limited to the laboratory, the weighbridge and the main office accounting system, allowing information to pass in both directions.

Production and operating conditions

To obtain optimum fuel usage a constant and uniform feed to the plant is needed. The aggregates should preferably be taken from the same place on the stockpile each time, avoiding the lower stock near to the ground and surface water from entering the loading shovel. This procedure will ensure that the minimum amount of water enters with the feed aggregates and provide uniform drying conditions. The dryer, burner and exhaust system operate more efficiently under stable conditions.

The dryer has a minimum capacity directly related to the effectiveness of the lifter pattern, the turn down ratio of the burner and the maximum inlet temperature permitted to the bag filter. The majority of filters in the UK are equipped with Nomex material for the bags. This material has a maximum operating temperature of approximately 200°C although higher peak temperatures can be accommodated. With a small aggregate feed to the dryer, the burner on minimum flame and a discharge temperature of say, 190°C, there will be occasions when too much heat is available and the inlet temperature probes to the filter will automatically switch off the burner fuel supply. The dryer angle can be lowered to try to overcome this by increasing the drum loading, but this will not resolve the problem when a small production rate is required. As a general rule, the minimum drying capacity for a plant equipped with a bag filter is approximately 50% of the capacities quoted for hot-rolled asphalts. This constraint does not apply to plants fitted with wet collection exhaust systems.

As the heated aggregates pass from the dryer to the mixer, there is an initial temperature loss of between 10°C and 20°C depending upon the required mix temperature. During long production runs, this temperature difference can be reduced to between 0°C and 5°C.

When drying and heating gravels, or marine gravels, for use in roadstone production it is important not to overheat the aggregate. Overheating can cause 'sweating' of the aggregate later in the process and this will eventually lead to stripping of the binder. For this reason, and as an aid to adhesion since crushed gravels still tend to be slightly rounded, the addition of 2% Portland cement or hydrated lime at the mixing stage is recommended.[2,3]

The efficiency of the vibrating screen on a conventional batch plant changes with different flow rates. Consistent feed gradings and constant flow rates are, therefore, preferred so that the aggregate separation into each hot stone bin compartment can be predicted. The continuous level indicators in each compartment are a useful check since, in practice, a consistent feed grading is difficult to achieve.

During production, care is taken not to contaminate hardrock mixes with limestone. However, the reverse situation can usually be accommodated. In a conventional asphalt plant, the hot stone bins are completely purged when the type of feed material changes. Purging can lead to considerable delays when changing specification. As each batch is manufactured individually with a batch heater, this situation does not arise. Continuous mixing plants can change specifications during production providing the same type of feed material is used, otherwise the plants are purged.

The ideal production schedule is to manufacture the lower temperature materials first and progress to the higher temperature specifications. In reality, this is not always possible and care has to be taken not to overheat a low temperature material that is required in the middle of a high temperature production run. Overheating can lead to oxidation and hardening of the binder and this may well change the mix characteristics which can have an adverse effect on the life of the pavement.

The paddle mixer has a recommended minimum capacity of 50% of the nominal capacity. Mixers do not function satisfactorily below this level as a homogeneous mix does not result and oxidation of the bitumen can occur. Prolonged mixing of a full or partial batch does not necessarily mean a more homogeneous mix. Entended mixing times can lead to binder oxidation and subsequent hardening.

To maintain the quality of the product, the equipment should be recalibrated on a regular basis. The feeders, the burner, the temperature measuring equipment and the load cell weighgear should all be inspected within a planned maintenance schedule to attain the quality assurance standards needed.

Mixed material storage

Mixed material silos are manufactured in both square and circular forms with bin capacities ranging from 30−200 t each. The top, the top doors, the sides, the conical section, and the discharge doors are all insulated with material which can be up to 150 mm thick. Heaters are usually attached to the discharge doors and the conical section and although the bin sides can be heated it is not common practice. The heating can be electric or indirect by a hot oil heater and the heating system usually allows the temperature of each individual silo to be controlled separately. The heat is applied purely to maintain the temperature by replacing lost heat and not as additional heat for a mix with a low temperature. Overheating of the material in the silo can cause the bitumen to strip from the aggregate.

Mixed material silos are available with grease, or oil, sealed doors at the inlet and discharge, with the facility to add an inert gas. This design feature provides extended storage times by preventing the mix coming into contact with the atmosphere and oxidizing.

Even without the sealing and the addition of an inert gas extended storage is possible. To achieve the best results the silo is completely filled. Dense materials store better over a longer period. For a twelve hour period the heat loss is minimal, say a maximum of 3°C and providing the outlets are large enough, the silos will discharge freely. If the material is to be stored for two to three days then a crust forms on the top and around the discharge outlet. This can be overcome by discharging one tonne of material at regular intervals, say every nine hours. Again, the temperature loss over these few days will be minimal, probably a maximum of 20C° due to the effectiveness of the heating and insulation. Hot-rolled asphalts and open-textured materials are not recommended for extended storage because of the difficulty in discharging from the silo outlets.

Sampling

Production must be monitored to ensure that the aggregate grading and bitumen content comply with the specification but not every batch can be tested. Samples are therefore taken at pre-determined intervals and these samples must be representative of the bulk material. Methods of sampling and achieving representative samples are given in BS 598: Part 100[10] for bituminous mixtures and BS 812: Part 102[8] for aggregates, sands and fillers. Sampling should only be performed by skilled personnel using well maintained and calibrated equipment.

Samples should be taken of all materials delivered to the plant i.e. aggregate, bitumen and filler. If the asphalt plant is located in the quarry with the aggregate plant then part of this requirement will have been fulfilled.

The aggregate samples from the hot stone bins are taken either from special sampling devices on the bins or directly through the paddle mixer. This tests the gradings within each bin and gives an indication of the efficiency of the screen. The results from these samples determine the amount to be weighed off from each bin during the weighing cycle. This exercise should be carried out at regular intervals to ensure aggregate grading control.

Samples of the mixed materials can be taken from the paddle mixer or drum mixer discharge and the mixed material silo discharge. The mixer and drum mixer discharge sample determines whether a homogeneous mix is produced and the silo discharge sample confirms that the material has stored favourably or indicates whether segregation or other deterioration has occurred. During production, samples are usually taken from the truck load of mixed material. The paddle mixer, drum mixer and silo discharge samples are taken during commissioning and calibration and subsequently only if problems occur.

Tests on the samples take time, as discussed in chapter 8 and a large quantity of material may have been manufactured and laid before the results are available. However, sampling is essential for calibration purposes and good quality control during production. It also gives confidence in the performance of the plant and, therefore, has to be undertaken on a regular basis.

References

1. BRITISH STANDARDS INSTITUTION. *Building and civil engineering terms: Concrete and plaster. Section 6.3 aggregates*. BSI, London, 1984, BS 6100: Part 6.
2. BRITISH STANDARDS INSTITUTION. *Hot-rolled asphalt for roads and other paved areas. Specification for constituent materials and asphalt mixtures*. BSI, London, 1992, BS 594: Part 1.
3. BRITISH STANDARDS INSTITUTION. *Coated macadam for roads and other paved areas. Specification for constituent materials and for mixtures*. BSI, London, 1993, BS 4987: Part 1.
4. BRITISH STANDARDS INSTITUTION. *Specification for air-cooled blast furnace slag, aggregates for use in construction*. BSI, London, 1983, BS 1047.

5. ALLIS MINERAL SYSTEMS (UK) LTD. Based on flow diagrams courtesy of Goodwin Barsby Division, Leicester.
6. MOORE J. P. Coating plant section. Reproduced by kind permission of Quarry Management, May 1992.
7. BRITISH STANDARDS INSTITUTION. *Road aggregates: Specification for single-sized aggregate for general purposes*. BSI, London, 1987, BS 63: Part 1.
8. BRITISH STANDARDS INSTITUTION. *Methods for sampling*. BSI, London, 1989, BS 812: Part 102.
9. BRITISH STANDARDS INSTITUTION. *Method for determination of particle size distribution. Sieve tests*. BSI, London, 1985, BS 812: Section 103.1.
10. BRITISH STANDARDS INSTITUTION. *Sampling and examination of bituminous mixtures for roads and other paved areas. Methods for sampling for analysis*. BSI, London, 1982, BS 598: Part 100.

5. Plant

5.1. Introduction

This chapter discusses the main items of plant involved in the laying of bituminous mixtures i.e. pavers, rollers and chipping machines. Lorries, cold planing machines, hot-boxes, infra-red heaters and miscellaneous plant involved in such works are also considered. If the correct plant, which must be properly maintained, is not used in the laying process the probability of the finished laid mat not meeting the required specification will inevitably increase. In any event it will make the paving team's job much more difficult if plant is deficient.

Regardless of the type of specification adopted by a client there are generally very few references to the plant engaged in laying operations. The two principal British Standards relating to the transport, laying and compaction of coated macadams and hot-rolled asphalt for roads and other paved areas are BS 4987: Part 2[1] and BS 594: Part 2[2] respectively. In the case of compaction plant the Department of Transport's *Specification for highway works*[3] gives a wide choice of compaction plant if a method type specification is to be adopted. The choice of plant for laying and compaction of bituminous materials is normally dictated by the surfacing supervisory staff with the experience of the laying squad sometimes exerting an influence.

In these times of a 'quality approach' to most construction works the costs of failure to achieve the desired result first time and every time in surfacing works can often be substantial. Once coated macadams and hot-rolled asphalts have been laid and compacted they cannot be re-worked so any failure to meet requirements related to line, level, compaction or surface texture will necessitate the removal of the full depth of layer and the work being carried out again.

It is therefore extremely short-sighted to use plant that is not suitable for a particular job or indeed plant that has not been properly maintained.

Each section of the chapter gives a description of the essential characteristics of the main plant items and describes how their performance influences the laying of hot bituminous materials. The chapter concludes by assessing the economic performance of plant in terms of choice of plant, running and maintenance costs on a life-cycle basis using the equivalent annual cost method.

5.2. Pavers

5.2.1. Introduction

Paving equipment has been in existence in one form or another for over 60 years. Early pioneers included Barber-Greene and Blaw-Knox. In 1932 the

Fig. 5.1. Adnun black top paver c. 1932. Photograph courtesy of Blaw-Knox Construction Equipment Ltd

latter introduced their first self-propelled paver which did not operate on forms and was used exclusively for laying bituminous mixtures and stone.[4] This machine was called the Adnun black top paver (Fig. 5.1) and comprised a self-propelled tailgate spreader with an oscillating, serrated screed based on a fixed screed configuration. The front of the machine was supported on two steerable wheels which rode on the underlying base. The rear of the unit was supported by two rollers which rode on the new, uncompacted mat and propelled the machine. The screed was mounted between the front and rear suspension so that any vertical irregularities encountered by the front wheels were reflected by a factor of only 25% at the screed. By 1938 Blaw-Knox had developed tracked laying pavers which exhibited many of the features prevalent in modern day pavers.

Throughout the 1940s a series of improvements were carried out culminating with the introduction of the first pneumatic, rubber-tyred asphalt finisher in the early 1950s see (Fig. 5.2).

Blaw-Knox pioneered technological advancements in the design and use of specialized paving equipment and continue to do so. Other manufacturers in the blacktop industry who have designed and developed high technology paving equipment include Barber-Greene, Bitelli, Ingersoll-Rand and Dynapac. Today the paving market is very competitive and each manufacturer believes their machines to have that extra edge over those of their rivals. Pavers are continually being improved and refined but the basic two elements of any paver remain the same namely the tractor unit and the screed.

Fig. 5.2. First pneumatic tyred paver c. 1954. Photograph courtesy of Blaw-Knox Construction Equipment Ltd

Modern pavers have various automated methods for controlling the flow of material through the paver and sophisticated automatic level controls to ensure that the correct mat profile is laid. Developments in technology have helped to reduce any poor workmanship which occurred as a result of operator error but a skilled paver crew remains fundamental to ensuring that good work is produced consistently (Fig. 5.3).

Fig. 5.3. Modern paver — a Dynapac 12000R. Photograph courtesy of Lothian Transportation INROADS

5.2.2. Types of pavers

There are three main types of pavers: fixed screed paver, slipform paver and floating screed paver. The floating screed paver can be further sub-divided into two categories: tracked or wheeled pavers.

Fixed screed pavers are normally used for laying a concrete pavement and have a screed which is fixed at the beginning of the paving operation and cannot be adjusted while the paver is in motion. The height of the screed and thus the thickness of the mat produced is determined by road forms upon which the paver runs and these also form the edge of the mat. Such a process is known as a 'concrete train' within the construction industry.

Slipform pavers differ from the concrete train in that they do not need road forms or any other means to contain the concrete. The edge of the mat is contained by long forms which travel with the machine and are connected to the sides of the machine at its varying laying widths.

Floating screed pavers are the type most associated with the laying of bituminous mixtures on roads. A floating screed paver may be fitted with a fixed screed or a hydraulically extendable screed which can lay varying widths. A fixed screed is preferential on high tolerance wearing course works such as motorways. Fixed screeds or hydraulically extendable screeds can be mounted on either wheels or tracks.

In the UK, the laying of hot bituminous mixtures is almost always carried out using a floating screed paver mounted on wheels. In France tracked pavers are used when laying the very thin wearing courses (20 mm) which are common there as the French consider that the tracks give the paver better stability. A comparison of the advantages of wheeled and tracked pavers is shown in Table 5.1

Pavers are sophisticated items of equipment which are required to perform a number of complex, inter-related tasks to transfer material initially arriving on a lorry into its final position in the pavement, adhering to very tight tolerances. In wearing course operations a paver is normally expected to lay to a tolerance of ± 6 mm and in some applications to ± 3 mm.

Figure 5.4 is a diagrammatic representation of a typical floating screed paver.

5.2.3. Wheeled pavers and their functions

The functions of a wheeled paver are as follows.

(a)   Tractor: the paver's tractor unit provides the power to drive all electrical, hydraulic and other functions. It may have one or two pairs of driving wheels at the rear and normally has two pairs of steerable wheels at the front.

Table 5.1. Advantages of wheeled and tracked pavers

Wheeled pavers	Tracked pavers
Readily transportable	Better control in poor ground
Moves quicker on site and between sites	Superior maneouverability
Do not mark mat unlike tracked pavers	Better tolerances on very thin layers

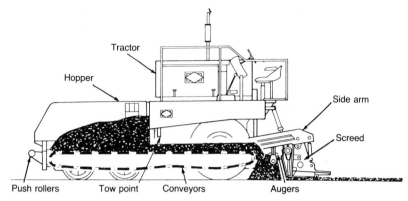

Fig. 5.4. Floating screed wheeled paver[4]

(b)   Hopper: lorries tip material into the hopper from which the conveyors carry the material to the augers at the rear of the paver.

(c)   Side arms: the side arms attach the screed to the tractor unit at the tow points. (Note that any movement at the tow points results in a movement of the screed. Hence the means of altering levels.) Switches at the rear of the paver allow the side arms to be raised or lowered, at the tow points, as required. The resultant level change at the screed is approximately 12.5% of that observed at the tow point.

(d)   Screed: this spreads and pre-compacts material received from the augers to the required finished mat profile width and depth. The screed may be a vibrating screed, a tamping screed or a combined vibrating/tamping screed.

(e)   Conveyors: these transport material from the hopper to the augers at a rate consistent with the rest of the paving operation and enable the operator to ensure that a constant head of material is present across the augers.

(f)   Augers: these spread and distribute material across the width of the screed. On many pavers the position of the augers can easily be adjusted when laying different bituminous mixtures or when laying different depths to maximize paver performance.

(g)   Push rollers: these allow constant contact between the paver and the delivery lorry and should be free to rotate fully at all times.

5.2.4. Summary of paving process

Material is discharged into the hopper of the paver. The lorry sits in neutral with its engine running (this also allows gentle use of the brakes if it is necessary to lay downhill) and with its rear wheels against the push rollers of the paver. The paver provides not only the motive power for itself but also that required to push the discharging lorry. The material is then transported from the hopper by the conveyors to the augers which distribute the material over the width of the screed. The screed then levels and pre-compacts the material to the desired profile.

The tractor unit and screed of the paver must meet a number of basic requirements when laying continuously: they must, in combination or in isolation, be able to

(*a*)   operate at speeds consistent with the material and tonnage required to be laid
(*b*)   produce an even and compact surface
(*c*)   impart initial compaction to mat
(*d*)   lay to required widths, depths, profiles, cambers, crossfalls
(*e*)   spread material without segregation or dragging
(*f*)   supply heat to the finishing screed.

The tractor unit and screed must meet these requirements whilst the laying process is in motion with a number of dynamic forces acting on the screed.

5.2.5. Factors affecting the performance of a paver's screed

Fig. 5.5 shows the main factors affecting the performance of a paver's screed (ref. 5.4). They include the forward motion of the paver (F1), the angle of attack of the screed (F2) and the amount (or head) of material in front of the screed (F3). All have an important influence on the finished quality of the laid mat.

In the laying process the weight of the screed will remain constant and can therefore be assumed not to influence the other factors, listed above, which may affect screed performance. It should be noted that when laying harsh bituminous materials mixtures such as stiff design mixes that in general terms a heavier screed will perform more efficiently than a lighter one. The vertical resistance of the screed in terms of the material being laid can also be assumed to be constant.

Forward motion will affect screed performance each time the paver stops or starts, if it is bumped by the lorry discharging material or the lorry leaves its brakes engaged and when the speed of the paver is changed. Poor maintenance may result in unwanted, random movement of the screed.

If the angle of attack of the screed is altered in the laying process there will be a change in both horizontal and vertical forces acting on the screed. This will affect the characteristics of the laid mat so no unnecessary changes should

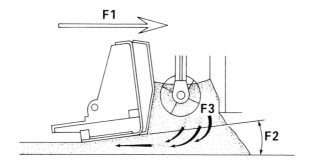

Fig. 5.5. Factors affecting the performance of a paver's screed[4]

be made. Poor maintenance of the paver may cause the screed to change its angle of attack independently of the operator with similar effects.

The head of material in front of the screed is the most critical of the three main factors which affect performance and must be kept as constant as possible across the length of the screed. Manufacturers and surfacing contractors agree that up to 90% of all mat deficiencies can be attributed to a failure to maintain a constant head of material across the screed. Many pavers now have automatic controls which continuously regulate the flow of material from the hopper along the conveyors through the augers to the finished laid surface.

5.2.6. Types of screed

Pavers used to lay bituminous mixtures have one of three types of screed fitted. These are

(a)   a vibrating screed
(b)   a tamping screed
(c)   a combination vibrating/tamping screed.

The three types are available for both fixed screeds or telescopic (extendable) screeds.

Vibrating screed

This type of screed relies on the natural feed action of the distribution of material to achieve the required levels and texture. This normally dictates that such screeds have a greater angle of attack than other types of screed as material is conveyed and distributed to the required level without any mechanical assistance such as that imparted by a tamping screed.

The frequency of vibration is determined by the speed of the vibrator which can be controlled by the paver operator. The vibration is provided by hydraulic motors driving shafts which have eccentrically loaded weights positioned on them and it is essential that these weights are set properly in phase to ensure proper action of a vibrating screed. Excess vibration may cause bitumen to

Table 5.2. Paver control functions

Factor to be controlled	Detailed description	Provided by
Control of head of material in front of screed	Control of flow of material from hopper via conveyor (and flow gates where applicable) to auger box Control of constant desired level of material across augers	Automatic sensor control devices to regulate the flow of material through the paver
Control of levels of mat	Control of mat levels on one or both sides of paver Control of mat level on one side of paver with slope level control on other side	Automatic level sensor control devices referencing levels on one or both sides of paver to reference level datum

rise to the surface leading to a laid mat of uneven density. The correct vibration speed for a particular screed and type of material will usually be found through experience. The frequency of vibration should be set low initially and increased until the correct finish is achieved.

Tamping screed

A tamping screed consists of a number of tampers (or blades) which strike the material being laid in front of the screed and compact it to the required level. The tampers operate with a vertical or inclined high amplitude oscillation at a comparatively low frequency. Tampers are normally set marginally below the main screed plate and only a small portion of the tamper unit is in contact with the mix which results in a degree of initial compaction on the material being laid. This is activated by the tamping action of the blades tucking material under the forward end of the screed. The tamper unit is normally followed by a static plate.

Combination vibrating/tamping screed

Combination screeds combine the advantages of both vibrating and tamping screeds with the tampers tucking the material being laid into the front of the screed and the vibrators compacting to high densities. The use of combined screeds provides a high degree of initial compaction which means that less compaction is required from the following rollers. Hence, the mat should not be disturbed as much by the subsequent rolling as would be the case with other types of screed. Levels are, therefore, more easily maintained.

Heating of screeds

All screeds need to be pre-heated, normally using propane gas, to the approximate temperature of the bituminous mixture being laid. This is to ensure that the material being laid will not adhere to the screed and will not be dragged when pre-compacted by the screed. This results in the mat having a closed appearance. The screed is the most important element of a paver and it must not be misused or it will inevitably affect the quality of the laid mat.

5.2.7. Automatic control devices used in paving operations

There are a number of important requirements of a paver which have to be achieved whilst the dynamic process of laying is in progress and many automatic control devices for ensuring that these important factors can be controlled as accurately as possible have been developed. The most important of these devices are shown in Table 5.2.

5.2.8. Maintaining a constant head of material across auger

Figure 5.4 shows the sequence of a typical flow of material from the paver hopper to its augers and Fig. 5.6 shows typical cross-sections of the effect of the head of material across the auger on the screed level. It should be noted that the angle of attack of the screed can be altered either at the tow point or at the screed.

While it is possible for an experienced skilled operator to maintain a constant head of material across the screed, the use of automatic control devices greatly

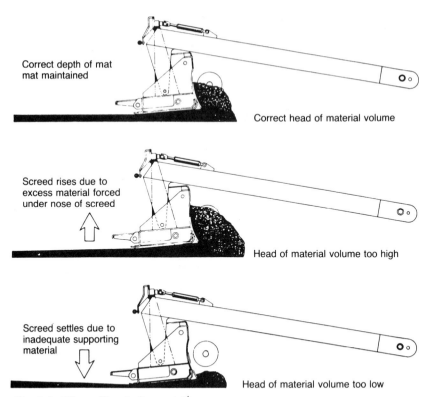

Correct depth of mat
mat maintained

Correct head of material volume

Screed rises due to
excess material forced
under nose of screed

Head of material volume too high

Screed settles due to
inadequate supporting
material

Head of material volume too low

Fig. 5.6. Effect of head of material[4]

assists this aim. The method of automatically controlling the feed and spread of material is as follows. Automatic control devices have to be located at the extremities of the auger box and positioned so that they 'cut-in' and 'cut-out' as required to maintain a constant material head. These may take the form of paddle-type switches or ultrasonic devices. The former may require some material be conveyed manually to determine, by trial and error, the optimum position of the paddle switches which maintain the correct head of material. The latter can be set and positioned to activate and de-activate when the material is a certain distance from the sensor thus requiring little, if any, manual operation to determine the optimum position.

Irrespective of the method of controlling the head of material across the augers there is clearly a need for control of material flow further back from the auger to ensure that the desired levels are maintained. Some pavers have the operation of their flow gates and auger/conveyor speed linked to the automatic sensors in the auger area to ensure a constant supply of material. Other pavers do not have flow gates but a further set of automatic sensors at the end of the conveyors to control fully auger/conveyor speeds and so the distribution of material across the auger.

5.2.9. Automatic level control devices

The use of automatic level control devices is essential in the laying of mats of high tolerances which may be as small as ±3mm and is preferable in all other circumstances. The use of such control devices are a natural extension of the principle of operation of a paver where the screed floats about the tow points and the raising or lowering of these points will produce corresponding changes of screed level and hence mat level. Automatic control devices simply maintain the screed tow points at a fixed height with respect to a datum and this results in the laying of a smooth finished mat. Automatic level control devices can be used for referencing from a datum in terms of height and also for monitoring the slope of the laid mat from the same referenced datum.

The main applications for automatic level control in surfacing operations are typically described as single joint matching, double joint matching or single joint matching with slope control.

Single joint matching

Longitudinal level control on one side of the paver is maintained with respect to a reference surface. The reference surface may be an adjoining lane laid previously, a line of kerb or a reference wire. The latter is established by levelling a line of pins at intervals of 20 m, 10 m or even 5 m as required. Automatic level control from the reference surface is by means of a grade sensor which can be mounted at various locations on the side arms of the paver. The grade sensor detects changes in vertical height from the reference surface and translates this information to an electronic control station on the same side of the paver which automatically raises or lowers that tow point and that side of the screed accordingly to lay the mat to the required levels. Level control on the other side of the screed is exercised manually.

Double joint matching

Longitudinal level control on both sides of the paver is carried out in exactly the same manner as for single joint matching but with an additional grade sensor and electronic control station on the opposite side of the paver. In double joint matching both tow points and both sides of the screed are automatically raised or lowered to lay the mat to the required levels.

A typical electronic control station is shown in Fig. 5.7 and grade sensor in Fig. 5.8.

Single joint matching with slope control

Slope control is essentially maintained with respect to gravity. Normally, a slope controller is mounted centrally on a cross-beam between the paver's side arms, the controller sensing the slope of the cross-beam with respect to gravity. Initially the slope controller is set for a specified slope, e.g. 1:40 or 2·5%, and the paver lays material to defined levels as for single joint matching on one side of the paver and controls the material depth on the opposite side of the machine relative to this by maintaining the required slope across the screed. A typical slope controller unit is shown in Fig. 5.9.

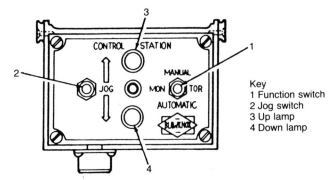

Key
1 Function switch
2 Jog switch
3 Up lamp
4 Down lamp

Fig. 5.7. Typical paver electronic control station[4]

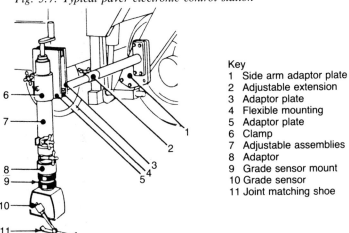

Key
1 Side arm adaptor plate
2 Adjustable extension
3 Adaptor plate
4 Flexible mounting
5 Adaptor plate
6 Clamp
7 Adjustable assemblies
8 Adaptor
9 Grade sensor mount
10 Grade sensor
11 Joint matching shoe

Fig. 5.8. Typical paver grade sensor[4]

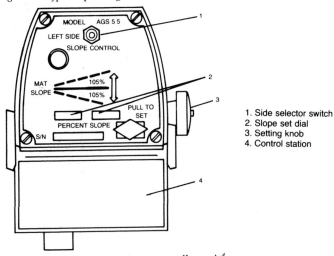

1. Side selector switch
2. Slope set dial
3. Setting knob
4. Control station

Fig. 5.9. Typical paver slope controller unit[4]

The procedures to be followed by the screedman/operator for the successful operation of an electronic level/slope control device are listed.

(a) Prepare paver for work.
(b) Lower paver screed onto a reference surface which mirrors depth and slope to be laid. Tow points are set while depth control station is set to manual.
(c) Depth control station is set to 'monitor'.
(d) Fit grade sensor-shoe; must trail, not follow, paver, by 45°.
(e) Fine tuning of grade sensor to be 'nulled' by raising/lowering until light indicators show all is OK. At this stage depth control station is set to 'automatic'.
(f) Mount slope controller and link to depth control station on opposite side of paver. Set dial to 0·00% initially.
(g) Slope controller to be 'nulled' by fine adjustment until light indicators show all is OK. Set depth control station to 'automatic'. Set slope controller to desired slope.
(h) Check proper operation of paver when both depth and slope control stations are set to 'automatic' visually ensuring that tow point movements are as set and specified.
(i) Fill hopper and during paving operations check depths and slopes at regular intervals. Adjust slope as required throughout.

This checklist emphasizes the ease with which these items of equipment can be applied to surfacing operations to ensure accurate and constant level control over the works. Each of these procedures is extremely straightforward and a paver can be properly 'set-up' by spending up to half an hour at the beginning of each paving run. Automatic level control devices greatly assist paving operations.

5.2.10. Selection of paving machine

The economics of purchasing paving equipment will be discussed in section 5.6. The selection of a paver for a contract will normally take into account such factors as width of mats to be laid, level tolerances and site considerations to name a few. Except for small surfacing operations such as those on lightly trafficked roads, in residential streets or in car parks, it is probable that a double drive wheel paver will be used. This gives much additional traction over a single drive wheel paver and this factor, combined with the additional power rating of double drive wheel pavers, means less unwanted movement of the screed when the surfacing process is in motion. Most modern pavers have a similar high specification with automatic material feed and level controls being the norm, rather than the exception. A comparison of two of the several double drive wheel pavers available is shown in Table 5.3.

Table 5.3 clearly shows that the specification for both the Dynapac and Blaw-Knox pavers differ only in the weights and widths of the screeds. All manufacturers provide data sheets on their range of pavers and this is one of the best ways of comparing specifications of different types of pavers. However, the best method of assessing the actual paver's performance is to watch them operate in similar conditions as those envisaged and analyse their comparative

Table 5.3. Comparison of paver models

	Dynapac 12000R	Blaw Knox BK191
Tractor unit		
Operating weight	15 400 kg	15 100 kg
Weight tractor	12 140 kg	12 650 kg
Weight screed	3260 kg	2450 kg
Number of drive wheels	2 pairs	2 pairs
Screed length	2.5–4.75 m	2.5–4.75 m
Screed width	300 mm	250 mm
Transportation		
Overall width	3.36 m	3.36 m
Overall length	6.70 m	6.36 m
Overall height	2.66 m	2.69 m
Travelling speed	0–18 km/h	0–20 km/h
Working speed	0–32 m/min	0–50 m/min
Engine		
Rated effect	82 kW (112 HP) at 2500 rev/min	76 kW (102 HP) at 2000 rev/min
Fuel tank capacity	160 l	160 l
Material distribution		
Hopper capacity	12 t	14.7 t
General		
Maximum depth of mat	Approx 300 mm	Approx 300 mm
Screed type	Combined tamping/vibrating	Combined tamping/vibrating
Automatic material feed control	Yes	Yes
Automatic level controls	Yes	Yes

performances in terms of outputs, finished product, operating costs, anticipated working life, availability of spares etc.

5.3. Chipping machines
5.3.1. Introduction

In the UK all heavily-trafficked routes and many carrying less traffic are surfaced with a hot-rolled asphalt wearing course with pre-coated chippings embedded in the surface to provide resistance to skidding. The specified configuration of the pre-coated chippings is principally contained in BS 594: Part 1[5] and in the *Specification for highway works*.[3] Except in certain limited circumstances the use of a mechanical chipping machine is mandatory. The original machine was designed by TRL with Bristowes, who still manufacture chipping machines today. Very few significant changes have been made since, although modern mechanical chipping machines are hydraulically rather than chain driven.

Fig. 5.10. Chipping machine. Photograph courtesy of Wrekin Construction

5.3.2. Operation of mechanical chipping machines

The mechanical chipping machine has a hopper into which pre-coated chippings are loaded. the hopper discharges chippings transversely across the width of the machine and the chippings are deposited via rotors and a rotating drum on top of the newly laid asphalt mat. The speed of the chipping machine must closely match the paver speed and preferably operate as close to the screed of the paver as is practical at all times to maximize the time available for compaction by the following rollers. A typical mechanical chipping machine is shown in Fig. 5.10.

It is necessary to be able to steer and control the speed of a mechanical chipping machine on both sides and these controls are duplicated with a control station at each end of the machine. This allows the machine to be manoeuvered with extreme precision at both sides which is an essential requirement in surfacing operations. The steering controls are shown in Fig. 5.11.

5.3.3. Control of rate of deposition of pre-coated chippings

The method of controlling the amount of pre-coated chippings being deposited is primitive in the sense that all that is measured directly from the mechanical chipping machine is the gap or gate setting which controls the amount of chippings deposited on the rotor arms and thereafter onto the new mat. However, if dry or trial runs are carried out it is possible to correlate fairly accurately the rate of spread of chippings which will be achieved by using chipping trays and balances as are used by monitoring technicians during the actual laying operation.

Gate settings are read from measuring gauges across the width of the chipping machine. It is essential that the chipping machine operator is conversant with

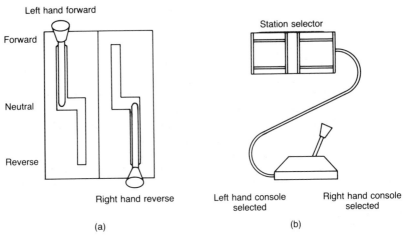

Fig. 5.11. Chipping machine steering controls

the chipping rate which results from the different gate settings. It must be noted, however, that each individual chipping machine will have its own characteristics and it is possible that the number of gauges across the width of the machine may have to be set to different readings to obtain a uniform chipping spread across the mat. The mechanism for adjusting the gate setting against the measuring gauge is shown in Fig. 5.12.

It is useful to prepare and carry a chart for a particular machine which gives an indication of required gate setting for different rates of spread of chippings. For example, a setting of 38 mm may give a rate of spread of 11 kg/m^2 while a setting of 57 mm may give a rate of spread of 15 kg/m^2. It is important to note that on inclines the rate of spread of chippings will not be the same as on a level surface. Account must be taken of this before wearing course

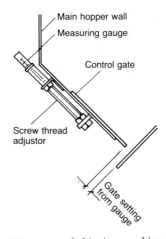

Fig. 5.12. Adjustment of chipping machine gauge setting

operations by adjusting the gate settings as necessary. The experience of the chipping machine crew plays an important role in this aspect.

5.3.4. Types of mechanical chipping machines

Mechanical chipping machines are available in a variety of widths with 12 ft (3.66 m), 18 ft (5.49 m) and 25 ft (7.62 m) being among the most common. In city streets or where safety fencing or such like is present immediately behind a kerbline it may be necessary to use a smaller chipping machine (say 10 ft (3.05 m) or less) to still allow passage of one lane of traffic without effectively closing off a complete single carriageway road. Runs which are 7.3 m wide and laid by a single or dual paver may require two smaller chipping machines (running on boards on one side) if a large 25 ft (7.62 m) type machine cannot be obtained. The chipping machines are normally fed by tractors or shovels and it is important that stockpiles of chippings are conveniently located to ensure that a chipping machine hopper is adequately supplied thus avoiding slowing or halting the laying operation.

One of the biggest problems with chipping machines is their slowness in moving from the end of one panel of finished wearing course mat back to the beginning of the next run. Their maximum working speed is approximately 15 m/min and the maximum travel speed is around 50 m/min. When a single kilometre of paving has been completed and the machine has to return to the start of the next run, it can take up to 30 minutes. The only safe, quick way to move such a machine is on its own trailer which requires a lorry with suitable air-line fittings to match those of the trailer.

5.4. Compaction plant

5.4.1. Introduction

Compaction is one of the key areas in surfacing operations which can, in many instances, either make or break a mat which has been perfectly laid by the paver. It may seem that rollers follow paving machines in a fairly random manner and at varying speeds to compact the mat but performance and durability of flexible bituminous pavements depends largely on the degree of compaction achieved during the construction phase. Compaction technology has improved in recent years with the use of vibrating rollers in surfacing operations becoming widespread. The Department of Transport has published recommendations for preferred methods of rolling to achieve optimum levels of compaction.[6] It specifies roller patterns for different types of rollers, training standards for roller drivers, the rolling length that can be undertaken at any one time, methods of checking that rolling patterns are achieved and emphasizes the need for adequate planning before surfacing commences.

5.4.2. Specification for compaction plant

The Department of Transport's *Specification for highway works*[3] identifies nine categories of compaction equipment which may be used to comply with method compaction specifications. These are

(a)   smooth-wheeled deadweight rollers (including vibrating rollers operating without vibration)

(b)   grid rollers
(c)   tamping rollers
(d)   pneumatic tyred rollers
(e)   vibrating rollers
(f)   vibrating plate compactors
(g)   vibro-tampers
(h)   power rammers
(i)   dropping-weight compactors.

Essentially this type of specification being a method specification classes the various types of compaction plant according to its mass. This is in many cases misleading, especially when considering vibrating rollers.

5.4.3. Compaction plant used in surfacing operations

The main types of compaction plant used in paver surfacing operations are smooth-wheeled deadweight rollers, pneumatic tyred rollers and vibrating rollers. In confined areas, however, vibrating plate compactors and power rammers are sometimes used. A brief description of items of compaction equipment is given below.

Three-point deadweight rollers

Compaction is achieved simply through the dead weight of the roller exerting pressure on the underlying surface. They are normally $8+$ to 12 t in weight but the rear wheels can be ballasted to provide a greater compactive effort of up to 15 t. Ballast can be water or sand. Since the latter is denser than the former the maximum weight is achieved using sand. Their speed of operation when compacting bituminous mixtures is usually around $1\frac{1}{2}$ –3 mph (2.4–4.8 km/h). The configuration of the three wheels of this type of roller gives a slow rate of coverage of a mat. For example, seven passes would normally be required for a 3.65 m wide mat to achieve 'half rear wheel overlap' as compared to three passes for a typical tandem, or vibrating roller. Weight distribution is such that the pressure exerted under the rear barrels is up to 30% greater than that imparted by the front barrel. This point should be borne in mind when rolling is undertaken using such plant. These rollers are sometimes used in pairs to increase mat coverage speeds.

Pneumatic tyred rollers

Compactive effort is dependent on the dead weight of the roller (usually between $10+$ and 35 t) and also on tyre pressures. They are often used as 'finishing' rollers to remove roller marks left by steel wheel drums and for surface sealing.

Vibrating rollers

The compactive effort in these machines emanates either from deadweight alone or by a combination of deadweight with vibration i.e. static and dynamic loading. They can conveniently be divided into three types

(a)   pedestrian operated vibrating rollers which have a single or a double drum (*b*)   lightweight vibrating tandem rollers where normally only the rear drum vibrates

(*c*)   vibrating tandem rollers with vibration on both drums.

Vibrating plate compactors

These items of compaction equipment are hand operated and provide compaction through vibratory action acting on a plate which is in contact with the material being compacted. These machines impart relatively low compactive efforts. Double drum pedestrian operated rollers or power rammers should preferably be used.

Power-rammers

These are easily transportable and lend themselves for use in the reinstatement of trenches or patches or in locations with awkward or limited working space. Compaction is provided by means of the rammer vibrating at a high frequency (450–600 strokes/min) while pounding the material which is being compacted.

In most works involving the laying of bituminous mixtures, deadweight rollers or vibrating rollers or a combination will be the main items of compaction equipment and these are now discussed in detail.

5.4.4. Choice between deadweight and vibrating rollers

There are a number of factors which may influence the decision on the best or most suitable item(s) of compaction equipment to be used to compact bituminous mixtures. These include the quantities of material to be laid, the width of mat, the depth of mat and the type of mix involved. Rollers must be selected to match and mirror anticipated paver speeds and anticipated rolling patterns to be adopted. A guide to the advantages and disadvantages for varying roller types is shown in Table 5.4.

In general terms, three-point, tandem or articulated three drum deadweight rollers are most suitable for compacting layers of depth up to 75 mm but the relatively large number of passes required for thicker layers makes their use inefficient compared to vibrating rollers. Furthermore, vibrating rollers cover the available compaction area more quickly than three-point deadweight rollers.

The three-point deadweight roller has traditionally been the hallmark of compaction plant on surfacing operations and even today their effectiveness in ironing out roller and other marks on a finished wearing course mat cannot normally be surpassed by any other type of compaction plant. Vibrating rollers are, generally, most suitable on deeper lifts and to ensure that the required number of passes are given in the minimum time. The sources of compactive efforts for different types of rollers are shown in Table 5.5

5.4.5. Deadweight rollers

There are three main types of deadweight roller, previously described as the three-point, the tandem (or two-wheeled) and the articulated three drum deadweight rollers. On standard 8–15 t three-point deadweight rollers, the static linear load (weight of roller divided by drum width) is normally between 50–80 kg/cm width and it should be noted that the pressure exerted by the

Table 5.4. Comparison of roller attributes

Roller type	Advantages	Disadvantages
Three-point deadweight roller	Excellent as a finishing roller	Slow coverage of mat means that more rollers are required to achieve reasonable productivity and avoid problems associated with poor compaction Differential pressures exerted by front/rear drums means inconsistent compaction Not suitable for deep lifts
Tandem deadweight roller or articulated three drum deadweight roller	Equal pressures exerted by front/rear drums means consistent compaction Coverage of mat quicker than 18−15 t three-point deadweight roller and equal to that of a similar width vibrating roller	Not suitable for deep lifts
Vibrating tandem	Coverage of mat quick and equal to tandem deadweight roller More efficient on deeper lifts requiring fewer passes than tandem deadweight, three-point deadweight roller or articulated three-drum deadweight roller	On wearing courses the rollers should be used in deadweight mode with little or no vibration. If in vibration mode, correct amplitude and frequency settings must be used. If settings are incorrect, rippling may occur in finished mat Not as good as a three-point deadweight roller for polishing surface

Table 5.5. Source of applied compactive effort for different types of rollers

Type of roller	Source of compactive effort
Deadweight roller	Weight of the machine distributed over the area of the drums
Vibrating roller	Weight of the machine distributed over the area of the drums Frequency of vibration of the barrels Amplitude i.e. height of eccentricity travelled by the drum from its central axis, of the barrels

Table 5.6. Typical frequency and amplitude settings on vibrating rollers for different applications (ref. 5.7)

Application	Frequency		Amplitude
	Hz	vpm	mm
Embankment construction	25–30	1500–1900	1.5–2.0
Sub-bases	25–50	1500–3000	0.8–2.0
Asphalt surfacing	33–50	2000–3000	0.4–0.8

front drum is upwards of 30% less than that imparted by the rear barrels. On articulated three drum deadweight rollers, the static linear load is normally equal on all three barrels. This is essential to ensure uniform compaction and avoid an unnecessarily awkward rolling pattern. Note that this is contrary to the differential pressures exerted by the drums of a normal three-point deadweight roller. A modern articulated three drum deadweight roller with a rolling width of 2.0 m can cover a mat width of 3.7 m in two parallel passes allowing for the necessary overlap. This is comparable with tandem deadweight rollers and vibrating rollers of similar widths.

5.4.6. Vibrating rollers

The effectiveness of vibrating rollers is more difficult to assess since it involves the weight which is constant, the amplitude which may be constant or variable depending on the model and the frequency of vibration which can be varied. The amplitude and frequency are fundamentally influential on how these type of rollers achieve compaction. Frequency can be described as the number of drum impacts or vibrations per second (Hz) and amplitude the maximum movement of the vibrating drum from the axis (mm). Numerous laboratory and field tests have been carried out to assess the performance of vibrating rollers and Table 5.6 gives an indication of typical frequency and amplitude settings for vibrating rollers for different applications.[7]

The figures for asphalt surfacing show high frequencies which help prevent surface rippling if impact spacing (the distance between each drum impact) is kept low. This can be achieved if roller speeds are kept relatively low as the combination of high frequency and low speed results in close impact spacing. The best way of altering the vibratory force exerted by a vibrating roller is to alter the amplitude setting. Essentially, with the choice of two amplitude settings the compacting effort can be adjusted to suit the different layer thickness and different types of material to be compacted. A thin asphalt layer (up to 50 mm) would normally require a lower amplitude setting, e.g. 0.4 mm, than that necessary for roadbase and basecourse compaction.

All vibrating rollers should have a system which automatically cuts out vibrations at speeds below a certain limit otherwise when the roller slows down to change direction the surface would be marked. It should also be noted that maintaining a constant speed is important in achieving a uniform degree of compaction. Some vibrating rollers have just one vibrating drum while others are capable of double drum vibration. The difference in compaction capacity between a vibratory tandem roller with two vibrating drums operating rather than one is in the order of 80% increase.

5.5. Lorries, cold planing machines, hot-boxes and infra-red heaters

5.5.1. Lorries

BS 594: Part 2[2] states that hot bituminous mixtures must be transported to site in suitably insulated lorries and be properly sheeted. This is to ensure that the material is protected from the elements and that there is the minimum loss of temperature in the mixture during transportation. A minimal amount of dust or sand on the floor of the vehicles is permitted to aid discharge of the mixture. The use of diesel for this purpose is not permitted. A lorry and driver carrying bituminous mixtures must satisfy a number of considerations. These include

(a)   Construction and Use Regulations
(b)   British/EC Drivers' Regulations
(c)   tachograph records
(d)   penalties for carrying overweight goods

The Construction and Use Regulations set out the maximum weights and dimensions of heavy goods vehicles. The maximum gross vehicle weights for the most common types of truck delivering bituminous mixtures to sites are shown in Table 5.7.

Maximum axle loadings are also specified (7.12 t for a steering axle and 10 t for a single axle) and it is important that the vehicle leaving a quarry meets all loading requirements. Penalties for overloading vehicles can be severe and the resultant delays to construction works will more than offset the benefit of carrying an extra couple of tonnes of material.

Lorries should have audible reversing devices and their geometry should be such that their overhang permits the paver's push rollers to be in contact with the lorry tyres when the body of the lorry is tipped upwards without causing damage to either the paver or the lorry.

Provision should be made on site for directing lorries to the required areas to ensure that in limited working spaces they are facing the correct way to discharge into the paver's hopper. The lorry and its driver are an integral part of the paver team and a good rapport between the rest of the squad and the drivers will help to achieve optumum surfacing results.

5.5.2. Cold planing machines

A common highway maintenance treatment is the removal of one or more layers of the pavement (normally 40 mm to 100 mm) prior to renewal of the

Table 5.7. Maximum weight and approximate capacities of trucks

Truck type	Maximum GVW, t	Approximate capacity, t
2 axle rigid	17	10
3 axle rigid	24	16
4 axle rigid	30	20

Fig. 5.13. Wirtgen 1000DC cold planing machine under test. Photograph courtesy of ALLROADS Technology

wearing course and/or basecourse. This is carried out to improve the skidding resistance and/or the riding quality of the surface.

The cold planer drives a drum (or a series of drums) which has high strength tungsten carbide picks (or teeth) attached to it. The high speed rotation combined with the cutting action of the picks, planes out the existing bituminous carriageway at a slow walking pace. The planed material is carried by means of a conveyor belt system to an accompanying lorry. Depending on the type of planing machine, the lorry may either be forward or reverse loaded. A typical cold planing machine is shown in Fig. 5.13.

These machines come in a variety of sizes with cutting widths varying typically from 0.3—4.2 m. The smaller machines e.g. those having a one metre cutting width, can normally plane out to a depth of 100 mm in one pass whereas the larger machines e.g. with a two metre cutting width, can plane out to a depth of around 300 mm in one pass. A comparison of the technical specifications of two cold planers is shown in Table 5.8.

One of the problems when using cold planing machines is that many of the smaller types scarify the existing road surface at a constant depth. Thus, any bumps or hollows will be mirrored in the planed surface. In such cases, fine material (called shaping) should be laid where necessary to remove any such irregularities.

On bigger machines, such as those compared in Table 5.8, full automatic level control of the depth of the cut is normally available. This is achieved by means of a reference grade sensor (e.g. levelling ski) in a manner similar to the automatic level control systems used on paving machines. The finished surface in this case should be smooth and regulated and ready for the laying of the basecourse and/or wearing course.

Table 5.8. Comparison of the technical specifications of two cold planers

	Wirtgen 2100 VC	Ingersoll Rand MT-7000E
Operating weight	39 525 kg	34 925 kg
Milling		
Milling width, max	2000 mm	1980 mm
Milling depth range	0−300 mm	0−300 mm
Number of teeth	148	145
Drum tip diameter	1080 mm	1082 mm
Operating/travel speeds		
Operating speed range	0−27 m/min	0−35 m/min
Travel speed	0−5.5 km/h	0−6.4 km/h
Conveyor system		
Belt width	1000 mm	800 mm
Power		
Output at 2100 rev/min	448 kW	447 kW

5.5.3. Hot-boxes

There are many occasions when the laying of bituminous mixtures cannot be carried out using a mechanical paver because access is restricted or the work is small. Laying by hand may be the only solution. The use of hot-boxes to store and lay bituminous mixtures should always be considered when patching or reinstatement work is to be undertaken.

A hot-box is essentially a dedicated or interchangeable unit which is mounted on a lorry chassis. When not in use as a hot-box the lorry may be used for normal haulage. A typical hot-box is shown in Fig. 5.14.

The main functions of the hot-box are to keep the bituminous mixture hot (by means of gas burners) and to allow controlled discharge of bituminous mixture by way of an auger followed by a chute at the rear end of the unit to the required location.

A 16 t GVW lorry will normally be able to carry an interchangeable hot-box with a capacity of approximately 8 t. Some hot-boxes have two compartments and can carry two different types of material.

The lorry is loaded at a mixing plant in the normal manner. Care should be taken to ensure that the plant can accommodate the additional height of the hot-box on the lorry beneath its loading hoppers as this is not always the case.

The gas-powered burners on the hot-box can normally keep material hot overnight if connected to a suitable power supply. However, the main benefit of hot-boxes is that the temperature of the material can be maintained at a higher value for much longer than would be the case if the material was kept in a sheeted lorry. This is important when carrying out repairs to small areas of

Fig. 5.14. Hot-box. Photograph courtesy of Proteus Equipment Ltd

patching. There is a much better chance of achieving consistent quality of work in such circumstances and also less material wasted due to it becoming unworkable through loss of heat. The one drawback in the use of hot-boxes is that an additional vehicle is normally required to remove any excavated material from proposed patching/reinstatement areas. However, for emergency use in urban areas they are far more efficient and cost effective than the majority of cold applied, expensive deferred-set materials. A great number of small/medium potholes can be made safe with 8 t of material.

5.5.4. Infra-red road heaters

It is not possible to re-work bituminous mixtures once laid and compacted except for the application of infra-red heat treatment. This method may be used to heat small areas, for example, where chippings have not been properly embedded, to a high temperature and thus, re-activate the bituminous mixture to allow re-compaction which in turn improves chip retention.

An infra-red heater unit is normally mounted on a lorry chassis and consists of an element which can re-heat the underlying surface to a temperature of up to 100°C.

The heating element is fuelled from gas supply tanks and all controls can be activated through a remote control box including forward/reverse vehicle movement, putting brakes on/off and raising/lowering the heater element. Infra-red heaters are very specialized items of equipment and few contractors own one. Hire of such machines is the best method of obtaining a fully serviced/operational heater with an experienced operator. Although their use in the surfacing industry is limited, infra-red heaters may provide a means of improving chipping embedment.

5.6. Economic performance of plant and surfacing operations
5.6.1. Introduction
This section investigates the principles of investment appraisal with respect to the purchase of major items of surfacing plant such as pavers.

5.6.2. Investment appraisal
Pavers and other items of surfacing equipment are very expensive items of plant. Typically, pavers can cost more than £100 000 (1993 prices). It is therefore essential to carry out some form of investment appraisal upon which the surfacing contractor purchasing plant can make the best economic decision. Clearly, before economics come into play it is necessary to prepare a technical specification listing the necessary or desirable features of the particular plant item. In the case of a paver such a list would include factors such as the maximum laying width, the types of screed available, the method of controlling material flow through the paver, the safety elements present, the ease of use by operator and back-up maintenance availability. At this stage a surfacing contractor may have several makes of paver model that meet the required technical specification. A useful technique of investment appraisal is called the net present value (NPV) or present worth (PW). This method adopts the principles of discounted cash flow (DCF) which takes the following important factors into account.[8]

(a)   The method allows for the fact that one pound today is worth more than one pound in the future since over a period of time any sum of money earns interest. Therefore, expected future returns i.e. profit, require to be discounted to reduce them to present day values.

(b)   The method helps overcome the problem of choosing plant which will yield in total the same expected profit over identical time periods but where the profits accrue at different times.

(c)   Discounted cash flow methods automatically take into account depreciation. Plant which attains a positive NPV implies that the discounted returns exceed the capital cost. At the same time, the plant is earning a rate of return which is higher than the rate of interest applied to the capital which funded the purchase.

(d)   An allowance for inflation can be incorporated into DCF calculations if required.

(e)   The calculation allows a surfacing contractor to make allowances for taxation regulations.

NPV or PW reduces all cash flows to a common base date either by discounting or compounding, or both, at an interest rate that can either be the cost of capital to the organization, or the return which an organization considers to be acceptable on its investments. Furthermore, the method takes account of annual maintenance costs (fixed or variable) and residual (scrap) value at the end of its estimated life. Taking all of these factors into account the NPV can be expressed as an 'equivalent annual cost'. This closely models the situation that would be encountered when purchasing a major item of plant.

The formula used in calculating the equivalent annual cost for an item of equipment or plant is as follows

$$\text{Equivalent Annual Cost} = (P\text{-}V)\frac{(1+i)^n i}{(1+i)^n - 1} + V_i + \text{annual maintenance costs}$$

where P = purchase price of the plant item
  V = residual value of the plant item
  i = rate of interest
  n = working life

The factor $\dfrac{(1+i)^n i}{(1+i)^n - 1}$

can be calculated easily or obtained directly from compound interest tables which have values for a range of interest rates. This particular factor is known as the uniform series that 1 will buy.

Where the annual maintenance costs vary, a further calculation is required.

The following examples illustrate the Equivalent Annual Cost principles using NPV technique.

Example 1.

$$
\begin{array}{ll}
\text{Cost of new paver} & = \text{£120 000} \\
\text{Lifespan} & = \text{5 years} \\
\text{Residual value} & = \text{£20 000} \\
\text{Maintenance costs:} & \\
\quad \text{Year 1} & = \text{£2000} \\
\quad \text{Year 2} & = \text{£5000} \\
\quad \text{Year 3} & = \text{£5000} \\
\quad \text{Year 4} & = \text{£7500} \\
\quad \text{Year 5} & = \text{£7500}
\end{array}
$$

The simplest method of analysing the problem is by assuming an annual maintenance charge of £2000 for the first three years inclusive and then discounting the additional maintenance charges back to year 0.

Method:

Assume an interest rate (i) of 14%.

For additional maintenance charges the compound interest factor is obtained from discount tables from the factor $1/(1+i)^n$ or the present worth of 1.

Year	Additional Maintenance Charge	Discount Factor	Discounted Cash Flow (DCF)
2	£3000	0.77	2310
3	£3000	0.67	2010
4	£5500	0.59	3245
5	£5500	0.52	2860
			10 425

$$14\% \text{ compounded over 5 years} = \frac{i(1+i)^n}{(1+i)^n - 1} = 0.29$$

Annual maintenance costs = additional maintenance charge + annual maintenance charge

$$= [(0.29 \times 10\,425) + 2000]$$
$$= £5023.25$$

Equivalent Annual Cost $= (P\text{-}V)\dfrac{(1+i)^n i}{(1+i)^n - 1} + V_i +$ annual maintenance costs

$$= [(120\,000 - 20\,000) \times 0.29] + [20\,000 \times 0.14]$$
$$+ 5023.25$$
$$= £36\,823.25$$

Example 2.

Cost of second hand paver	= £65 000
Lifespan	= 3 years
Residual Value	= £5000
Maintenance Costs	

Year 1 = £7500
Year 2 = £7500
Year 3 = £7500

$$14\% \text{ compounded over 3 years} = \frac{i(1+i)^n}{(1+i)^n - 1} = 0.43$$

Equivalent Annual Cost $= (P\text{-}V)\dfrac{(1+i)^n i}{(1+i)^n - 1} + V_i +$ annual maintenance costs

$$= [(65\,000 - 5000) \times 0.43] + 5000 \times 0.14 + 7500$$
$$= £34\,000$$

The methods illustrated in examples 1 and 2 permit an economic decision to be made by an organization by relating purchase, maintenance and residual costs to a common base date using the equivalent annual cost method.

References

1. BRITISH STANDARDS INSTITUTION. *Coated macadam for roads and other paved areas. Specification for transport, laying and compaction.* BSI, London, 1993, BS 4987: Part 2.
2. BRITISH STANDARDS INSTITUTION. *Hot-rolled asphalt for roads and other paved areas. Specification for the transport, laying and compaction of rolled asphalt.* BSI, London, 1992, BS 594: Part 2.
3. DEPARTMENT OF TRANSPORT. *Specification for highway works.* HMSO, London, 1992, **1**.
4 BLAW-KNOX CONSTRUCTION EQUIPMENT CO LTD. *Paving Manual.* Blaw-Knox Construction Equipment Co Ltd, Rochester, Dec., 1986.
5. BRITISH STANDARDS INSTITUTION. *Hot-rolled asphalt for roads and other paved areas. Specification for constituent materials and asphalt mixtures.* BSI, London, 1992, BS 594: Part 1.

6. DEPARTMENT OF TRANSPORT STANDING COMMITTEE ON HIGHWAY MAINTENANCE. *Preferred method 8 — road rolling.* Cornwall County Council, Truro,

7. DYNAPAC. *Compaction and paving. Theory and practice.* Dynapac.

8. PILCHER, R. *Principles of construction management.* McGraw Hill, Maidenhead, 1976, 2nd edn.

6. Laying operations

6.1. Introduction

There is little point in carefully designing bituminous mixtures, both in terms of composition and thickness, for use in road construction if the materials are not laid properly. This chapter examines the laying of bituminous mixtures from a practical standpoint i.e. from the surfacing contractor's perspective.

The first section deals with the most important consideration on any building or construction site, safety. The rest of the chapter considers surfacing work, with sections on the initial estimate, programming, ordering, transportation, laying, rolling and compaction, quality control and after care.

6.2. Safety

Construction sites, especially congested roadworks, are particularly dangerous places to work. It is vitally important that everyone concerned is aware of his personal responsibility for the safety of himself and his colleagues. One means of achieving this is for surfacing contractors to formulate and adhere strictly to a site safety policy.

6.2.1. Site safety policy

A site safety policy should address the following issues

(a) Section 7 of the Health and Safety at Work Act, which places strict legal requirements on all construction staff

(b) wearing of safety equipment including safety helmets and reflective jackets or vests

(c) access through the site which must not be restricted by careless parking of vehicles. Marshalling areas for delivery vehicles and stand-by plant should be stipulated

(d) site speed limits need to be fixed and enforced

(e) the disposal of waste

(f) the reversing of vehicles, which needs limiting and must involve the use of banksmen

(g) overhead power cables and other hazards; locations of buried utilities need signing

(h) the danger of children and pets on sites; these usually arrive in vehicles.

6.2.2. Site safety officer

Each site should have a site safety officer who may also act as a traffic officer on smaller sites since traffic management and organisation are often major factors in controlling site safety.

6.2.3. General safety

Each surfacing gang should have its own safety representative and someone trained in administering first aid. This need not necessarily be the same person.

His responsibilities would include day to day implementation and monitoring of the company safety rules. In addition, he may be responsible for seeing that at the end of the working day all plant is parked and sheeted correctly, cabs are locked, plant immobilized wherever possible and flashing beacons are in position and working on all items of plant.

Fuel gas (LPG) bottles should be stored in a locked compound. Each nuclear density meter should be stored in its own special secure box or cabin and, when in use, a strict code of practice followed, including the prominent displaying of the nuclear symbol and unauthorized or 'badged' personnel kept at least two metres away from such equipment at all times.

The use of jack hammers or asphalt saws for cutting of joints should only take place when the operative has the appropriate protective clothing, including ear muffs or plugs, goggles and masks. Dangerous locations such as where there are overhead or buried cables should be suitably fenced off by scaffold poles, goal posts or bunting with a clear indication of the voltage. The tipping of delivery vehicles under high voltage cables must never be allowed and the supervision of tipping lorries should be under the control of an experienced banksman at all times, with particular attention being given to the reversing of vehicles and the setting-up of turning points and cleaning-out areas. The most common cause of serious accidents, and occasionally deaths, is due to the reversing of contractor's plant and delivery vehicles. Audible reverse warning systems should be fitted to all large items of plant.

Horse-play, children, animals and drink are taboo on any surfacing site. Site speed limits should be strictly enforced. Mobile plant must only be operated by a competent, certified driver and no unauthorized person should be permitted to ride on such plant, in particular the paver, when travelling up and down the site.

6.2.4. Precautions for paver work

The following points, as appropriate, should be observed by anyone involved in paver work.

(a) The driver should always walk around the paver and check for obstructions and equipment in the hopper before mounting. Deck plates must be kept free of oil, grease, obstructions or other hazards. Safety guards must always be in position. The driver must ensure that all controls are in neutral or a safe position before starting the engine and he should always look around and signal before moving off. The wash down system, which uses diesel, should not be operated while the screed is hot.

(b)   No personnel should walk between the paver and a reversing lorry or enter the hopper while the engine is running.

(c)   All personnel must wear protective clothing, know the locations of fire extinguishers and be aware of other traffic at all times.

(d)   Extendable screeds should not be operated without checking for personnel of other obstructions.

(e)   Machines must be braked properly when parked or during servicing and maintenance.

(f)   Appropriate warning measures must be employed at all times.

(g)   Servicing and maintenance must not be carried out while the engine is running and batteries must be disconnected when the paver is undergoing electrical servicing. Hydraulic pressure should be released before carrying out hydraulic servicing.

(h)   During transportation on low loaders, pavers must be adequately secured.

6.3. Estimating for work

Roads contracts are normally awarded to the contractor who submits the lowest tender. This must allow him to produce a pavement of the specified quality at a cost which produces a return on capital at the end of any contract. An examination of the estimating process highlights many of the practices which an efficient surfacing contractor will adopt if these aims are to be met.

Contracts essentially consist of hand laying work, machine laying (paver) work or a combination of both types of operation.

6.3.1. Main components of the estimate

The important elements of any estimate are

(a)   labour: number, skills and time
(b)   plant: types, combinations, time
(c)   material: specification, laid thicknesses
(d)   sundry expenses: traffic safety and control, supervision, setting out
(e)   overheads and profit.

6.3.2. Labour

The sizes of surfacing gangs vary according to the type and size of contract. However, for estimating purposes, Table 6.1 shows the compositions which can be used in various combinations to cover most eventualities.

To calculate an average daily gang cost, the number of men employed is divided by the anticipated average daily tonnage laid to produce a mean cost of labour/t laid.

6.3.3. Plant

The types and uses of the various kinds of available plant is covered in detail in chapter 5. The items of plant which may be required for a machine laying contract are shown in Table 6.2.

Ancillary plant, small tools and sundry material should also be included, where appropriate. Some of the more common items and their functions are given in Table 6.3.

Table 6.1. Composition and usage of surfacing gangs*

Gang	Operative skills	Plant	Usage	Remarks
4 man hand-laying gang	Foreman, raker, roller driver, general operative	Crew bus, roller	Footways with good access	Often forms chipping gang in 12 man machine laying gang
6 man hand-laying gang	Foreman, 2 rakers, roller driver, loader driver, general operative	Crew bus, roller, shovel loader	Footways with poor access or where larger outputs are available	Often forms chipping gang and string men in 14 man machine laying gang
6 man machine-laying gang	Foreman, paver driver, screwman, raker, roller driver, loader driver/general operative	Crew bus, paver, 2 rollers, shovel loader	Small roads, housing estate roads, roadbase on larger jobs where level control is supplied	Basic machine laying gang
8 man machine-laying gang	Foreman, paver driver, screwman, 2 roller drivers, raker, loader driver, general operative	2 crew buses, paver, 2 rollers, shovel loader	Roadbase on larger jobs where level control is not supplied or where second roller is required for high output	
10 man machine-laying gang	Foreman, paver driver, 2 screwmen, 3 roller drivers, raker, loader driver, general operative	2 crew buses, paver, 3 rollers, shovel loader	Basecourse on larger jobs and motorways	
12 man machine-laying gang	2 foremen, paver driver, 2 screwmen, 2 roller drivers, chipping machine driver, 2 chippers, loader driver, general operative	2 crew buses, paver, 2 rollers, chipping machine, shovel loader	Wearing course on re-surfacing jobs and housing estates	
14 man machine-laying gang	2 foremen, paver driver, 2 screwmen, 3 roller drivers, chipping machine driver, 2 hand chippers, 2 loader drivers, general operative	2 crew buses, paver, 3 rollers, chipping machine, 2 shovel loaders	Wearing course on motorways	

*Subject to variation according to site conditions, output and material types.

Table 6.2. Plant for machine-laying work

Item	Detail
Pavers	One paver plus a standby if the contract size warrants spare capacity. It may be sensible to have one paver with an extending screed plus one fixed screed.* Should an extending screed machine be unacceptable then it would be usual to have two fixed screeds built up to different widths.
Rollers	Two three-point deadweight rollers, both equipped with cutting wheels and scarifiers, with water sprinklers and mats. One large vibrating roller plus one smaller vibrating roller or vibrating plate for difficult, small areas.
Chipping machine	With weigh scales and trays for checking calibration.
Loading shovel	Lifting offcuts from laid mats and feeding chipping machine. Also equipped with compressor, jack hammers and hoses and asphalt cutters.
Specialist plant	Mechanical road sweeper, for use in initial sweeping of existing surfaces before application of tack coat, and for use in picking up joint offcuts and surplus pre-coated chippings. Cold planing machine for use in preparing tie-ins at extremities of site, such as start and finish lines and side road entrance. Infra-red heater for use in minor remedial areas, such as the removal of roller marks, re-embedment of chippings or rectifying small areas of under-compaction.

*Extending screed pavers are not normally used to lay hot-rolled asphalt wearing course. However, they are very useful when laying roadbase or basecourse in varying widths to suit joint patterns or when laying in excavated areas, bellmouths, etc. where frequent changes of width of mat are necessary.

On larger sites, it is often the case that the full area of surfacing is not made available at one time. The surfacing contractor should, when sub-contracting to a main contractor, ask for a simple programme which shows when particular areas will be released so that he can include an allowance for the particular costs associated with moving on and off site. As with the gang labour cost, similarly the daily cost of plant is divided by the average daily tonnage output to produce a cost/t for plant.

6.3.4. Materials

The material content of the estimate usually accounts for around 60% of the total cost and on a major motorway contract, may account for as much as 90% of the total cost.

While the average daily tonnage output is extremely significant, an equally important figure in any surfacing estimate is the material superage figure i.e. the area covered/t of material. Different materials have different compositions and therefore different coverages. The properties of the geological aggregate used in any mix will also influence the superage.

Superage figures for any combination of specification and aggregates might vary by as much as 10%. The volume per unit weight is likely to be in the range $0.4-0.45 \, \mathrm{m^3/t}$.

$$\text{Superage in } \mathrm{m^2/tonne} = \frac{1000 \times \text{volume/Unit weight } (\mathrm{m^3/t})}{\text{Thickness (mm)}}$$

Table 6.3. Ancillary plant and small tools used in surfacing operations

Item of plant	Use in surfacing operations
Emulsion sprayer	Usually driven by an integral petrol engine (used for spraying a layer of material ready for the next coat) where there is doubt whether the existing layer will bond with the mat to be laid. A hand-operated standby is often kept in reserve, particularly on larger contracts.
Bitumen boiler	Heating of bitumen to be used on joints, around street furniture, etc. Bitumen pots are also required.
Chipping boards	Placed and moved with the chipping machine to ensure that the channels i.e. the extreme sides of the mat, are kept free of chippings for the purposes of allowing a free flow for surface water run off and for aesthetic reasons.
Shovels	Different types for shovelling materials at edges of mats and start and finish of runs and also for hand chipping.
Tampers	Melting bitumen in joints ensuring all voids are filled, known as sweating and also hand compaction of materials in confined spaces and around street furniture.
Rakes	Hand spread (dressing) asphalt in channels, at joints and around street furniture.
Tool heaters	Heat all tools which come into contact with materials to prevent adhesion between the materials and the tools.
Mechanical sweepers	Used for sweeping carriageway before next layer, usually where roadbase or basecourse has been used for traffic.
Core drilling rig	Used for taking cores in laid materials usually to check on achieved compaction levels.
Fuel tanker	Fuel for plant.
Water tanker	Water for rollers to prevent barrels sticking to hot materials.
Tapes and steel rules	Check depths and widths to be laid and during and after laying.
Chipping trays/balance	Check that the desired rate of chipping is achieved on chipped wearing courses.
Sundries	A supply of tack coat, enough for at least two days work. Bitumen pot with burner and supply of gas. Spare bottles of LPG for the paver. Large drum of diesel, for topping up plant when fuel tanker breaks down or is late and also for cleaning the paver and the tack coat sprayer at the end of the day. Supply of spray road marking paint for delineating mat widths, depths, chainages. Paver driving lanes and spot levels. Supply of bitumen and turks head brushes for joint painting. 80 lb breaking strain cat-gut line. At least two block-up staffs or dipping stands. A 2 m rule. A 3 m straight edge for use in cutting back cross joints and checking levels across wearing course joints.

e.g. Taking volume per unit weight as $0.425 \, \text{m}^3/\text{t}$

(a) 100 mm of roadbase supers at

$$\frac{1000 \times 0.425}{100} = 4.25 \, \text{m}^2 \text{ per tonne}$$

(b) 60 mm of basecourse supering at

$$\frac{1000 \times 425}{60} = 7 \, \text{m}^2 \text{ per tonne}$$

(c) 40 mm of wearing course supering at

$$\frac{1000 \times 425}{40} = 10.6 \, \text{m}^2 \text{ per tonne}$$

The above are average figures. For a large contract more accurate figures should be used taking account of the anticipated degree of compaction or air voids content, the specific gravity of the aggregate to be used and the percentage binder content of the mix. Experience with local materials is the most reliable basis for applying accurate superages.

Sample calculations are given below.

Superage of DBM basecourse.

(a) Data

Relative densities	2.72 — aggregate
	2.51 — sand
	2.65 — filler
	1.00 — bitumen
Specification	BS 4987: Part 1: 1987[1]
Centre of specification	61% aggregate
	33.5% sand
	5.5% filler
	100%

Binder is 4.7% of total mix.

(b) Calculation

Aggregate proportion $= 100\% - 4.7\%$
$= 95.3\%$

So, to take account of binder content, the proportions of aggregate, filler and sand above should be reduced, proportionately, such that they total 95.3%

$$
\begin{aligned}
\text{Aggregate} &= 61\% \times 0.953 &= 58.2\% \\
\text{Sand} &= 33.5\% \times 0.953 &= 31.9\% \\
\text{Filler} &= 5.5\% \times 0.953 &= 5.2\% \\
& & 95.3\%
\end{aligned}
$$

Theoretical maximum density $= \dfrac{100}{\dfrac{58.2}{2.72} + \dfrac{31.9}{2.51} + \dfrac{5.2}{2.65} + \dfrac{4.7}{1.0}}$

$= 2.453 \, \text{t/m}^3$

So, the theoretical maximum density (voidless) of the mix is 2.453
Assume 5% air voids in laid 20 mm DBM basecourse

$$\begin{aligned}\text{Compacted Density} \quad &= 2.453 \times (1.00 - 0.05) \\ &= 2.330 \, \text{t/m}^3\end{aligned}$$

Effectively 2.33 Mg of DBM basecourse occupies $1 \, \text{m}^3$

$$\begin{aligned}\text{So 1 Mg of DBM basecourse occupies} \quad &\frac{1}{2.33} \, \text{m}^3 \\ &= 0.43 \, \text{m}^3\end{aligned}$$

$$\begin{aligned}\text{At 60 mm thickness, superage} \quad &= \frac{0.43}{0.06} \\ &= 7.2 \, \text{m}^2/\text{t}\end{aligned}$$

So, each tonne of this material at 60 mm thick will cover $7.2 \, \text{m}^2$.

Superage of chipped hot-rolled asphalt wearing course.

(a)   Data
Relative densities   2.72 — aggregate
                         2.51 — sand
                         2.65 — filler
                         1.00 — bitumen
Pre-coated chippings — same aggregate as bituminous mixture aggregate

Specification   BS 594: Part 1: 1992[2]
Centre of specification   34% aggregate
                          56% sand
                         10% filler
                         100%

Binder is 8.3% of total mix.
(b)   Calculation

$$\begin{aligned}\text{Aggregate proportion} \quad &= 100\% - 8.3\% \\ &= 91.7\%\end{aligned}$$

So, to take account of binder content, the proportions of aggregate, filler and sand above should be reduced, proportionately, such that they total 91.7%

$$\begin{aligned}\text{Aggregate} &= 34\% \times 0.917 &= 31.2\% \\ \text{Sand} &= 56\% \times 0.917 &= 51.3\% \\ \text{Filler} &= 10\% \times 0.917 &= \underline{9.2\%} \\ & & 91.7\%\end{aligned}$$

Relative density proportions in the mix

$$\begin{aligned}\text{Theoretical maximum density} \quad &= \frac{100}{\dfrac{31.2}{2.72} + \dfrac{51.3}{2.51} + \dfrac{9.2}{2.65} + \dfrac{8.3}{1.0}} \\ &= 2.371 \, \text{t/m}^3\end{aligned}$$

So, the theoretical maximum density of the mix is 2.371 t/m³
Fig. 6.1 represents a layer of chipped hot-rolled asphalt wearing course
For the purposes of this calculation only, if the chippings are treated

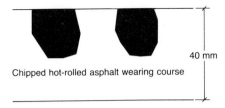

40 mm

Chipped hot-rolled asphalt wearing course

Fig. 6.1. Physical model of chipped hot-rolled asphalt wearing course

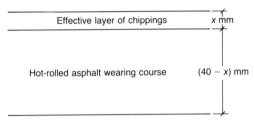

Effective layer of chippings x mm

Hot-rolled asphalt wearing course $(40 - x)$ mm

Fig. 6.2. Mathematical model of chipped hot-rolled asphalt wearing course

as a homogeneous layer of aggregate at the top of the wearing course then the volume occupied by the chippings can be discarded. This is shown in Fig. 6.2

Rate of spread of pre-coated chippings   = 12 kg/m^2
Relative density of pre-coated chippings = 2.72
Density of pre-coated chippings   = 2.72 Mg/m^3
i.e. 2.72 Mg occupies 1 m^3
so 12 kg occupies $\dfrac{12}{2720}$ m^3 = 4.4118 × 10^{-3} m^3

So the effective thickness of chipping aggregate i.e. x in Fig. 6.2 is 4.4 mm
Effective asphalt thickness is 40 mm − 4.4 mm
                                  = 35.6 mm
Assume 6% air voids in laid hot-rolled asphalt wearing course
Net effective Relative Density   = 2.371 × (1.00 − 0.06)
                                  = 2.228 t/m^3
Effectively 2.228 Mg of hot-rolled asphalt wearing course occupies 1 m^3

So 1 Mg of hot-rolled asphalt wearing course occupies $\dfrac{1}{2.228}$ m^3

                                  = 0.45 m^3
At 35.6 mm thickness superage   = $\dfrac{0.45}{0.0356}$
                                  = 12.7 m^2/t

So, each tonne of this material at 40 mm thick will cover 12.7 m^2

Obviously, the values of the parameters will have a significant effect on the calculation and the resultant superage. Since the cost of bituminous mixtures is such a large proportion of the surfacing contractor's tender, superage inaccuracies exceeding 10% are too high to be acceptable. However, this factor seldom receives the same attention as the gang size or the tonnage output despite the fact that it warrants close consideration.

6.3.5. Calculation of surfacing tender rates
Increasingly estimates are produced by computer using spreadsheet programs. However, the basic system is illustrated in Table 6.4.

6.4. Pre-planning and programming
As is the case with all construction processes, adequate time must be spent on pre-planning and programming to ensure that the finished product meets the requirements of the specification and that operations are carried out in an efficient, cost-effective manner. Adverse weather conditions, stop/start and frequent interruption of the continuous laying process, coupled with a lack of attention to temperature control, compaction and poor site preparation are the most common reasons for poor laying and premature failure of bituminous surfaces. Programmes often become quickly outdated and should be reviewed regularly. Table 6.5 covers points to be considered.

The first contract programme should be prepared during the tender period.

Table 6.4. Calculation of surfacing rates*

Specification	40 mm DBM roadbase	28 mm DBM basecourse	14 mm hot-rolled asphalt wearing course
Area (m^2)	5000	5000	5000
Thickness (mm)	100	60	40
superage (m^2/t)	4.25	7	10.6
Material tonnage (t)	5000/4.25=1176	5000/7=714	5000/10.6=472
Daily output (t/day)	500	350	235
Labour cost (£/t)	650/500=1.30	650/350=1.86	1000/235=4.25
Plant cost (£/t)	300/500=0.60	300/350=0.86	500/235=2.13
Material cost (£/t)	18.00	20.00	24.00
Sundries (£/t)	50/500=0.10	50/350=0.14	75/235=0.32
Sub-total cost (£/t)	20.00	22.86	30.70
Add overheads and profit say 10%	2.00	2.29	3.07
Cost/m^2	22.00/4.25 £5.18	25.15/7 £3.59	33.77/10.6 £3.18

*Figures are typical values but care should be taken to ensure that the values employed for a particular material/situation/contract are accurate. The estimate should not be confused with the programme and must never be used as a substitute for a proper programme.

Table 6.5. Points to be considered in compiling a surfacing programme

Topic	Detail
Site preparation	Any preparatory work to be carried out by the surfacing contractor, usually done by the main contractor.
Labour	Numbers and skills requirement.
Plant	Numbers and types including standby plant.
Specialist plant	Cold planers, infra-red heater, core drill rig, mechanical sweepers, etc.
Materials	Full specification including BS table numbers, aggregate nominal size and percentage, binder type and percentage content. Total tonnage plus daily output and hourly rate of supply which should be maximized at all times if the average used in the estimate is to be achieved. Full delivery address, together with limitations, if any, on points of entry to the site and access road. Special requirements: minimum batch tonnage in cases where there is more than one source of supply, laying special pre-coated chippings, early delivery to allow for testing.
Notices	Notices to Engineer's Representative, main contractor and other sub-contractors.

It will show the phasing of the works, traffic management schemes, etc. to facilitate the most economic price for the work. Re-programming on a daily, weekly or sectional basis, should be a continual process throughout the contract with all relevant staff providing an input.

6.5. Pre-start inspection

Surfacing works, usually the last operation on most road contracts, are invariably inadequately planned by the main contractor. This often results in impossible demands being made upon the surfacing contractor. No plant or

Table 6.6. Pre-start checklist

Item	Detail
Sub-base/planed area	Checks on the surface to be covered including: the absence of soft spots in the sub-base; dipping of the sub-base or existing surface; ensuring that the sub-base has been properly shaped to give the correct cambers, crossfalls, moving crowns; adequate compaction of the sub-base; adequate cleaning and sweeping of planed areas; ability to support surfacing plant and delivery vehicles.
Water supply	Suitable water supply either stop-tap box or water bowser for filling up rollers.
Access	Site entrance, road signs, ramps, etc., paving machines, rollers and delivery vehicles must have reasonable access to the area to be surfaced and access must be able to support surfacing plant and delivery vehicles.
Cleanliness	All accesses and surface to be overlaid to be free of mud, debris, etc.
Safety	Inspection of site safety rules and any hazards.

materials should be ordered until the site has been thoroughly inspected by the surfacing contractor (usually a surfacing supervisor and a foreman) together with a representative of the main contractor and the clerk of works or resident engineer.

Points requiring particular attention would include those listed in Table 6.6.

6.6. Ordering plant, equipment and materials

6.6.1. Ordering plant and equipment

Once an order is placed or a firm acceptance of a tender is received, the items of plant and equipment, typically as shown in Tables 6.2 and 6.3, should be ordered internally or externally when hiring specialist items. Provisional orders are commonplace for specialized items of plant or materials such as chipping machines and pre-coated chippings where advance notice of hire is required. Table 6.7 gives a surfacing operations checklist which includes plant considerations.

6.6.2. Ordering materials

Areas available for work

Obviously the area should be as large as possible. Should any isolated areas be unavailable, for instance at a bridge approach, then a patch of 50 m or 100 m (the actual size of the patch is not important) should be left unsurfaced. Any attempt to surface numerous small areas in such circumstances invariably leads to unsatisfactory results both in terms of variability in level control and finish.

Areas are not considered available until

(a) The sub-base is trimmed, rolled and passed by the engineer's staff.

(b) Any overlay or planed off area is thoroughly cleaned and passed by the engineer's staff.

(c) All the required setting out is in place. This should be clear and unambiguous. Colour codes should be used if more than one crossfall is required using pins and tapes (further reference to this subject is made later in this Chapter). All references to old or spurious setting out must be removed. The surfacing gang should be thoroughly acquainted with the system used.

(d) All earthworks, kerb laying, pipe laying and manhole or gully construction is complete. Street furniture (gullies, access covers, manholes etc.) should be set at levels which are no higher than the top of the first layer of bituminous mixture which is to be laid.

(e) All unnecessary site traffic has been diverted.

(f) Site access is in a clean and hard condition since delivery vehicles can quickly turn a site into a mud bath. The access must be clear and not blocked by cranes, concrete pumps etc.

(g) All temporary sub-base or hardcore access ramps into the excavation area have been removed and replaced with timbers or hand-laid bituminous mixtures.

Supply of materials from the mixing plant

Once the work area is available the required tonnage can be calculated and an order placed. The tender rates are now largely irrelevant. The surfacing

Table 6.7. Plant, contractual and environmental surfacing checklist

Environmental	Weather to be expected
	Exposed location?
	Air and wind temperatures
	Have previous problems been encountered in the area?
	Schools, populated areas: inform them if possible
Contractual	Noise levels
	Restricted working hours
	Compaction specifications: method or end product?
	Laying specifications (BS 594, BS 4987, contract specification)
	Has site been checked to make sure that it is ready for surfacing?
Paver	Plan sequence of paver mat runs
	Size/type of paver required: fixed width/telescopic?
	Foreman and gang availability
	Source/s of material
	Width/length of runs
	Mat profile
	Crossfall, camber or crown?
	Automatic level control required?
Rollers	Number and type of rollers for each layer of construction
	Availability of water supply for sprinklers
	Cutting devices on rollers to trim joints
Chipping machine	'Dry run' to set chipping rate
	Preferred locations for chipping machine to be delivered
	Spanners for changing gauge setting
	Width of mat runs
	Obstacles, fences, width restrictions
Bituminous materials	Specifications of materials to be laid
	Minimum and maximum delivery and rolling temperatures
	Distance of site from mixing plant
	Contact name and telephone number of mixing plant
	Method of placing orders (when and by whom)
	Has proper amount been ordered?
	Are enough chippings present?
	Who checks material coverages (superages) on site to adjust material orders as necessary?
Other plant: Emulsion sprayer Lorries	Quantity of emulsion needed per layer (if applicable)
	Are they properly insulated?
	Are they properly sheeted/double sheeted?
	Is material covered until just before use?
Tractors	For feeding chipping machines
	For cutting 'cold' joints
	Number of compressor guns/power of tractor
	Compressor hoses, etc.
Bitumen boiler	For painting bitumen on longitudinal/transverse joints
	Quantity and type of bitumen required
Hand tools and other	Rakes, shovels, tampers, safety equipment (gloves, hard shoes, hard hats, high visibility jackets, etc.)
Tool heaters	For heating rakes, shovels and tampers to prevent adhesion of material

Table 6.7 continued

General	Procedures and responsibilities for daily maintenance of all plant
	Supply of fuels/oils to all items of plant
	Gas supply for screed heating
	Storage of plant overnight
	Forward planning of low loader movements to next job
	Other provisions for workforce if weather not suitable for laying
	Steel rulers, 30 m tape, string lines, metre rule, paint, etc.
	First Aid equipment/fire extinguisher

contractor should aim for maximum output given the production capability of the source of supply, the nature of the site and the size of the laying gang.

Tonnages are calculated from the area, depth and density of the material being laid. This is the easiest part of the ordering process. Timing and rates of laying are equally important if a good working relationship is to be established between the laying gang and the supplier.

Rates of laying are determined not only by paver speed but also by the width and depth of the mat to be laid and access restrictions.

Laying at a constant speed of 200 m/h and assuming a density of 2.6 t/m^3 will give the following outputs

DBM roadbase	5 m wide × 125 mm depth =	300 t/hr
DBM roadbase	3 m wide × 100 mm depth =	150 t/hr
DBM basecourse	3 m wide × 60 mm depth =	85 t/hr
Hot-rolled asphalt wearing course	3 m wide × 40 mm depth =	60 t/hr

The outputs quoted above should be regarded as the maximum on all but the very largest motorway type contracts. Poor access for delivery vehicles, or reversing of lorries onto the paver over long distances will reduce these figures considerably. Increasing the speed of the paver is not a solution because if the paver lays too fast the screed will drag and the mat will have a poor finished appearance.

In calculating the daily tonnage requirement and the timing of deliveries, allowance should always be made for the following contingencies which invariably occupy the first hour of the working period. These are that

(a) the paver, rollers etc. must be fully fuelled, moved onto site and positioned ready for laying to begin
(b) the paver screed may have to be built up
(c) the screed must be heated to the laying temperature
(d) joints must be cut, prepared and painted with bitumen
(e) rollers must be filled with water
(f) the chipping machine must be calibrated and checked
(g) tack coating must be completed
(h) driving lines and levels must be marked out.

Material deliveries should be specified at intervals which allow sufficient time for the previous load to be correctly laid. A discussion with the foreman is always beneficial, especially if there is hand-laying work to be undertaken.

Where possible, the surfacing contractor should hold delivery vehicles on site for the minimum time particularly early in the working day. Small, awkward areas should be left until later in the day when the need to return lorries to the mixing plant for another load is less urgent. Material deliveries should fit in with meal breaks and one hour should be left at the end of the shift to allow for rolling off and preparation of the site for a prompt start on the next shift.

Finally, before orders are placed with the supplier, the specification should be written down and checked. The required tonnage and the timing of deliveries should be made clear to the supplier. The order should consist of, where appropriate

- (a)   the name of the surfacing contractor
- (b)   the name of the person placing the order
- (c)   contact telephone number
- (d)   contract description
- (e)   delivery locations
- (f)   lorry sizes
- (g)   tonnages of each of the materials
- (h)   full material specification/s
- (i)   start times (morning and after lunch, say)
- (j)   delivery times or frequencies and any other special instructions.

The order is normally placed over the telephone to the supplier's order clerk. He should be asked to repeat it back to the person placing the order to avoid mistakes or misunderstandings. Whenever possible, a faxed confirmation of the order should be sent. On a major surfacing contract such orders have a substantial financial value so there can be no room for error.

Efficient turn around of delivery vehicles will assist the supplier and help foster good relations. This will lead to good service from the mixing plant. Normally a half hour site waiting time is built into the prices of supply by the mixing plant. Where this is exceeded, a charge for the extra period (standing time) is normally incurred. However, it is often useful for the surfacing contractor to examine all delivery times to ensure that he has not incurred standing charges as a result of late delivery by the mixing plant. It may also be the case that delivery vehicles have, on average, been emptied in substantially less than the half hour allowed and that the mixing plant has had some benefit from the efficiency of the surfacing contractor.

Gang output and motivation

Surfacing gangs are specialists in the laying of bituminous mixtures. They are seldom capable of, or inclined towards, coping with the unexpected or undertaking other works such as preparation of sub-base, kerb laying or other specialist work which should have been completed ahead of the laying process. They perform at maximum efficiency when everything runs smoothly and the laying process can continue uninterrupted between meal breaks.

If sites are ready to receive bituminous mixtures and the correct equipment is available on time and in good working order and material is supplied at a steady rate without interruption, then a gang will produce consistently good quality work at a rate of productivity which will ensure the profitable completion of contracts and the minimum of remedial works. Failure to comply with these basic requirements means that morale and motivation suffer.

It is most important not to expect too much of a gang. If it is found that extra capacity is available then rather than order more material for the same day it is often wiser to discuss the tonnage with the foreman and increase the order for the following day. Such an approach is less likely to adversely affect the morale of the gang and should lead to better overall productivity.

If the order is changed during a working day then it is wise to confirm the total tonnage for that day to avoid the possibility of duplication if, say, the alteration has already been communicated directly from site.

Ordering the exact tonnage of material necessary to complete surfacing up to a closed joint, such as excavation areas, patches or tie-ins, calls for considerable skill and experience. Wherever possible, it is advisable to organize the day's work so that there is an adjacent area, such as a footway or central reserve, where excess material can be laid. Where this is not possible, an extra 1 or 2 tonnes should be ordered. The superage obtained in the area of work should be monitored carefully so that the coverage can be accurately anticipated. It is wise to liaise closely with the mixing plant on the exact weight of the last load.

6.6.3. Surfacing operations checklist
Table 6.7 summarizes actions and considerations necessary before surfacing work can start.

6.7. Transportation
All bituminous mixtures should be mixed and delivered within the limits specified in the contract documents so optimum compaction is possible. Good transportation practice is summarized in Table 6.8.

Fig. 6.3 shows a delivery vehicle discharging into a paver. The lorry is clean, well insulated and sheeted and reversing under the control of a banksman. The sheeting on the lorries is removed only as far as is necessary to facilitate discharge. The 3 point deadweight roller is tight behind the paving operation. The vibrating roller is working to a pre-determined pattern.

6.8. Laying
6.8.1. Setting out
Before laying can begin, all necessary setting out should be completed. The gang foreman will usually organize this operation whilst the paver crew are setting up the paver, filling rollers with water, checking sprinklers, clearing off the wheels, calibrating the chipping machine, checking stockpiles of chippings, working out the pattern of delivery of chippings from the stockpile to the point of laying and heating tools and bitumen.

A laying pattern should be worked out which maximizes the lengths of runs thus avoiding the need for joints and hand-laying as far as possible. The following points should be remembered when working out a laying pattern

Table 6.8. Good transportation practice

Topic	Detail
Delivery vehicles	Hand-laying operations invariably require smaller trucks than materials placed by pavers. Consequently, delivery vehicles should be capable of carrying the maximum weight of material consistent with site conditions and the nature of the operation. Four-wheeled lorries are only appropriate where access is difficult or hand-laying is being undertaken.
Insulation	Delivery vehicles should be insulated and sheeted (preferably double sheeted) to minimize heat loss in transit or while awaiting discharge.
Vehicle cleanliness	The inside of the body should be clean with the minimum of dust or sealing grit applied to facilitate discharge. Diesel treatment to the tailboard mechanism is worthy of particular attention providing none comes into contact with the material.
Rate of delivery	The rate of delivery should be controlled to provide continuity of laying. As indicated previously, clear instructions should be included as part of the order to the supplier. Normally bituminous mixture deliveries are specified as beginning at a particular time and continuing at a rate of 15–20 t every 5–10 min, as appropriate. When laying chipped hot-rolled asphalt wearing course, two or three loads are often ordered for the start time so that continuity can be maintained.
Site address	Clear instructions should be shown on the delivery note, together with access points, turning locations, marshalling areas and limited weight or access routes. Many motorway contracts are subject to restricted access routes which must be specified.
Arrival on site	Immediately upon arrival on site the driver should report to the marshalling point or to the foreman or other authorized person. Failure to do so may delay temperature checks or sampling procedures and cause delay.
Discharge	Excessive reversing distances can be extremely dangerous. If necessary, turning points should be provided. Reversing in dangerous locations or in the vicinity of other operations while backing onto the paver must always be under the supervision of an experienced banksman.
Sheeting removal	Sheeting should only be removed as far as is necessary to facilitate discharge. Hand-laying does not normally warrant complete removal. Such removal should not take place until immediately before tipping.
Hand-laying	Hand-laying operations necessitate controlled discharge to avoid the possibility of excess material being tipped. Failure to do so involves unnecessary effort and/or material becoming cold before it is properly compacted.
Cleaning out	Once the load has been tipped, the lorry should draw fully away from the paver and cleaning out or tipping of excess material restricted to a designated cleaning out point on site.

(*a*) avoid hand work and bleeding out (i.e. hand spreading where the laying width is greater than the screed width)

(*b*) the paver runs and widths should be planned to give as many long runs as possible avoiding joints and hand-laying at the beginning and end of runs.

Fig. 6.3. Lorry deliveries to a paver. Photograph courtesy of Wrekin Construction

Driving lines should be marked on the existing surface or lower layer. These assist the paver driver to maintain a constant line and assist roller drivers when cutting long joints in hot-rolled asphalt wearing course. Fig. 6.4 shows typical joint patterns and laying widths for a large car park.

In Fig. 6.4 hand-laying is limited to the start of panels 1 and 2, the start and finish of panel 3 and at the radius. Joint cutting is reduced to the ends of panels 1 and 2. The direction and order of laying panels 1, 2 and 3 gives maximum cooling time for run 2 before the paver rides up onto the laid material in panels 1 and 2. This would avoid hand-laying for a paver length at the end of runs 4 to 7.

An alternative method would be to lay panels 3, 4, 5, 6 and 7 before 1 and

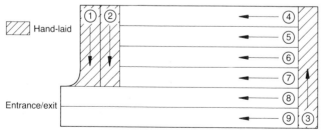

Fig. 6.4. Surfacing a car park

2. This would avoid the paver operating on the fresh material of 1 and 2 but it introduces hand-laying and joint cutting at the ends of 4, 5, 6 and 7. On a hot day with a thick wearing course, (40 mm or more), this method would have to be adopted since runs 1 and 2 would be too hot to take the weight of the paver.

All situations have to be assessed individually by an experienced person to achieve the best results with the least work and therefore minimum cost. A temporary hardstanding of hardcore can be laid using an ordinary excavator around the car park entrance/exit to avoid damage to the formation. Level control is normally exercised by means of steel pins or timber stakes or coloured tape on steel pins or timber stakes set along both sides of the carriageway (at a maximum of 10 m centres). These are set in relation to the datum levels for the contract. The top of the pins or coloured tape would be fixed at common heights above the finished surface levels, say 300 mm above roadbase finished surface levels, 240 mm above basecourse finished surface levels and 200 mm above wearing course finished surface levels.

It is essential that the pins or stakes are driven into firm ground, or preferably set in concrete, otherwise they are likely to become loose and the bituminous courses will be laid to incorrect levels. Dipping below a cat-gut line, held tight across the carriageway between two such pins or stakes, provides a simple but accurate means of level control. The use of chalk on kerbs to denote crown lines and stop points, crown line off sets and block heights etc. is recommended.

Block heights should be underlined to ensure that figures are read correctly. Chainages should be marked 'CH X' to avoid confusion between levels and chainages.

6.8.2. Use of block-up staff

The vertical alignment of a carriageway is designed to a high degree of accuracy and requires close control when the road layers are placed.

On major carriageway repairs where diversions or contraflows cannot be used, level control can be a problem. The requirement to maintain a free-flow of traffic means that only half the carriageway is accessible. Accurate level control requires the establishment of control levels along the centre of the road. This is normally done with a painted cross or masonry nail with an associated block-up height in millimetres. Block-up staffs make it possible to hold a line at the exact height required and thus give perfect level control. Good quality staffs are made from the best materials and will withstand years of wear and tear on site. This makes them cost effective against standard rules which are initially cheaper but have a shorter life.

The major components of a block-up staff are 1.04 m hardwood staff with an armoured foot, a heavy duty brass rule and a stainless steel sliding cursor which locks in place (Fig. 6.5).

The lower edge of the cursor is set at the appropriate level on the staff. It is firmly locked in this position using a thumbscrew on the back. This screw acts on a captive stainless steel pressure plate and so cannot harm the wood. An 80 lb breaking strain mono-filament nylon fishing line is then pulled under the locating tab below the cursor (Fig. 6.6). The other end of the line is secured by looping over a setting out pin at the top of marking tape or hooked over

Fig. 6.5. Block-up staff. Photograph courtesy of TBS Engineering

a kerb, as required. The existing surface can now be dipped from the line to calculate the next layer thickness. Fig. 6.6 illustrates a number of errors to be avoided when setting or checking levels. These are

(a) if the line breaks the chainman may fall into the line of traffic
(b) the operative cannot watch for site traffic running through the line as he must look at the rule to hold a constant level
(c) wrapping the line around fingers is dangerous and can cause injury if site traffic snags the line
(d) the operative is using a standard rule which is too thin and will be unstable

Fig. 6.6. Setting levels in a live carriageway incorrectly

Fig. 6.7. Setting levels in a live carriageway correctly

(e)   the graduations on a standard rule wear off quickly when handled with soiled hands, and they may also be hard to read

(f)   this method relies on the operative being able to read the rule correctly and set the level accurately

(g)   after use checking involves repeating the process

(h)   it is very uncomfortable to remain in this position for long

(i)   if the line is dirty, too thick, tied in knots or has tapes on it then it may sag

(j)   many operatives are not confident when measuring.

Compare this with the method illustrated in Fig. 6.7, which shows the following points.

(a)   The line is now being held at the top of the staff. Tension is maintained by setting up 0.5 m short of the datum and the operative moving the staff sideways with his foot. This maintains constant line tension and there is no danger of injury due to line breakage.

(b)   The operative is now free to be alert to any danger from site traffic, etc.

(c)   The line is wrapped round a short stick which is easy to hold or let go if site traffic catches the line.

(d)   The staff is strong and stable.

(e)   The rule can be set by a competent person and given to the operative.

(f)   After use the setting can be checked before being altered for the next chainage.

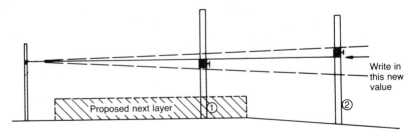

Fig. 6.8. Offsetting control levels

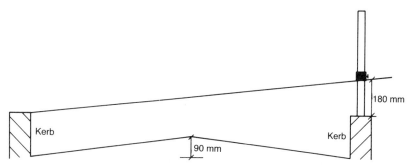

Fig. 6.9. Setting a level where there is a crown line at the centre

(*g*) When being set the staff can be held at the most convenient distance to suit the operator's eyesight.

(*h*) Since the line is held round a right angle bend at the cursor and braced against the top of the staff it is easy to keep the tension constant.

(*i*) The upright stance is comfortable for long periods of time and makes the chainman more visible to site traffic and improves safety.

(*j*) The recommended line is the best balance of strength and weight. Using a warning board is advised.

Figure 6.8 shows how to offset a block height at a point which will be covered by the next layer. To offset a block height from position 1 to position 2, the top of the cursor is set on the first staff at the original height. The second staff is set up at the new offset location. The line is pulled in the normal manner and the cursor on the second staff slid up or down until the line coincides with the top of cursor 1. Cursor 2 is locked in position. The second staff is rocked slightly to check that the line just clears the top of cursor 1 and adjusted, if necessary. The location of the second staff is marked on the ground and the height of cursor 2 recorded.

Figure 6.9 shows how to set out levels where there is a crown line at the centre. The cursor is set at twice the value of the crown height. the staff is then used to block up on top of the right hand kerb. The line is pulled taut between the top of the left hand kerb and the cursor. The left half of the road can now be dipped. The procedure is reversed for the right half of the road.

Figure 6.10 shows how to set out levels where there is an obstruction such as a crash barrier.

If the tapes on the pins are set at a control level of, say 200 mm and the line cannot be pulled because of site obstructions such as piles of pre-coated chippings, barriers or timber baulks, then the staff can be inverted and the cursor set on top of the pin. The staff is now slid through the cursor until a convenient reading is against the top of the tape, say 800 mm. This height should always be kept to a minimum to avoid errors caused by the staff not being vertical. If the height is above 0.5 m then a spirit level should be employed. The work can now be dipped using the new control of 800 mm + 200 mm = 1000 mm.

Figure 6.11 shows a laying gang dipping the top of a chipped hot-rolled asphalt wearing course.

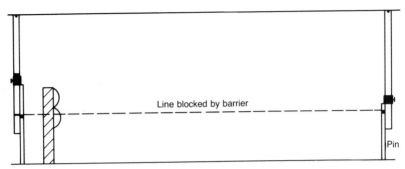

Fig. 6.10. Setting out over an obstruction

6.8.3. Hand-laying

Hand-spreading and raking of bituminous mixtures is a highly skilled and physically demanding operation and it is rare to find a good machine crew who are equally skilled at hand-laying.

Good hand-laying practice can be summarized as follows.

(a)   Wherever possible, material from the lorry should be tipped in small heaps, exactly where the material is to be raked. Large heaps cause unnecessary carrying and spreading of the material and compacts the bottom of each heap under its own weight resulting in humps in the finished mat under the roller.

(b)   Unnecessary raking and shovelling causes segregation of the material.

(c)   Tipping and raking should be completed speedily to minimize heat loss and enable rolling to begin as quickly as possible.

(d)   At the start of rolling, joints should be treated first since all materials cool first at the joint. Alternatively, the joint may be hot ironed as raking progresses.

(e)   Wearing course joints should be raked past the actual joint and onto the existing surface so that any segregated aggregate may be scattered back onto the mat with a flick of the rake. Failure to do this can result in a concentration of coarse aggregate at the joint resulting in plucking and premature failure of the mat.

(f)   Best results are achieved by spreading the material as accurately as possible allowing 30% extra thickness for rolling down followed by a flashing over the mat with the roller. After this initial compaction any small irregularities can be rectified before final compaction.

(g)   Ease of access for the delivery of materials to the point of laying will increase the speed and efficiency of the laying operation.

The following exercise provides a means of evaluating gang size.

Examples of gang size calculation

Site A — a footway job. Most sites have easy access to a standard 1.8 m wide footpath by way of the adjacent roadway, although increasingly there are grass verges between the footpath and the carriageway. Such a site lends itself to the use of a tractor shovel loader and a small vibrating tandem roller.

Fig. 6.11. Dipping to check the finished road levels. Photograph courtesy of Wrekin Construction

A six man gang could reasonably lay 60 t/day with deliveries at 8 am, 10.30 am and 1 pm to allow for tea breaks and lunch. This assumes 10 t of bituminous basecourse per man day laid at 40 mm thick, with one man driving the roller, another man driving the tractor and the remaining four men handling the material.

Each load should be tipped at a location near to the point of laying, normally in an adjacent bellmouth on previously laid basecourse material. This gives manoeuvring space for the tractor loader. Any slight damage to the existing basecourse will be covered by the wearing course in the finished work. Tipping hot material onto an existing wearing course marks the surface and must be avoided.

Sixty tonnes of material will cover approximately 366 linear m of footpath 1.8 m wide and 40 mm thick. Each shovel load is around $\frac{1}{3}$ t. Allow 20 s to load and a further 20 s to tip. Speed of tractor is 10 mph (4.5 m/s).

Average distance travelled by the loading shovel = 366/2 × 2 = 366 m
Number of trips to move 60 t of material = 60 × $\frac{1}{3}$ = 180
Time per round trip = 366/4.5 + 40 = 2 min approximately
Time to move 60 t = 180/60 × 2 = 6 h total

The following timetable will operate during a typical day from 7.30 am to 4.30 pm.

7.30 – 8.00	Set up site
8.00 – 10.00	Lay first load
10.00 – 10.30	Break
10.30 – 12.30	Lay second load

12.30− 1.00	Lunch break
1.00− 3.00	Lay third load
3.00− 4.30	Roll off, tidy site, etc.

The timetable shows that the rate of laying depends on the tractor's ability to supply 60 t in the 6 h laying period. In reality, delays in delivery of materials, break downs and site problems may slow down the day and it would not be unusual to finish rolling off as late as 6.00 pm. It is also evident that although the output per man would be rated at 10 t, since only four men are actually handling the material, the output is actually 15 t/man. It would seem therefore that when laying more than 366 m away from the tipping point, it is not possible to feed enough material in 6 h to keep four men working. The options are to reduce the number of men or increase the capacity to deliver.

Site B — similar to site A but with access problems. On a site where there is no convenient tipping point or no easy access from the road to the footpath then the use of a dumper should be considered in addition to the loading shovel.

In this instance the six man gang would be equipped with a small vibrating tandem roller, a tractor loading shovel and a 3 t dumper. Again, consider laying 60 t of material at 40 mm thickness with one man driving the roller, another man driving the dumper but also loading the dumper with the tractor and the remaining four men handling the material.

During the day, the dumper only has to make 20 journeys. However, it may take 360 s to load and to unload based on 20 s for each manoeuvre of the tractor or dumper. Assuming an average trip of 400 m, the time taken to deliver the 60 t would be 4 h whereas an average trip of 800 m would take 6 h for the same 60 t laid by four men handling 15 t/man with the only additional costs for the day being the charge for the dumper.

The temptation to cut laying costs by reducing the number of men in the gang from, e.g., six to four men, although showing a saving of approximately £150, would usually mean reducing the output from 60 to 40 t. The effect of this would not be an overall saving in the cost of laying since the overhead cost of supervision, the crew bus, roller, tractor and/or dumper remain the same.

The number of men actually laying is now reduced to two, compelled to work 33 % harder with each man laying 20 t. While this may be possible for a short period, it does tend to produce stress and strain, days off work and men looking around for easier ways of making a living.

Conclusions

(*a*)   On sites where delivery distances exceed 800 m, an additional dumper and driver could increase the delivery distance up to 1600 m from the tipping point. Employing one extra man and dumper would double the efficiency of a six man gang working 800−1600 m from the point of tipping.

(*b*)   Gang resources must be adjusted to take account of variation in site conditions. However, it is seldom economical to reduce the size of gang from six down to four and the standard six man gang operates most efficiently when tipping is no more than 400 m from the point of laying.

Table 6.9. Comparison of 4 man and 6 man gang costs

4 man gang costs	Item	6 man gang costs
4 × £75 = £300 1 × £30 = £30 1 × £30 = £30 1 × £30 = £30 £40 £40	Labour Roller Tractor Van Fuel Supervision	6 × £75 = £450 1 × £30 = £30 1 × £30 = £30 1 × £30 = £30 £60 £40
Total cost = £470		Total cost = £640
Laying 40 t/day	Output	Laying 60 t/day
470/40 = £11.75/t	Unit cost	640/60 = £10.67/t

Recommendations
(a) In all cases, explore the possibility of increasing gang resources and thus increasing unit output.
(b) Calculate gang output on the basis of the output per man actually laying material, rather than the total number of men in the gang.
(c) Programme the minimum resources for delivery at the point of laying with 15 t/man day for each man actually handling the material being a reasonable and sustainable output.
(d) Try to programme work so tipping points are available within 400 m of the point of laying.
(e) Use a dumper where the delivery distance is between 400 and 800 m and two dumpers plus one extra man where delivery exceeds 800 m.

Cost calculation
The calculation in Table 6.9 relates to a site where the material is delivered to the point of laying and illustrates the false economy of reducing labour from six men to four men.

Table 6.10 shows the benefit of using an extra dumper and driver when the point of tipping is greater than 800 m from the point of laying.

Table 6.10. Comparison of 6 man and 7 man gang costs

6 man gang costs	Item	7 man gang costs
6 × £75 = £450 1 × £30 = £30 1 × £30 = £30 1 × £20 = £20 1 × £30 = £30 £30 £40	Labour Roller Tractor Dumper Van Fuel Supervision	7 × £75 = £525 1 × £30 = £30 1 × £30 = £30 2 × £20 = £40 1 × £30 = £30 £60 £40
Total cost = £630		Total cost = £755
Laying 30 t/day due to distance	Output	Laying 60 t/day
630/30 = £21/t	Unit cost	755/60 = £12.58/t

6.8.4. Machine laying

The types and method of operating pavers is covered in detail in chapter 5. However, some simple basic rules are summarized as follows

(a)   As the paver increases in speed the thickness of the laid mat will decrease. As the paver reduces in speed so the thickness of the laid mat will increase.

(b)   When laying on a radius, the inside of the screed travels slower than the outside with a corresponding thickening on the inside and thinning on the outside of the mat.

(c)   Hot material passing under the screed will lay thicker. Cold material passing under the screed will lay thinner.

(d)   As the speed of the tampers increase so the thickness of the mat will increase. Failure of the tampers will result in the mat laying thinner.

(e)   A short-fed screed will lay thinner. Bleeding out or a build-up of material in front of the screed will result in a thickening of the mat.

(f)   Variation in the consistency of the material flowing under the screed will result in a variation in the thickness laid e.g. stiff or soft mixes.

(g)   A hot screed will produce a thicker mat. A cold screed will produce a thinner mat.

Even allowing for these variables, an experienced screwman, working regularly with the same machine, will produce consistently high quality work. Variations in these parameters will require considerable adjustment by the screwman so conditions should be as consistent as circumstances allow.

It has to be appreciated that each time the screwman changes the layer thickness, as well as increasing the indicator reading to allow for say 15 mm extra, he must add a percentage to take into account that the material will reduce in thickness as it compacts, perhaps a further 5 mm making 20 mm in all on the screw indicator. He must then wait for perhaps three paver lengths for the level change to take effect.

Figure 6.12 shows this effect.

The floating screed paver is capable of taking out frequent small irregularities in the surface to be covered. Each successive mat will reduce these bumps and hollows in the base producing a more level surface but it is not possible to lay a mat where finished levels vary markedly over short distances (Fig. 6.13). Often, surfacing contractors are asked to plane off a 40 mm hot-rolled asphalt wearing course and replace it with the same material. The rolling straight edge is then applied to the finished road. If the original road had bumps or hollows these will exist in the new surface (albeit to a lesser degree because of the nature of pavers). If they are in excess of the rolling straight edge requirements, the new surface will appear not to comply with the specification. Surfacing contractors should be wary of undertaking work where this may be the case and may request a rolling straight edge before beginning planing operations. Patching should be undertaken after the planing operation is completed to ensure that any sizeable irregularities (less than 10 mm in the case of 40 mm hot-rolled asphalt wearing course) are removed. Patching tends to be expensive because of the high labour and plant element compared with the low material content. However, doing the work again is substantially more expensive.

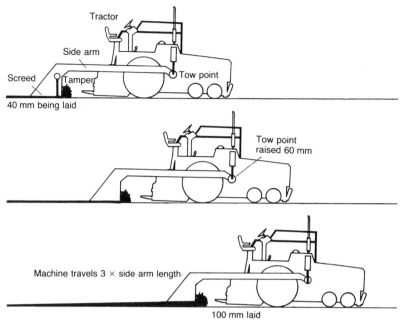

Fig. 6.12. *Time lag for screed changes to take effect*

Laying bituminous mixtures on a gradient

Start at the bottom of the gradient and lay uphill. Laying uphill means that the roller driver only has to use his brakes to stop at the low end of each pass where the material is coolest and, therefore, stiffest. At the high end of each pass the roller will naturally roll to a standstill and, consequently, does not displace the material which would result in a poor finish. When laying chipped hot-rolled asphalt wearing course any displacement causes bald patches with few pre-coated chippings and low texture depth, as the laid mat is stretched before the hump being formed. In heavy rain, downhill laying may be necessary to prevent water build-up in front of the screed. This requires extreme care. Fig. 6.14 illustrates why bituminous mixtures should be laid uphill. Fig. 6.15 shows the effect of laying chipped hot-rolled asphalt wearing course downhill.

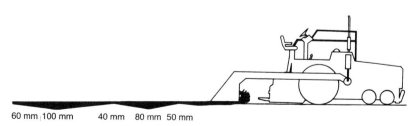

Fig. 6.13. *Overlaying an existing surface which has poor vertical geometry*

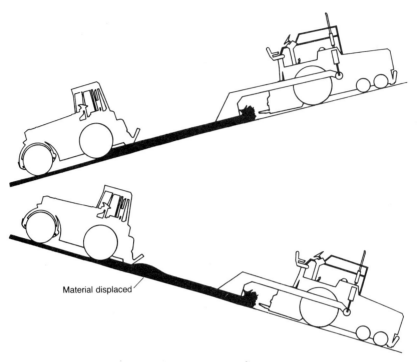

Fig. 6.14. Laying bituminous mixtures on a gradient

Laying on crossfalls

Figure 6.16 shows why great care must be taken when laying bituminous mixtures on a surface which has a crossfall. Laying thick layers of bituminous mixtures is to be avoided if possible in the circumstances shown in Fig. 6.16(b). Low density and loss of level control will be experienced on the low side free edge. In wet or very windy weather, laying may have to be suspended as the risks of losing the whole mat are too great.

Use of paving machines

The correct use of pavers is a complex process. Training and experience are vital. Table 6.11 lists points essential for the correct use of these machines.

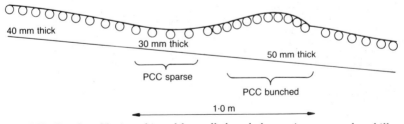

Fig. 6.15. Results of laying chipped hot-rolled asphalt wearing course downhill

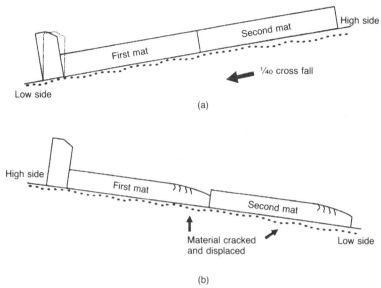

Fig. 6.16. Laying on a crossfall

Chipping machines

Figure 6.17 shows the chipping density being checked and Fig. 6.18 illustrates a cross-section of the hopper and control gate of a chipping machine.

When adjusting chipping machines alterations should not be made on the basis of readings from small trays (300 mm square). BS 594[3] recommends the

Fig. 6.17. Checking the rate of spread of pre-coated chippings. Photograph courtesy of Wrekin Construction

Table 6.11. Use of pavers

Remember
Safety comes first. Observe all safety precautions
A paver should not stop when laying to obtain best results
Flow control gates should be used
The paver should go to the truck and not vice versa
Material cools on the flat surfaces of the hopper which should be folded after at least every second truck load
A large permutation of speeds is available because of variable engine speed, a large gearbox and/or hydrostatic drives
The screed need only be 3–8°C above the temperature of the material being laid
It may take up to 5 tow lengths for a mat thickness correction to become fully effective. Use two tow lengths as a rule of thumb
Only two men should touch the controls — the paver driver and the screwman
The tampers should work in opposition to each other
Tampers are important and should be set correctly: 1 mm for fixed screeds, 2 mm for telescopic screeds, above the surface of the screed plate
Tyre pressures should be kept equal on both sides of the machine
The auger is for spreading material and not mixing it and the material should be correctly mixed before it reaches the paver
The floating screed will react to different temperatures in material and, therefore, these should be kept consistent
A roller will mirror an uneven contour on deep lifts due to differential compaction
The paver engine has to be checked for the correct amount of lubricating, fuel oils and water before starting
The oil levels in the hydraulic system must be correct before starting
The paver should be cleaned and lubricated at the end of the working day and not the following morning
Engine oil pressure and battery charging rates should be monitored when working
The bolts holding the screed to the side arms must be checked for tightness, especially after adjustment of the turnbuckles
The bottom tip of the auger should be 50–70 mm above the finished surface of the mat for best results
Over-vibration is harmful to the machine and to the mat
If you get high amplitude, low frequency beats, speeding up either the tamper speed or the vibrator speed will eliminate these

Table 6.11 continued

For the highest densities you work in the lowest traction gear or speed
The parking brake has to be applied before starting the engine
All gears must be in neutral before starting the engine
The auger clutches must be checked to ensure that they are disengaged before starting the engine
The traction clutch should be disengaged or the traction speed control lever should be in neutral before starting the engine
All control switches must be in the off or neutral position before starting the engine
If standing for a period, put gears in neutral and engage the traction clutch to reduce wear
The maximum aggregate should be less than $\frac{2}{3}$ of the mat thickness
Vibrators should always be set in line (in phase)
Material must not be dropped in front of the paver
The flow gates must not be wide open unless necessary
Fold the hopper after every second load
Auger boxes must not be over filled
Do not pave in too high a gear or speed bearing in mind the supply of material available
The lorry must not be permitted to hit the paver
Work with the augers at the correct height if height adjustment is available
Do not overheat the screed
Do not turn the air off when extinguishing the flame in the screed heater
Support the screed adequately and evenly when work is finished otherwise it may cool down with a twist in it
Park the screed with even support
Have the hoist cables or rams supporting the screed weight except when lifting the screed from the mat
Do not make any sudden movement on controls, i.e. main clutch, pump speed control, brakes, or excessive lock on front wheels except in cases of emergency
Do not attempt to move the tampers until the screed is heated
Do not idle the engine for long periods
Do not run the engine with a faulty injector
Do not permit the fuel to become contaminated with dirt or water
Take anti-frost precautions, remembering also the tyres if these are hydro-filled
Check the engine for abnormally low or high oil pressures

Table 6.11 continued

Keep the various reservoirs i.e. engine fuel, hydraulic oils, etc. topped up
Examine all fuel, air and oil filters regularly
Do not run the engine with abnormally high water temperature
The screed control buttons and switches should not be operated needlessly
The screed must not sit stationary on the laid mat with the burners operating
The deckplates must not be lifted while the engine is running due to danger from moving parts
Remove electrical leads from battery and alternator before carrying out any welding on paver
No one must stand in the hopper while the conveyer is running
Fit safety pins to auger clutch levels before travelling machine
Exhaust air reservoir before travelling machine if air-assisted clutches are fitted
Switch off the sensor unit before lifting screed
Drain the air reservoir (where fitted) daily after use
Have the engine running before moving hydrostatic control levers
Collect all accessories before moving paver from working site
If tyre pressures are reduced for extra traction, the tyres must be re-inflated before the paver is travelled

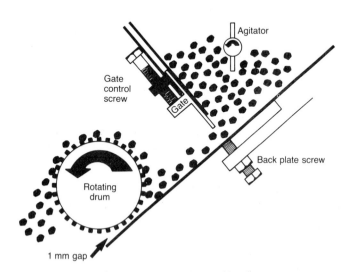

Fig. 6.18. Cross-section of a chipping machine hopper

use of ten number 300 mm square trays in conjunction with a small spring balance. Providing that the density is reasonably uniform across the width of the chipping machine, it is easier and equally accurate to use a tray 707 mm × 707 mm (i.e. area 0.5 m^2) in conjunction with a 0–25 kg set of scales i.e. (15.25 kg on scales − 9 kg tray) × 2 = 12.5 kg/m^2.

The actual rate of spread is best established by trials on site as varying wearing course thickness, material stiffness and type of pre-coated chippings will all affect the rate of spread given by a particular gate opening. The rate achieved should be checked every 100 m of single lane carriageway. The chipping machine will not run for long without needing some adjustment.

(a) Changes in gradient affect the rate of spread. Where the drum is before the backplate the machine will chip more heavily downhill and more lightly uphill. This situation is reversed with the hydraulic drive machines which usually have the drum behind the hopper backplate.

(b) Very often the chipper gate control has some slack in its mechanism and this will have to be allowed for when making adjustments.

(c) The back plate is easily damaged when driving over rough ground or kerb. If there is a gap between the back plate and drum, any eccentricity in the drum will allow chips to drop through on each revolution and create a wash-board effect on the road surface.

(d) Small blockages will produce a line of low rate of spread which must be hand chipped. An iron bar to dislodge small clumps of pre-coated chippings etc. is an essential part of the chipping machine driver's kit.

(e) The most important factor related to erratic chipping is the condition of the pre-coated chippings. Pre-coated chippings should always be wet before use except where the temperature is at or below 0°C. It is not uncommon on site for stockpiles to be watered before laying begins. The chipping machine may chip well initially but then it may start to chip lightly or erratically. The reason is that the wet pre-coated chippings have been used up and dry ones have begun to stick in the machines. The solution is not to increase the chipping rate but to adjust the poor areas by hand-chipping and watering the pre-coated chippings as soon as possible.

(f) The calibration tray should be checked before being placed for pre-coated chippings and excessive water, or asphalt adhering to the back of the tray as these would distort the result.

(g) Pre-coated chippings can easily jam in one of the many sprockets on the drive mechanism. This will produce a lurching motion which will throw rows of pre-coated chippings off the drum.

6.9. Rolling and compaction

The type and number of rollers to be used will be influenced by

(a) material specification
(b) layer thickness
(c) speed of laying
(d) compaction requirements.

The importance of using skilled, experienced roller drivers and the need to roll to a pre-determined pattern for equal compactive effect across the full width of the mat, cannot be over-emphasized.

At the start of rolling the temperature of the mat should be at the maximum permitted within the specification allowing full time for optimum compaction before the minimum rolling temperature is reached and avoiding waves in the mat.

The main factors affecting the time available for rolling are

    (a)   the thickness of the mat
    (b)   temperature of the mat at start of rolling
    (c)   wind speed.

Special attention should be given to the rolling of joints. The cross joint should be treated immediately the paver has moved away. The long joint should be rolled progressively as the paver moves forward and before rolling the adjacent mat.

The basic rules for rolling are

    (a)   follow the paver as closely as possible
    (b)   compact the joint first
    (c)   start compaction at the low edge of the mat
    (d)   turn off vibrations before reversing when using vibrating rollers
    (e)   change rolling speeds smoothly
    (f)   run forwards and backwards in the same rolling lane
    (g)   change rolling lanes on the cold side — avoid lane changes where the material is hot
    (h)   run in parallel rolling lanes and reverse at another section away from these adjacent rolling lanes
    (i)   keep the drums sufficiently wet to avoid picking up hot material but not more than necessary
    (j)   do not let the roller stand on the hot mat.

6.9.1. Rolling patterns

Figure 6.19 illustrates good rolling practice. This is based on the following points.

    (a)   Rolling patterns should start at the joint and move across the mat with overlapping passes.
    (b)   Each run should advance across and forward along the mat as the paver progresses. In practice this operation is not easy with as many as three rollers on the mat.
    (c)   The rolling pattern should be agreed between the foreman and roller drivers before laying begins.
    (d)   Rollers should approach the paver on the first run then reverse away from the paver in a straight line on the return run.
    (e)   The move over to the second pass should be made away from the paver on the cooler compacted material to avoid undue displacement of the uncompacted material.

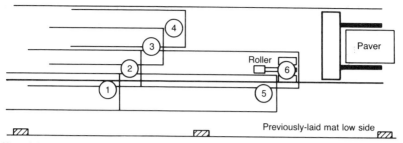

Fig. 6.19. Rolling of bituminous mixtures

(f)   Shorter runs give better compaction with the first roller on hot material immediately behind the paver.

(g)   Short runs have the disadvantage of frequent stops and manoeuvres on hot material however, which produces irregularities in the finished work.

(h)   Runs of 30—50 m and one roller for every 50 t/h laid give good compaction for DBM laid in 100 mm thick layers.

The Department of Transport Standing Committee *Preferred Method 8*[4] gives good guidance on all aspects of rolling although it suggests that low air temperature substantially reduces the time available for rolling. As this and other chapters have emphasized, the single most important variable in this respect is the wind factor.

6.9.2. Correlation between layer thickness and density of the compacted mat

The introduction of the Percentage Refusal Density test (PRD) and the use of dense bitumen macadam roadbases in varying thicknesses dependent on site conditions and design life, has led to a need to establish the optimum layer thickness at which the contract PRD can be achieved.

Although the standard UK specification does not require ongoing monitoring many clients require checking as part of a modified specification. Densities are often measured by nuclear density meters. Their principle of operation is illustrated in Fig. 6.20.

Contracts completed over a twelve month period showed that when laying 28 to 40 mm dense bitumen roadbase or heavy duty roadbase failure to achieve PRD can be expected when the layer thickness is less than 70 mm. When layers exceeded 85 mm they usually met the PRD specification. Optimum thicknesses were 95—100 mm. (Fig. 6.21).

6.9.3. Factors affecting rolling technique and the achievement of optimum mat density

When placing bituminous surfacing materials in a contained situation such as a trench reinstatement or when laid against a kerb foundation, good compaction is relatively easy since the material cannot spread under the compactive effect of the roller.

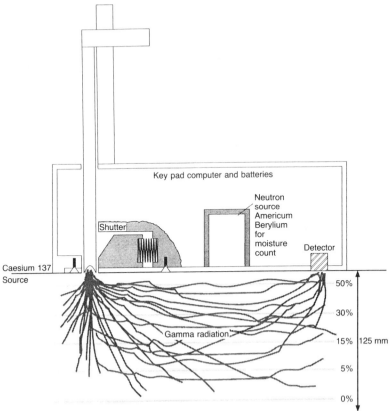

Fig. 6.20. Principle of operation of nuclear density meters

If the material is not contained the only forces that prevent the material from spreading are the tensile strength that the mat gains as the material cools and the friction between the mat and the underlying surface.

The tensile strength and friction are greatly reduced by too much compactive effort which can easily crush the point of contact between adjacent pieces of aggregate in the mix and this is likely when using limestone mixes. When this happens the bitumen bond is broken at the point of contact where the bitumen is absorbed by the dust created. The effect of this is a loss of tensile strength which does not improve as the material cools, resulting in the mat spreading under the effect of the roller rather than compacting.

Once the tensile strength of the mix is lost then the only force acting to keep the mat together is the friction between the fresh material and the underlying surface. There will be no bond if this layer comprises sub-base or DBM using limestone aggregate which has been rolled white i.e. the aggregate at the surface has been crushed and a fine powder is present. Similarly, there will be no bond if the underlying surface is wet or very cold.

When any or all of these conditions exist then a crazing pattern may appear on the mat surface and density will drop off alarmingly. Once this has happened

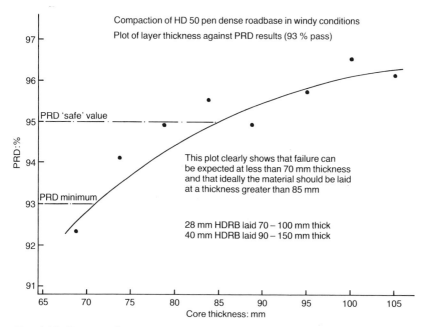

Fig. 6.21. Densities from nuclear density meters

there is no way to correct it. This situation is called over-rolling and was very common in the early days of the use of heavy vibrating rollers.

To avoid this the initial compaction should be kept to a minimum until the mat has gained tensile strength then final rolling can be followed by polishing off to close up the surface of the mat. It is a matter of judgement based on experience to achieve a good result.

The following factors will influence the result

(a)   material type
(b)   gradient, crossfall and edge restraint, if any
(c)   type and number of rollers used
(d)   surface condition — wet or dry, surface temperature and cleanliness of surface.

When rolling a fragile material in adverse circumstances it is good practice for the roller driver to keep away from the low-side, free edge, after the first pass and to roll off at the minimum rolling temperature or lower. Since this material is cut away, the relatively low level of compaction is unimportant.

6.10. Quality control and checks

Factors which influence the quality of the finished product are varied and include the following

(a)   the gang
(b)   plant
(c)   weather
(d)   site conditions
(e)   the material.

6.10.1. The gang

The use of well-trained and experienced operatives under the control and guidance of a competent and enthusiastic foreman is of prime importance. Paver drivers, roller drivers, tractor drivers and chipping machine operators should all be accredited under the CITB or the QPTB schemes. The morale of the gang and the attitude of the site staff are extremely important. The temptation to push the work along too fast should be avoided. There is no point in over-ordering materials which results in lorries standing on site going cold while a tired and dispirited gang try to lay cold materials in the dark.

The best approach is to assess the gang's likely output and place a reasonable order taking into account all the circumstances. The job will then run smoothly, remedial works will be kept to the minimum and valuable time will not be diverted away from planning and organization to troubleshooting.

6.10.2. Plant

All plant should be well maintained and under the control of a competent operator. Manufacturer's operating instructions and safety rules should be followed at all times.

The temptation to save on plant costs is understandable. The use of the additional dumper (referred to under hand-laying), the need for an extra roller when laying high tonnages or having standby plant such as a spare paver, roller and chipping machine on a motorway type contracts are all factors which have a direct effect on the quality of the finished mat.

Some breakdowns are unavoidable. On a critical job, the standby plant must be tested regularly to ensure that it is viable for use immediately in the event of plant failure. A spare roller is no use if it has a flat battery or seized water pump.

6.10.3. Weather

The weather causes more disruption and poor quality work than the sum total of most other problems. Since the UK does not suffer from the extremes of weather found in some parts of Europe, bituminous mixtures are regarded as all-weather materials and so receive less care and attention than some other construction materials.

Much can be done to combat the effects of poor weather.

(a)   If frost is forecast for the next day, laying a 100 mm thick layer of roadbase ensures that the mat will be frost free and ready to be over-laid the following morning.
(b)   On a sunny winter day the surface temperature of the road is often three or four degrees above the reference air temperature used by the engineer's staff. The use of thermometers at the point of laying

can often confirm this and mean an earlier start for operations and avoid keeping material standing in lorries.

(c)   A cross-wind has a far more detrimental effect on bituminous mixtures than either low air temperature or light rain. A still, dry atmosphere with air temperatures below freezing may well be ideal laying conditions. Although later the same day, as the air temperature rises and a wind blows up, may be the time that the specification permits laying to commence the wind will have the effect of substantially reducing the time available for compaction.

(d)   Laying in wet or very damp conditions should be avoided if the low side edge of the mat is unsupported. In these circumstances the material may slide sideways under rolling, due to lack of adhesion to the surface below and the lack of edge restraint.

(e)   Weather centres such as those located at local airports are extremely good at forecasting likely weather conditions up to two to three hours ahead which should be enough time to plan for the delivery of materials from the supply source to the site.

(f)   It is relatively easy to insulate the sides of the hopper of the paving machine with 40−50 mm hardwood planks. This greatly reduces heat loss from material while in the hopper. Indeed, most paver manufacturers offer this as an extra. All pavers should be manufactured with insulated hoppers as standard practice.

(g)   Tack coating existing surfaces the night before laying may avoid delays the following morning waiting for the surface to dry off or for frost to thaw.

(h)   Tighter control of water sprinklers on the wheels of rollers may be necessary in cold laying conditions.

The list is not exhaustive but emphasizes that time spent in anticipation of problems is more profitable than that spent on expensive remedial works.

6.10.4. Site conditions

Reference has been made to the need for a thorough site inspection before starting work. All too often this is not done and the consequences can be catastrophic.

6.10.5. The material

Bituminous mixtures are usually manufactured to the highest standards of quality and consistency. However, they should be given a visual inspection and temperature check before despatch to site. Other precautions to avoid the likelihood of sub-standard material being laid include the following

(a)   Well-maintained, insulated lorries should be used for the delivery of bituminous mixtures.

(b)   Where possible, but always in the case of hot-rolled asphalt wearing course, the material in the lorry should be double-sheeted.

(c)   Material temperatures should be checked immediately upon arrival on site and lorries should not then stand for long periods.

(d)   Segregation of the mix should be avoided by ensuring that loose aggregate from the side of the load in the lorry is well mixed in the hopper of the paver. Similarly, avoid a build up of material at the extremity of the feed screws as this can also be a cause of segregation.

(e)   Cold material should not be allowed to accumulate in the corners of the hopper. Remove by raising hopper sides as required.

(f)   Attention must be paid to the nominal stone size of the mix in relation to the thickness of mat to be laid and the texture of the finished mat.

6.10.6. Other factors

There are other factors which may affect the quality of a bituminous mat.

(a)   A well-trained surfacing gang which will take pride in its work with each member taking personal responsibility for his own operations. The golden rule is — Do it right, do it once and each man to his own job.

(b)   Attention to detail is paramount. Premature failure and subsequent remedial works are usually preventable by careful work in the initial stages.

(c)   Appointing one member of the gang, usually the foreman, to oversee and check back behind the paving operation is vital to ensure quality control.

(d)   The temptation to meddle should be avoided at all costs. Set up the paver, start the paving train and feed it at a steady rate with consistent material and little can go wrong. Frequent re-adjustment to the screed of the paver should not be necessary.

(e)   Frequent checking of the following will ensure that quality does not suffer
  (i)   material consistency and temperature
  (ii)   weather conditions — temperature, rain and most important of all, wind and wind chill
  (iii) a steady flow of material and elimination of stop/start mentality
  (iv)   attention to detail and pre-planning with all preparatory work completed well ahead of the surfacing operation
  (v)   level control, mat thickness and mat temperature
  (vi)   efficient rolling patterns: the important of compaction cannot be over-stated; once the mat has cooled then there is nothing to be done but remove it and replace it; Nuclear Density Meters (NDMs) are now commonplace on all large surfacing contracts
  (vii) in the case of chipped hot-rolled asphalt wearing courses then regular use of chippings trays and weigh scales and cleaning laid runs each day are advisable
  (viii) use of only the best plant available and an effective plant maintenance system.

(f)   The surfacing, and in particular the wearing course, should be the last operation on the contract. Conflicting priorities often mean works are not always completed in sequence. The temptation to use bituminous road layers as a convenient 'work bench' or temporary haul road should be avoided and adequate protection provided if it is necessary to use the surface in this state.

Table 6.12. Tolerances in surface levels of pavement courses[5]

Road surfaces	± 6 mm
Basecourse	± 6 mm
Upper roadbase in pavements without basecourse	± 8 mm
Roadbase other than above	± 15 mm
Sub-base under concrete pavement surface slabs laid full thickness in one operation by machines with surface compaction	± 10 mm
Sub-bases other than above	± 10 mm
	± 30 mm

6.11. Aftercare

6.11.1. Checks confirming compliance with specification

Once the wearing course had been laid, the following checks should be made before the road is opened to traffic.

(*a*)   Line and level — the finished surface should be checked for level using the same control levels used during the laying process. A common problem arises if such check levelling is conducted by the engineer's staff, independently of the main contractor and the surfacing sub-contractor. Lines pulled between two level pins must be taut. Often, apparently incorrect levels are the result of measuring below a slack line. This is easily avoided by conducting an agreed tri-partite dipping exercise similar to that conducted on the sub-base before surfacing. The standard UK Specification[5] gives the tolerances shown in Table 6.12 for levels of individual layers.

(*b*)   The standard UK Specification[5] requires that sudden variations in longitudinal profiles should not occur. This can be checked using the rolling straight-edge (Fig. 6.22).

This apparatus measures the number of surface irregularities and

Fig. 6.22. Rolling straight edge. Photograph courtesy of Lothian Transportation

Table 6.13. Maximum number of surface irregularities[5]

	Surfaces of carriageways, hard strips and hard shoulders				Surfaces of lay-bys, service areas, all bituminous basecourses and upper roadbases in pavements without basecourses			
Irregularity	4 mm		7 mm		4 mm		7 mm	
Length, m	300	75	300	75	300	75	300	75
Category A roads	20	9	2	1	40	18	4	2
Category B roads	40	18	4	2	60	27	6	3

the UK Specification[5] limits the number to those shown in Table 6.13.

Roads are categorized A or B according to traffic speed with category B having a traffic speed of less than 50 km/h.

The rolling straight-edge has a bell which rings when a pre-set depth of irregularity is reached or exceeded. A gauge at the front of the machine shows the depth of irregularity. The level at which the bell rings is normally set at 4 mm. All rings are recorded as being equal to or greater than 4 mm but less than 7 mm, or as being equal to or greater than 7 mm but less than or equal to 10 mm, or as being greater than 10 mm. No irregularities above 10 mm are permitted.

(c)   Levels across the mat and across joints are measured by use of a 3 m straight-edge and wedges. The specification[5] states that the maximum allowable difference between the surface and the underside of the straight-edge, when placed parallel with, or at right angles to, the centre line of the road at points decided by the engineer shall be

(i)   for pavement surfaces   3 mm
(ii) for basecourses   6 mm
(iii) for upper roadbase in pavements without basecourse   6 mm
(iv) for sub-bases under concrete pavements   10 mm

(d)   Texture depth is measured by one of two methods; manually by use of the sand patch method or by laser using the mini texture meter. BS 598: Part 105[6] supports the sand patch test for the resolution of contractual problems and recommends the laser texture meter for screening large areas quickly to find any suspect areas. Both test methods are referred to in the DTP *Specification for highway works.*[5] The minimum 1.5 mm sand patch texture depth is required on roads where 85% of the traffic travels at a speed of more than 50 km/h. Full details of these tests are contained in chapter 8.

(e)   Density can be measured by one of three methods
(i)   comparison of the density achieved on site, as determined from cores, with a standard laboratory compactive effort on uncompacted material i.e. percentage Marshall
(ii) comparison of the site density of cores with the maximum theoretical density to determine the percentage air voids
(iii) comparison of the site density of cores removed from the road

Table 6.14. Quality checklist

Materials specifications	Sampling of materials at pre-determined frequencies, testing of samples to ensure compliance with client's specification
Precise location of each load of material deposited	Properly identified mat runs, reference to chainages (contract chainages)
Characteristics of mat	Width Depth (planned and actual, back up with dip sheets)
Environmental data	Air temperature, wind speed and a general description of weather conditions
On site testing of finished mat	Surface regularity Transverse straight edge Surface texture
Off site testing of bituminous materials	Grading Binder/filler content Density (PRD, etc.)
On site signing and safety	Daily record verifying that everything on site is safe and fit for purpose
Any deviations from client's specification	Must be referred to client for consideration

pavement with the density of the same cores compacted to refusal in the laboratory i.e. percentage refusal density (PRD).

Screening of the compacted mat is carried out by use of a nuclear density gauge.

6.11.2. Quality checklist

Table 6.14 shows a typical Quality Assurance checklist for use on bituminous pavement contracts.

6.11.3. Economic analysis of operation

It is usual, at the end of each phase of the work and on completion of the contract, to conduct a detailed analysis of actual cost and performance against the figures produced for the estimate. This exercise serves the dual purpose of checking that optimum performance has been achieved on site and that estimating policy and technique is in line with optimum site performance. It is extremely important that the surfacing contractor keeps adequate records at all stages of the operation.

The foreman would usually complete a standard daily report showing

(a) labour — number and names of men and hours worked
(b) plant — type and number of items together with working time, standing time and breakdowns
(c) materials — summary of total tonnage of each material, and a list of ticket numbers and weights of each load

(d)   work completed — surface area and thickness of each area completed, together with ancillary works such as sweeping, planing, tack coat, etc.

(e)   non-productive time — plant breakdown, weather, waiting for materials, delays due to other contractors, public utilities, etc.

(f)   superage for each material laid.

The supervisor should keep a more general daily diary detailing site instructions and variations, with small diagrams showing locations of tie-ins, made-up areas, etc. The weekly cost and valuation sheet may be completed by either the supervisor or the quantity surveyor after reference to both the foreman's and supervisor's daily diary and returns, the weekly time sheets showing the actual cost of labour, the weekly plant returns showing the actual cost of plant and the invoices for material supplies.

All these costs, together with an agreed percentage addition to cover site and head office overheads are then compared with a fully priced, detailed measure of the completed works, comprising measured works, dayworks, contra-charges, variations and additional works, etc. Many major surfacing contractors use a simple computer spreadsheet program which compares these actual costs and values with daily outputs, superage figures, costs of remedial works, material wastage, labour and plant costs, etc. with the figures used in the estimate.

Typical methods of recording surfacing work progress on a daily and summary basis respectively are shown in Tables 6.15 and 6.17. These are normally mounted on a computer spreadsheet in the format shown in Tables 6.16 and 6.18.

6.11.4. Remedial works

On the ideal site, inclement weather would not occur, material supplies would arrive on time and comply with the middle of the specification, plant would never break down, operatives would work at peak performance and promises made by others would always be kept. Unfortunately the ideal site does not exist and remedial work is necessary.

Typically, an allowance of $0.5-1\%$ is added to the total cost at the estimating stage to cover average remedial costs. As a guide to likely costs, one t of hot-rolled asphalt wearing course may be laid for between £25 and £30 whereas removal and replacement would cost between £100 and £150/t as traffic management charges add heavily to the second figures.

Most contract documents and specifications make little reference to remedial methods, simply stating that failure to comply with the terms of the specifications necessitates removal and replacement, not only of the suspect areas of material but also the adjacent areas which make up the minimum acceptable patch size of full mat width. Clearly, removal of laid materials must be avoided if possible. When gauging the need for remedial work the engineer's staff should weigh up the benefits of replacement. Often the replaced material is little better, the surface is scarred and extra joint lengths are left for maintenance by the client.

The increasing use of end result and performance-related contracts will allow for more appropriate methods of dealing with minor deviations from

Table 6.15. Daily surfacing diary

Item	Daily surfacing diary
General	
Contract	Details of contract location
Job Number	Number to reference contract
Supervisor	Name of supervisor/foreman on contract
Date	Date when material was laid
Weather	Brief description e.g. dry, cold and windy
Material Order	Material order for particular date
Material Cost	Cost of each type of material/t delivered
Waiting Time Cost	Cost of lorry bringing material/h (first $\frac{1}{2}$ hour on site incl)
Gang Cost	Cost of labour and plant costs for gang/h
Supplier	Material supplier
Labour and plant hours	
Labour	Number of hours worked daily by each gang member
Plant hours	Number of hours worked daily by each item of plant
Labour cost	Number of hours worked × daily 'all in labour rate'
Plant cost	Number of hours workd × various hourly plant rate
Gang cost	Gang cost = labour cost + plant cost
Material	
Load/t	Weight of individual load of material
Ordered time	Ordered times of all material loads
Time in	Actual delivery time of all loads of material
Time out	Actual time of completion of discharge of all loads of material
Contractor cost	Cost that is incurred by contractor for waiting time over and above that allowed (normally first $\frac{1}{2}$ hour on site) calculated from order time/time in and time out
Quarry cost	Cost incurred by contractor due to supply delay by quarry calculated by multiplying unit gang cost by difference between order time and delivered time
Summary of daily surfacing costs	
Total cost	Cumulative costs of labour, plant, materials, supervision and waiting time costs
Superages	
Estimated coverage	Estimated m^2/t for materials laid
Actual coverage	Actual m^2/t for materials laid
Estimated outputs	Estimated output in terms of t/day
Actual outputs	Actual output in terms of t/day

specification such as extended guarantees or reduced payments. In the current climate acceptable remedial methods must be negotiated between the parties.

Infra-red road heaters, when operated by an experienced foreman in conjunction with an experienced raker and roller driver, can markedly improve an area where pre-coated chippings were not fully embedded or where they had been lost from the mat due to chilling. Similarly, small areas of total chip

Table 6.16. Example of daily surfacing spreadsheet layout

Contract	A6106 Re-alignment	Date	5 November 1993
Job Number	HY211/023/RNH	Material cost	£21.00/t
Supervisor	R McLellan	Waiting time cost	£15.00/h
Supplier	Quality Asphalt Ltd	Squad cost/h (incl. plant)	(£840 + £592)/8 = £179.00
Weather	Dry, mild, no wind	Material type/order	30% HRAWC/175 t

Labour	Hours	Plant	Hours	Rate	Cost
C. Glackin	8.00			10.50	84.00
L. Adam	8.00			9.50	76.00
O. Redding	8.00			8.50	68.00
B. Holly	8.00			8.50	68.00
C. Berry	8.00			8.50	68.00
R. Orbison	8.00			8.50	68.00
I. Hayes	8.00			8.50	68.00
S. Cropper	8.00			8.50	68.00
A. Green	8.00			8.50	68.00
E. Floyd	8.00			8.50	68.00
T. Turner	8.00			8.50	68.00
B. King	8.00			8.50	68.00
		BK 191 paver	8.00	25.00	200.00
		Crew bus — 1	8.00	5.00	40.00
		Crew bus — 2	8.00	5.00	40.00
		Wallis SA DW Roller	8.00	4.00	32.00
		Hamm DV 06 — 1	8.00	5.00	40.00
		Hamm DV 06 — 2	8.00	5.00	40.00
		Bristowes Mk 6 Ch M/c	8.00	15.00	120.00
		JCB Loadall shovel	8.00	10.00	80.00
Total labour cost (A)	£840.00	Total plant cost (B)			£592.00

Material	Weight	Order time	Arrive	Departed	Waiting cost	Standing charge
30% HRAWC	15.87	8.30	8.20	8.50	0.00	0.00
30% HRAWC	16.24	8.30	8.20	9.25	6.25*	0.00
30% HRAWC	16.25	9.30	9.20	9.50	0.00	0.00
30% HRAWC	15.95	10.00	9.50	10.20	0.00	0.00
30% HRAWC	15.96	10.30	10.25	10.55	0.00	0.00
30% HRAWC	15.83	11.00	10.45	11.25	0.00	0.00
30% HRAWC	15.87	11.30	11.25	11.50	0.00	0.00
30% HRAWC	15.92	2.00	1.40	2.30	0.00	0.00
30% HRAWC	16.25	2.30	2.30	3.00	0.00	0.00
30% HRAWC	16.10	3.00	3.55	4.25†	0.00	164.08
30% HRAWC	15.90	3.30	4.20‡	4.40	0.00	0.00
Total weight	171.14			Totals		
Material cost (C)	£3698.94		Total waiting cost		£6.25	
			Total standing charge			£164.08

*25 minutes at £15.00/h

†Squad had to wait from 3.00 pm to 3.55 pm for material to arrive — no production, therefore charge to mixing plant

‡No charge to mixing plant since previous load had left at 4.25 pm and last load was on site at 4.20 pm — waiting costs and standing charges must be considered globally

Summary of daily surfacing costs/charges

Labour cost	£840.00
Plant cost	£592.00
Material cost	£3698.94
Supervision cost	included
Waiting cost	£6.25
Total cost	£5137.19

Mixing plant charge
£164.08 (records kept for consideration at the conclusion of contract may be used to offset against waiting time)

Superages

Estimated coverage	11.0 m²/t
Actual coverage	10.8 m²/t

Outputs

Estimated output	175 t
Actual output	171.14 t

Table 6.17. Summary surfacing diary

Item	Summary surfacing diary
General	
Cost of material Gang cost Cost of waiting time	Cost of material/t delivered Cost of labour and plant for 8 hour day Cost of waiting time/h for lorry (after allowance, normally first $\frac{1}{2}$ hour)
Output	
Estimated output Actual output % difference Cost profit/loss	Estimated output for 8 hour day Actual output for 8 hour day % difference between estimated and actual outputs Expressed as a cost produced by multiplying % difference by additional or reduced time needed to achieve estimated output
Coverage	
Estimated coverage Actual coverage % difference Cost profit/loss	Estimated coverage in m^2/t for materials Actual coverage in m^2/t for materials % difference between estimated and actual coverages Expressed as a cost produced by multiplying % difference by material cost to quantify loss (through additional material used) or profit (through additional area covered)
Waiting time	
Estimated waiting time Actual waiting time % difference Cost of waiting time	Estimated waiting time/load (excluding allowance, normally first $\frac{1}{2}$ hour) Actual waiting time/load (excluding allowance, normally first $\frac{1}{2}$ hour) % difference between estimated and actual waiting times Expressed as a cost produced by multiplying % difference by cost of waiting time/h
Summary	
Hours worked Cost difference Cumulative cost difference	Number of hours actually worked in a day Difference between estimated and actual costs ($+ =$profit, $- =$loss) Summation of cost variance/day ($+ =$profit, $- =$loss) taking outputs, coverages and waiting time into account
Complete Contract	
Total outputs Coverage Waiting time Work hours Cost difference	Estimated against actual expressed as a percentage — profit/loss Estimated against actual expressed as a percentage (for each material) — profit/loss Estimated against actual expressed as a percentage — profit/loss Cumulative hours worked Total cost difference for project

Table 6.18. *Example of summary surfacing spreadsheet layout*

Contract: A6106 Re-alignment Job number: HY211/023/RNH			Supervisor: R. McLellan Supplier: Quality Asphalt Ltd							Cost of material: £21.00/t Gang cost (inc. plant): £179.00/h Cost of waiting time: £15.00/h				
Day	Date: D/M/Y	Material	Estimated output: t	Actual output: t	% difference	Cost: profit/loss	Estimated coverage	Actual coverage	Difference: %	Cost: profit/loss	Estimated waiting time	Actual waiting time	Hours worked	Cost difference	Remarks
Friday	5/11/93	30% HRAWC	450	450	0	0	11	11	0	0	0	0	8	0	Everything okay
Monday	8/11/93	30% HRAWC	450	400	−11.1	−158.95	11	11	0	0	0	0	8	−158.95	Output reduced
Tuesday	9/11/93	30% HRAWC	450	400	−40.7	−583.41	11	11	0	0	0	0	12	−583.41	Output reduced, i.e. 450 t in 12 h not 8 h
Wednesday	10/11/93	30% HRAWC	450	550	+22.2	+318.22	11	11	0	0	0	0	8	+318.22	Output increased
Thursday	11/11/93	30% HRAWC	450	450	0	0	11	10	−9.1	−859.10	0	0	8	−859.10	Coverage reduced
Friday	12/11/93	30% HRAWC	450	450	0	0	11	12.5	+13.6	+1288.64	0	0	8	+1288.64	Coverage increased
Monday	15/11/93	30% HRAWC	450	400	−40.7	−583.41	11	10	−9.1	−859.10	0	0	12	−1442.51	Output reduced, coverage reduced
Tuesday	16/11/93	30% HRAWC	450	450	0	0	11	11	0	0	0	680 mins	8	−170.00	Waiting time costs incurred
Wednesday	17/11/93	30% HRAWC	450	550	+22.2	+318.22	11	12.5	+13.6	+1288.64	0	0	8	+1606.86	Output increased, coverage increased
Thursday	18/11/93	30% HRAWC	450	400	−11.1	−158.95	11	10	−9.1	−859.10	0	900 mins	8	−1243.05	Output reduced, coverage reduced, waiting time costs increased

loss or pot-holing may be treated by means of invisible patching. This involves re-heating the full depth of wearing course around the suspect area, removal of the affected material by shovel and placement of fresh material. Careful raking of the treated patch followed by hand chipping and careful rolling completes the operation producing a reinstatement invisible to the inexperienced eye once traffic has run on the material for a few days.

References

1. BRITISH STANDARDS INSTITUTION. *Coated macadam for roads and other paved areas. Specification for constituent materials and for mixtures.* BSI, London, 1993, BS 4987: Part 1.
2. BRITISH STANDARDS INSTITUTION. *Hot-rolled asphalt for roads and other paved areas. Specification for constituent materials and asphalt mixtures.* BSI, London, 1992, BS 594: Part 1.
3. BRITISH STANDARDS INSTITUTION. *Hot-rolled asphalt for roads and other paved areas. Specification for the transport, laying and compaction of rolled asphalt.* BSI, London, 1992, BS 594: Part 2.
4. DEPARTMENT OF TRANSPORT STANDING COMMITTEE ON HIGHWAY MAINTENANCE. *Preferred method 8 — road rolling.* Cornwall County Council, Truro.
5. DEPARTMENT OF TRANSPORT. *Specification for highway works.* HMSO, London, 1992, **1**.
6. BRITISH STANDARDS INSTITUTION. *Sampling and examination of bituminous mixtures for roads and other paved areas. Methods of test for the determination of texture depth.* BSI, London, 1990, BS 598: Part 105.

7. Compaction

7.1. Introduction

The laying of bituminous mixtures involves a number of processes which can all profoundly affect the life of the final pavement. Each is important but it is the compaction process which commands particular attention as failure by the laying contractor or the engineer to understand and adequately address the key issues may lead to a substantial reduction in the life of the road. Compaction is the densification of material by the application of pressure initially from the paver tamper and subsequently from the rollers. As the density of a bituminous mixture increases so does the strength of the material.

This chapter sets out general observations about compaction and justifies these by reference to published work on the effect of compaction on pavement performance. It also covers rolling methods, compaction plant and other factors, including environmental aspects. It examines the current compaction specification and considers the practical application of compaction technology.

7.2. Importance of compaction

7.2.1. General observations

Bituminous mixtures are a blend of aggregates and a binding agent. Depending on the material type, these aggregates can range from coarse materials through fine aggregate to filler. The binding agent is normally a bitumen which is obtained by the distillation of crude oil.

In terms of common UK usage, bituminous mixtures can be conveniently subdivided into bitumen macadams and hot-rolled asphalts. Bitumen macadams are described as continuously-graded since they consist of a number of aggregate sizes. Hot-rolled asphalts are called gap-graded because the aggregate sizes within the mix are not continuous. (See chapter 1).

Fig. 7.1 illustrates a number of points about the resistance to compaction (compactability) of dense bitumen macadams and hot-rolled asphalts. DBMs derive their strength from internal friction and mechanical interlock between the pieces of aggregate present in the mix. It is these mechanisms which provide the resistance to compaction. Hot-rolled asphalts derive their strength from the sand/filler/bitumen mix (mortar). Both DBM and hot-rolled asphalts are laid down and compacted at relatively high temperatures. This ensures that the binder is sufficiently fluid to act as a lubricant between the aggregate particles, reducing the internal friction of the mixtures and assisting in achieving good aggregate interlock.

Fig. 7.1. Core of hot-rolled asphalt wearing course and DBM basecourse

Therefore, as the mix temperature rises, a bituminous mixture is more easily compacted. In the case of DBM this is because the binder is less viscous and acts as a lubricant between the aggregate particles. In the case of asphalts this is because the bitumen has not set and the asphaltic mortar has little strength until the binder hardens.

Figure 7.1 also illustrates how variations in the composition of these materials can have a profound effect on their compactability. If, for example, a material has a higher stone content or a harder binder then this will reduce its compactability. This variability is summarized graphically in Fig. 7.2.

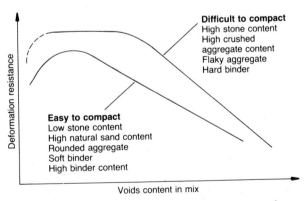

Fig. 7.2. Variation in the compactability of bituminous mixtures due to changes in composition[1]

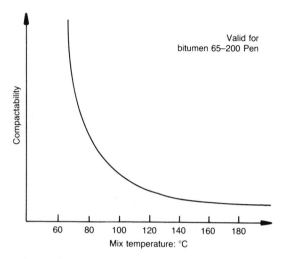

Fig. 7.3. Effect of temperature on the compactability of bituminous mixtures[1]

Figure 7.3 shows how the compactability increases as the temperature rises.[1] The reasons given explain why asphalts are more easily compacted than DBMs.

7.3. Published work

Proof that compaction of bituminous mixtures is a singularly important topic comes from the large volume of published work on the topic. Most literature concentrates on DBMs for the reason that these materials predominate in roadbases and basecourses. Asphalts contain more bitumen (by far the most expensive constituent in bituminous mixtures) than DBMs and are therefore more costly. Since asphaltic roadbases and basecourses have higher fatigue strength but lower dynamic stiffness than the equivalent DBM, the extra expense of asphalts is more than compensated by the increased lifespan of the pavement.

Nevertheless, the vast majority of roadbases and basecourses are DBMs. The stiffness of a carriageway is derived from these layers, which explains why research is directed towards them. Much of the published research work relates to asphaltic concrete which is a material used in the USA and mainland Europe. It is continuously-graded and, therefore, results for this material will, in the main, be valid for DBM.

It is convenient to split compaction technology research into four categories

   (*a*)   the effects of compaction on pavement performance
   (*b*)   rolling methods
   (*c*)   compaction plant
   (*d*)   other factors affecting compaction.

7.3.1. Effects of compaction on pavement performance

The compositions of roadbases and basecourses are not designed at present. The Marshall test is normally used to design the most economic hot-rolled asphalt wearing course mix for a given set of constituent materials. This is done using a cylindrical mould which contains the specimen under test. The sample is compacted by a standard hammer for a specified number of blows and then subjected to various tests from which certain parameters are ascertained.

The wheel tracking test measures the amount of deformation by running a standard wheel backwards and forwards across a compacted sample of a bituminous mixture. Full details of both this test and the Marshall test are given in chapter 8.

Jacobs[2] carried out laboratory work on the compaction of both hot-rolled asphalts and asphaltic concrete using both the Marshall test and the wheel tracking test apparatus. The relationship that he found between deformation resistance and the achieved level of compaction is reproduced in Fig. 7.4 and illustrates how the stiffness of a bituminous mixture increases as the degree of compaction rises.

Jacobs concluded that hot-rolled asphalts are more easily compacted than asphaltic concretes (and therefore DBMs). This substantiates the earlier general observation. He also found that the resistance to deformation increased as the degree of compaction rose. In the case of asphaltic concretes the deformation increased once void contents fell below 3.5%. No such phenomenon occurred with hot-rolled asphalt but it does not cover void contents of 2.5% or less. The change in resistance to deformation was less marked in hot-rolled asphalts than in asphaltic concretes. His general conclusion was that compaction control was more important in dealing with asphaltic concretes (DBMs) than with hot-rolled asphalts.

Lister and Powell[3] examined compaction levels and the relationship between compaction and pavement performance in DBMs. Among other areas of

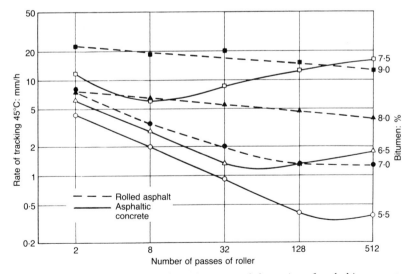

Fig. 7.4. Effect of compaction on the resistance to deformation of asphaltic concretes and rolled asphalts[2]

examination they plotted values of void content and/or voids in the mineral aggregate (VMA) against deflection (values were adjusted to include hot-rolled asphalts) and Young's Modulus (i.e. the stiffness) (Fig. 7.5). They also plotted void contents against mean track depth for different bitumen contents.

The voids content is the volumetric proportion of air between the coated aggregate particles to the total volume of the compacted mixture. The voids in the mineral aggregate are the proportion of air voids and the amount of binder not absorbed into the pores of the aggregate particles compared to the total volume of the compacted mixture. The methods of determination of these parameters can be found in chapter 8. Lister and Powell concluded that worthwhile extensions in pavement life could be obtained if compaction is increased.

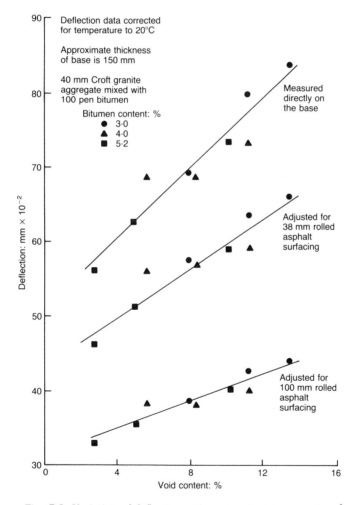

Fig. 7.5. *Variation of deflection with void content in a roadbase*[3]

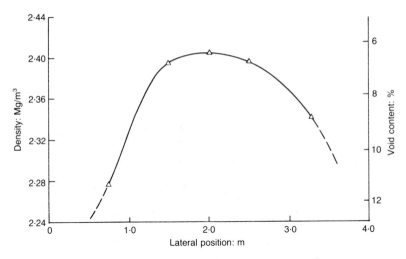

Fig. 7.6. Variation in compaction across a lane width of DBM[4]

7.3.2. Rolling methods

Leech and Selves[4] examined the rolling of DBMs. They state that there is substantial variation in the achieved levels of compaction across the lane width (Fig. 7.6).

In any road there is a tendency for the wheels of traffic to use the same narrow strips thus concentrating the loading on these areas. These are known as the wheel paths and are the part of a road which are likely to show wear or signs of failure first. Their location varies according to factors including the road width. For a single two lane carriageway having a width of 7.3 m they are centred around 0.9 m and 2.7 m from the nearside kerb. The inner wheel path is narrower because vehicles tend to maintain a standard distance from the kerb so this part of the road is stressed to the highest level at the greatest frequency. As can be seen from Fig. 7.6, the maximum compaction level, and therefore the greatest density, is achieved at the centre of the lane where there is only sporadic loading. This comes as no surprise as the traditional method of rolling is from the centre out towards the edges. Thus the compactive effort is concentrated in the middle of the lane and this is where the density is maximized.

Leech and Selves[4] also examined existing roads to find where densification had occurred under the action of traffic and looked at modifying the rolling pattern to equalize the compactive effort across the lane. They concluded that the level of compaction and density achieved across the width of the lane was typically 3% lower in the nearside wheel path than in the centre of the lane. They also confirmed that there was little or no additional densification as a result of the action of traffic except where there was poor compaction at the time of construction. Therefore thorough compaction is essential at the time of construction. Modified rolling was found to reduce the disparity in density across the lane width.

Fig. 7.7. A three-point deadweight roller. Photograph courtesy of Lothian Transportation INROADS

7.3.3. Compaction plant

Initially, compaction is imparted to the bituminous mixture by the screed of the paver. As discussed in chapter 5 the screed may be a tamping screed, a vibrating screed or a combination (tamping/vibrating) screed. Little information is available on the degree of compaction imparted by screeds in roadbases and basecourses. Hunter[5] carried out some testing on chipped hot-rolled asphalt wearing course, after laying but before rolling, which returned average air voids of 11%. Normally the target for asphalt is around 4% (this is a traditional figure for which no published justification can be found). Apparently, the rollers need only induce a further 7% reduction in the voids to achieve full compaction. However, one of the effects of the impregnation of the pre-coated chippings is to de-compact the asphalt in the upper part of the layer. Most published work in this area relates to roller performance.

Traditionally, three-point deadweight rollers have been used to compact both DBM and chipped hot-rolled asphalt wearing courses. A typical example of such a roller is shown in Fig. 7.7.

During the last ten years or so, vibrating rollers have become standard on surfacing operations. A typical tandem vibrating roller is shown in Fig. 7.8.

Powell and Leech[6] examined the use of vibrating rollers in relation to a number of trials. Some trials were carried out at the Military Vehicles and Engineering Establishment. The results are shown in Fig. 7.9.

Powell and Leech[6] concluded that vibrating rollers are capable of achieving 3% more compaction than conventional 8–10 t deadweight rollers which could result in up to 30% increase in the life of a pavement. BOMAG[7] publishes an extensive compendium of information on this subject. Chapter 5 discusses plant, including the available types of roller, in detail.

Fig. 7.8. A tandem vibrating roller. Photograph courtesy of Bomag (GB) Ltd

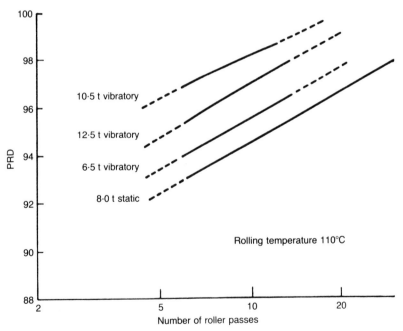

Fig. 7.9. Comparison of compaction levels achieved by vibrating and deadweight rollers[6]

7.3.4. Other factors affecting compaction

Other factors which affect the degree of compaction achieved are related to the material itself and the environmental conditions under which it is laid. It is convenient to think of heat in a hot bituminous mixture as a lubricant. In a soil as the amount of water increases it becomes easier to compact (up to the optimum moisture content). Any soil compacted around its optimum moisture content will achieve the maximum dry density, by definition. Similarly, the more heat that exists in a bituminous mixture when it is compacted, the greater will be its final density for a given compactive effort (again, up to the optimum heat content). As soon as the material is mixed it begins to lose heat. The temperature at which it is mixed at the production plant is governed by BS 594: Part 1[8] for hot-rolled asphalts and BS 4987: Part 1[9] for DBMs. It is transported to site in insulated vehicles ready for laying. Chapters 5 and 6 give more detail on this subject.

It is rare for a bituminous mixture to be over-compacted although it can occur with asphaltic concretes and, therefore, DBMs.[2] As far as the laying contractor is concerned, material which is being compacted has to be as hot as possible within the limits of the specification. If it is too hot it may squeeze out under the roller but this is a rare occurrence. UK roads specifications place no maximum limit on the degree of achieved compaction. Generally, the material should be as hot as the specification allows so that adequate compaction can be effected with ease.

Subject to specification requirements, failure by the contractor to maximize compaction will not result in immediate failure in roadbases or basecourses. However, failure of a hot-rolled asphalt wearing course to retain the pre-coated chippings is likely to occur soon after the road is opened to traffic if the material has not been compacted at temperatures high enough to ensure bonding of the pre-coated chippings with the parent asphalt layer.

Lister and Powell[3] carried out scanning of cores to ascertain how density varies through a layer of a bituminous mixture. Fig. 7.10 shows their findings

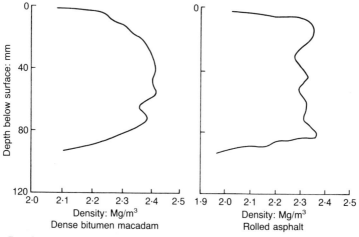

Fig. 7.10. Typical density/depth profiles[3]

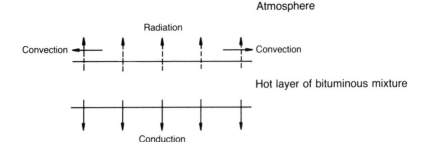

Fig. 7.11. How heat is lost from a hot bituminous mixture as it is laid

for a DBM and a hot-rolled asphalt used in roadbase or basecourse. This illustrates how the temperature of a bituminous mixture affects the final level of compaction achieved. Since the upper and lower portions cool faster than the centre of a layer the final densities at the top and bottom of the layer are correspondingly lower.

Unchipped and chipped materials cool by the same mechanisms but the chipped analysis is much more complex due to the effects of the addition of pre-coated chippings to the latter.

The cooling of unchipped bituminous mixtures

Figure 7.11 illustrates the parameters which affect the rate at which heat is lost from a hot bituminous mixture. The mechanisms by which heat is lost from the material are

(a) thermal conductivity, which is the transfer of heat through the hot material into the base layer and then into the layers below
(b) convection, which is heat carried away by the wind
(c) radiation, which is heat transmitted out into the atmosphere.

The rate at which heat is lost by these mechanisms is influenced by the following parameters

(a) air (ambient) temperature
(b) wind speed
(c) emissivity of the surface i.e. how efficiently it radiates heat
(d) absorption coefficient of the surface i.e. how efficiently it absorbs heat from the sun
(e) amount of solar radiation i.e. how much heat comes from the sun
(f) depth, thermal conductivity, density, specific heat and temperature of the hot layer
(g) depth, thermal conductivity, density, specific heat and temperature of the first underlying layer

(h)   depth, thermal conductivity, density, specific heat and temperature of the second underlying layer — it is unlikely that there would be any effect below this level.

There have been a large number of studies analysing the rate of heat loss from a layer of a hot bituminous mixture in order to establish the time available for compaction. Four papers on this subject are related. Corlew and Dickson[10] carried out the pioneering work. This was continued by Jordan and Thomas[11] and thereafter by Daines[12] and Hunter.[13]

These papers relate to methods which calculate temperatures in the layer after a short period of time. These values are then used to calculate the temperatures after another equal time increment. This process is repeated until the values for the required time periods are computed. As many discrete calculations are involved such analytical methods are ideally solved by computer.

The layer is considered infinite in terms of the width which is not an unreasonable assumption. Temperatures along any horizontal line have, therefore, the same value for any given time after laying. This means that the temperatures vary one-dimensionally.

A method which computes the temperature values at the current time by using those ascertained at the last time increment is called an explicit method. It can be shown mathematically that a more accurate method can be obtained by using only the temperature values at the current time. This is known as an implicit method.

Corlew and Dickson[10] used an explicit method to compute temperatures. Jordan and Thomas[11] used an implicit method but obtained the first estimate of temperatures at the next time level with an explicit method thus effectively making the whole analysis explicit. Their analysis runs on a mainframe computer. Daines[12] uses Jordan and Thomas' computer analysis so that too is explicit. Hunter[13] used a wholly implicit method to derive temperatures. This is a more accurate means of computing temperatures and, thus makes possible an examination of the relevance of each of the parameters involved in the flow of heat in a hot bituminous mixture as it is laid.

Daines[12] concludes that wind speed and air temperature have a substantial effect on the rate of cooling. As will be shown, the former does but the latter does not. Hunter[13] wrote a computer program using a wholly implicit method which computes temperatures in a bituminous mixture and allows them to be interrogated by the user. Temperatures at one minute intervals are stored during computation. The program runs on an Acorn/Archimedes micro-computer in less than a minute.

Hunter ran a large number of examples using a variety of parameters.[14] Variation of each of the parameters in turn allowed the relative effect of each of these factors to be ascertained. On the basis of the results he concluded that changes in the various parameters affecting the rate of cooling of unchipped bituminous mixtures have the following effects

(a)   wind speed has a major effect on the cooling rate
(b)   layer thickness has a major effect. Although thicker layers lose heat at a higher rate, the percentage loss is lower

(c)   ambient temperature has little effect on the cooling rate
(d)   thermal conductivity of the hot layer has a major effect
(e)   density and specific heat of the hot layer have little effect
(f)   incident solar radiation has little effect on the cooling rate
(g)   the underlying layer temperature, thermal conductivity, density and specific heat have little effect on the cooling rate.

Although the thermal conductivity of the material is an important determinant in the rate of heat loss, it is unlikely that there are any practical options available to alleviate this. The major finding is that the amount of heat loss does not vary markedly for substantially different air temperatures. Section 6.1 and note (b) of BS 594: Part 2[15] goes some way to recognizing this fact. The major variable is convection loss i.e. incident wind. The importance of these findings will become apparent when the specification is considered later.

Cooling of chipped bituminous mixtures

The cooling of a chipped layer is much more difficult to analyse. The functions of the roadbase, the basecourse and the wearing course have been covered in chapters 1, 2 and 3. Premature failure of any layer will manifest itself in different ways in that roadbase and/or basecourse failure shorten the life of the pavement whereas wearing course failure is likely to be more immediate. Nevertheless, failure of any material in a pavement is little short of a disaster and must be avoided. In order to minimize the possibility of wearing course failure it must be adequately compacted. The same mechanisms which govern the cooling of unchipped layers also apply to chipped materials such as chipped hot-rolled asphalt wearing course.

In relation to the temperature in a chipped hot-rolled asphalt wearing course Daines[12] states 'it is estimated that the average asphalt temperature is reduced by 15C°'. No basis for this estimate is given and it seems highly unlikely that one value will apply for all the different sources of materials that exist and also for the vast range of circumstances which may occur during any laying operation.

Hunter[13] devised a mathematical model which accurately analyses heat loss during the laying of chipped hot-rolled asphalt wearing course. Work by Hunter and McGuire[16] used this model to analyse temperature changes (Figs 7.12 and 7.13).

In terms of heat retention, chipped hot-rolled asphalt wearing course does not have much in its favour. It is a thin layer, normally 40 mm, but specification tolerances usually allow this to be as little as 35 mm. Immediately after it is laid, cold pre-coated chippings are rolled into the material. It is these chippings which complicate the cooling process. Since the chippings are three-dimensional, so too is the variation in temperature. Thus the analysis is also three-dimensional.

Hunter[13] wrote a computer program which computes temperatures in the asphalt and allows them to be interrogated by the user. Temperatures at one minute intervals are stored during computation. The program operates on an Acorn/Archimedes in around four minutes and calculates a matrix of temperatures in chipped hot-rolled asphalt wearing course at one minute

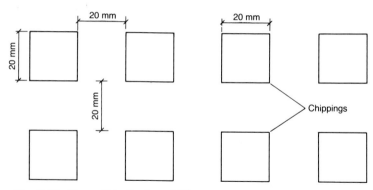

Fig. 7.12. Plan of idealized model[16]

intervals for 30 minutes after placement.

As with the unchipped bituminous mixtures Hunter ran a large number of examples using a variety of parameters.[14] Again, variation of each of the parameters in turn allowed the relative effect of these to be ascertained. On the basis of the results he concluded that changes in the various parameters affecting the rate of cooling of chipped bituminous mixtures have the following effects

(a)   wind speed has a major effect on the cooling rate
(b)   minor changes in layer thickness have a major effect because the nominal thickness is so small
(c)   ambient temperature has little effect on the cooling rate

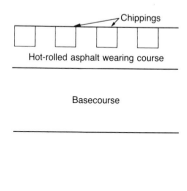

Fig. 7.13. Section through idealized model[16]

(d) thermal conductivity of the hot layer has a major effect
(e) density and specific heat of the hot layer have little effect
(f) incident solar radiation has little effect on the cooling rate
(g) the temperature, thermal conductivity, density and specific heat of the underlying layer and pre-coated chippings have little effect on the cooling rate.

As in the unchipped analysis, the incident wind speed is of fundamental importance. Note that ambient temperature has a minor effect. Thermal conductivity is a major factor in determining the rate of heat loss but it is unlikely that this fact will be of any practical use.

Pre-coated chippings are added to hot-rolled asphalt to provide frictional resistance for braking vehicles. They are normally 20 mm single size aggregate coated with bitumen of the same type and hardness (penetration) as that used in the manufacture of the hot-rolled asphalt wearing course. The rate of spread of pre-coated chippings is normally dictated by the specification as 70% of the shoulder-to-shoulder coverage achieved by following the method given in BS 598: Part 108.[17] Full details of this test are given in chapter 8. Examination of this method confirms that the more cubic the chipping the greater will be the required rate of spread. The chipping density is commonly around 13 kg/m^2 but values exceeding 15 kg/m^2 can occur. Hunter and McGuire[16] examined the effect of higher chipping densities on the rate of heat loss in chipped hot-rolled asphalt wearing course. They compared the heat content in the upper 20 mm of the layer for pre-coated chipping densities of 12 kg/m^2 and 15 kg/m^2. Results are shown in Table 7.1 and illustrated in Fig. 7.14.

This graph assumes that chippings are fully embedded five minutes after

Table 7.1. Temperature values in a 40 mm thick chipped hot-rolled asphalt wearing course layer with different chipping densities

Distance from surface mm	Temperature* at 12 kg/m^2	Temperature* at 15 kg/m^2
0	70	53
5	75	57
10	80	61
15	83	65
20	85	70
25	85	72
30	83	74
35	80	73
40	75	70

* Temperature at 1 min after chipping impregnation.

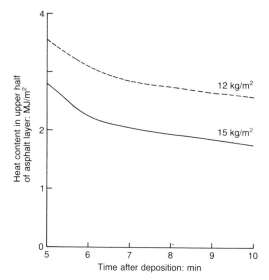

Fig. 7.14. Heat contents in a 40 mm thick chipped hot-rolled asphalt wearing course layer with different chipping densities

the asphalt is deposited and shows that the heat loss in the first minute after chipping impregnation is greater for the material with the higher chipping density. The material with the lower chipping density has more heat five minutes after laying than the material with the higher chipping density has after one minute.

Hunter[13] also modified his program to compute temperature values for a 35 mm layer. A comparison could then be made between the temperature profiles in a 35 mm thick layer and those in a 40 mm thick layer (Fig. 7.15).

The temperatures are around 10C° lower for the 35 mm layer. Hence, the reason why the thinner the asphalt layer the more difficult it becomes to lay satisfactorily. Some authorities specify a 45 mm or 50 mm thick hot-rolled asphalt wearing course which has substantial benefits when compaction takes place in marginal conditions.

7.4. Compaction specification

The general specification commonly applied to road construction in the UK is the *Specification for highway works*[18] with its accompanying Notes for Guidance.[19]

This specification normally refers to the relevant British Standards for bituminous macadams and hot-rolled asphalts[8,9,15,20] except where its express provisions override these.

The Notes for Guidance on the *Specification for highway works*,[19] is not normally mentioned in contract documents and is, therefore, at first sight, non-contractual. It is, nevertheless, a document which dictates the way that specifications are prepared. This raises the status of this document to one which is quasi-contractual which means that it is, in effect, a contract document.

trials in accordance with the requirements laid down in BS 598: Part 109[22] or by the contractor producing evidence of independent trials which satisfy the engineer that the proposed vibrating roller achieves a state of compaction which is equivalent to that obtained by using an 8 t deadweight roller. In relation to the former BS 598: Part 109[22] sets out a trial area which is split into two areas longitudinally. One area is compacted by an 8 t deadweight roller and the adjacent area is compacted by the proposed vibrating roller. Specified cores are taken and the densities obtained in accordance with the procedure given in BS 598: Part 104.[23] The mean density of the proposed vibrating roller must be at least the same as that achieved by a deadweight roller. The achieved coverage obtained by multiplying the width of the vibrating roller by the speed of travel must also at least equal that of the deadweight roller. In reality, most contractors would be able to obtain information from manufacturers of vibrating rollers in the form of evidence from independent trials, e.g. Bomag.[7] This contains details of independent trials carried out by TRL, the Military Vehicles and Engineering Establishment and various county highway authorities which might remove the need for the trials cited above. The same clause also states that if the degree of compaction achieved is to be determined in accordance with Clause 901.19, then the provisions above do not apply and the contractor may use any plant to achieve the required compaction above the specified minimum rolling temperatures. Items of plant used for compaction are described in chapter 5.

Normally, the degree of compaction achieved in dense roadbases and basecourses is tested in accordance with Clauses 901.19 and 927.1. The former clause requires that a trial area is compacted at least three days before DBM from each source is laid. The material is laid in a trial length of 30–60 m and at a width and thickness approved by the engineer. Cores are then taken and tested for PRD. Compliance is achieved if the average density is 93% of the refusal density as defined in BS 598: Part 104.[23] Basically the test consists of determining the bulk density of the trial cores. These cores are then heated and re-compacted to refusal in a laboratory. The bulk density should be at least 93% of the refusal density. Full details of this test are given in chapter 8.

Since no density testing is specified during the actual laying operation the engineer must ensure that the conditions and methods under which DBM in roadbases and basecourses is laid are at least as favourable as those which occurred during the compaction trial.

There are no requirements to achieve specific compaction levels in hot-rolled asphalt wearing course other than that stated in Clause 901.13 to achieve 'adequate compaction'. What constitutes 'adequate compaction' is not defined.

Clause 901.20 relates to the addition of pre-coated chippings. They are added to hot-rolled asphalt wearing course to provide skid resistance for vehicles. Full details of the materials used to produce pre-coated chippings and their functions are given in chapter 1. The pre-coated chippings must be applied by a mechanical spreader (chipping machine) except for specified instances such as in confined spaces. Full details of these machines are given in chapter 5.

The required texture depth is generally specified at 1.5 mm. It is a common misconception that texture depth is achieved by the pre-coated chippings sitting

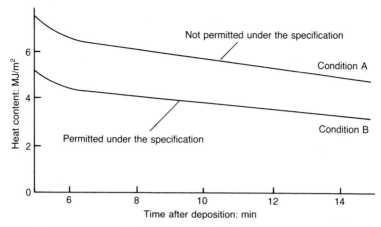

Fig. 7.16. Heat content of chipped hot-rolled asphalt wearing course under varying physical conditions[24]

above the finished surface. This is quite incorrect. Microtexture is the roughness of the applied chippings. Macrotexture is the roughness of the asphalt/chipping mixture. These terms are discussed fully in chapter 8. Texture depth comes from the combination of both microtexture and macrotexture. Checking of texture depth is discussed fully in chapter 8.

In UK roads there are no upper limits on achieved compaction levels. The tendency may be for the contractor to over-compact in an attempt to ensure that the specified minimum performance is attained. Over-compaction is rare, but the engineer is advised to monitor achieved compaction levels during construction to avoid the possibility.

The limitations of the specification[24] are shown in Fig. 7.16. This compares heat contents with time after laying for two specific sets of circumstances. Condition A is for temperatures of air, chippings and substrate of − 10°C and a wind speed of 0 mph. Condition B is for temperatures of air, chippings and substrate of 0°C and a wind speed of 5 mph. The paradox is that the condition which results in the material having the higher heat content would not be allowed under the Specification while the condition which results in the lower heat content would be permissible. Clearly the emphasis in the Specification must change.

7.5. Practical application of compaction technology

To maximize the time available for compaction, material should arrive on site at the maximum temperature permissible. Micro-chip control of mixing plants allows accurate control of the proportions, mixing times and temperatures. The laying contractor should insist that bituminous mixtures are mixed at the high end of the temperature ranges. It is far better to have a delay in commencing compaction because the material is too hot than to try to achieve adequate compaction when the material is too cold for that to be possible. The production process is discussed in detail in chapter 4.

Chapter 6 details the measures taken during transportation and laying to ensure that the heat loss during these phases is minimized.

Rolling has to be carried out by skilled operatives. It is of vital importance to provide thorough training in effective compaction practice.

Despite incorrectly stating that air temperature and base material temperature are important parameters affecting the time available for compaction, *Preferred Method 8 — Road Rolling*[25] gives excellent guidance for engineers, supervisors, foremen and roller drivers on all aspects of rolling hot bituminous mixtures. It covers the planning of surfacing operations including evaluation and flow charts, sample calculations, methods for checking rolling patterns and a separate section specifically for roller drivers which covers all aspects of rolling from the constitution of bituminous mixtures through good rolling practice to daily maintenance.

The application of pre-coated chippings to hot-rolled asphalt wearing course means that there are special problems in ensuring adequate compaction of this material. Fig. 7.17(a)−(d) illustrates the compaction of hot-rolled asphalt wearing course with pre-coated chippings.

The paver imparts initial compaction to the hot-rolled asphalt wearing course. The chippings are dropped onto the top of the laid material. As the pre-coated chippings are pushed into the laid mat, they de-compact, i.e. de-densify, the upper asphalt material. This asphalt will cool rapidly due to the de-densification and its physical contact with the cold pre-coated chippings. As the chippings continue to be forced into the material, this interstitial asphalt is re-compacted. It is vital that the asphalt is re-compacted as quickly as possible. If the final density is inadequate, the material will not be able to resist the shearing force of traffic tyres and abrade. This abrasion, in turn, exposes the chipping to a higher level of the shearing force than that experienced by embedded chippings and, as a result, some of those will be removed from the asphalt. This phenomenon is called 'fretting' (Fig. 7.18). Other problems, failures and their causes are covered in chapter 10.

Fretting invariably occurs shortly after the road is opened to traffic. The affected depth increases until equilibrium is reached i.e. when the hot-rolled asphalt matrix has enough stiffness to resist the abrading effect of traffic. Fretting usually stops after a day or two of exposure to traffic. Its occurrence is serious and may lead to the layer being planed out and replaced or perhaps surface dressed (sprayed with a bituminous layer and dry or coated chippings spread and compacted into the binder). Either eventuality is serious. The former means further delays for traffic while the latter means that the surface will not be of the expected quality. (Surface dressings are an excellent maintenance measure but are undesirable on a new or renewed road.) Both options probably involve substantial expense.

It is relatively easy to tell by visual inspection whether chipped hot-rolled asphalt wearing course has been properly compacted. The interstitial asphalt has a smooth surface. If that area is rough then that may mean there has been inadequate compaction and the material may fail.

In time, technology will undoubtedly take the guesswork out of compaction. Some rollers have a meter in the operator's cabin which shows, by visual inspection, the degree of compaction achieved as the roller passes over the

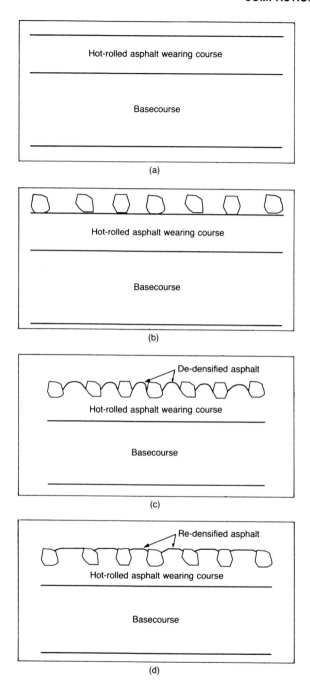

Fig. 7.17. The stages of compacting chipped hot-rolled wearing course: (a) hot-rolled asphalt wearing course laid on basecourse (b) chippings placed on top of hot-rolled asphalt wearing course (c) chippings partially embedded (d) chippings fully embedded

Fig. 7.18. The fretting of chipped hot-rolled asphalt wearing course

material being compacted. The same system can be extended to record information about the levels of compaction achieved on a carriageway. Such systems have a liquid crystal display which indicates

(a) the average compaction value for the entire area being compacted
(b) the average compaction level for each lane
(c) the number of passes which have been given to each lane.

Until these refinements become commonplace, engineers and contractors will have to continue to exercise great care when controlling the compaction of bituminous layers. There is no substitute for an efficient surfacing gang. An experienced foreman will know if the material is being effectively and uniformly compacted simply by its appearance. In the case of unchipped bituminous mixtures there are two key items of plant: the paver and the roller. When chipped hot-rolled asphalt wearing course is being laid there is an additional key item or plant: the chipping machine. Having delivered the material at the highest possible temperature, in all the circumstances, the items of plant must be co-ordinated to ensure that compaction of roadbases and basecourses or compaction/chipping embedment takes place at the earliest possible time to maximize the effect of the available compactive effort.

References

1. KIRSCHNER R. and KLOUBERT H. *Vibratory soil and asphalt compaction.* Bomag-Menck GmBH, Boppard, 1988.
2. JACOBS F. A. *Properties of rolled asphalt and asphaltic concrete at different*

states of compaction. Transport Research Laboratory, Crowthorne, 1977, Suppl R288.

3. LISTER N. W. and POWELL W. D. *The compaction of bituminous base and base-course materials and its relation to pavement performance.* Transport Research Laboratory, Crowthorne, 1977, Suppl R260.

4. LEECH D. AND SELVES N. W. *Modified rolling to improve compaction of dense-coated macadam.* Transport Research Laboratory, Crowthorne, 1976, LR 274.

5. HUNTER R. N. *An examination of the fretting of hot-rolled asphalt wearing course.* Heriot-Watt University, Edinburgh, MSc thesis, 1983.

6. POWELL W. D. and LEECH D. *Compaction of bituminous road materials using vibratory rollers.* Transport Research Laboratory, Crowthorne, 1983, LR 1102.

7. BOMAG. *The compaction of bituminous materials.* Bomag (Great Britain) Ltd, Larkfield. Bomag publishes an extensive compendium of available information on this subject.

8. BRITISH STANDARDS INSTITUTION. *Hot-rolled asphalt for roads and other paved areas. Specification for constituent materials and asphalt mixtures.* BSI, London, 1992, BS 594: Part 1.

9. BRITISH STANDARDS INSTITUTION. *Coated macadam for roads and other paved areas. Specification for constituent materials and for mixtures.* BSI, London, 1993, BS 4987: Part 1.

10. CORLEW J. S. and DICKSON P. F. Methods for calculating temperature profiles of hot-mix asphalt concrete as related to the construction of asphalt pavements. *Proc. Tech. Sess. Asphalt Paving Technology,* 1966, **35**, 549−50.

11. JORDAN P. G. and THOMAS M. E. *Prediction of cooling curves for hot-mix paving materials by a computer program.* Transport Research Laboratory, Crowthorne, 1976, LR 729.

12. DAINES M. E. *Cooling of bituminous layers and time available for their compaction.* Transport Research Laboratory, Crowthorne, 1985, RR 4.

13. HUNTER R. N. *Calculation of temperatures and their implications for unchipped and chipped bituminous materials during laying.* Heriot-Watt University, Edinburgh, PhD thesis, 1988.

14. HUNTER R. N. *Cooling of bituminous materials during laying.* Institute of Asphalt Technology, Staines, 1986, **38**, 19−26.

15. BRITISH STANDARDS INSTITUTION. *Hot-rolled asphalt for roads and other paved areas. Specification for the transport, laying and compaction of rolled asphalt.* BSI, London, 1992, BS 594: Part 2.

16. HUNTER R. N. and McGUIRE G. R. *Are texture and density in hot-rolled asphalt wearing courses incompatible?* Institute of Asphalt Technology, Staines, 1988, **40**, 29−31.

17. BRITISH STANDARDS INSTITUTION. *Sampling and examination of bituminous mixtures for roads and other paved areas.* BSI, London, 1990, BS 598: Part 108.

18. DEPARTMENT OF TRANSPORT. *Specification for highway works.* HMSO, London, 1992, **1**.

19. DEPARTMENT OF TRANSPORT. *Notes for guidance on the specification for highway works.* HMSO, London, 1992, **2**.

20. BRITISH STANDARDS INSTITUTION. *Coated macadam for roads and other paved areas. Specification for transport, laying and compaction.* BSI, London, 1993, BS 4987: Part 2.

21. DEPARTMENT OF TRANSPORT. *Rolled asphalt wearing course — cold weather working.* DTp, 1980, Advice Note HA 10/80.

22. BRITISH STANDARDS INSTITUTION. *Sampling and examination of bituminous*

mixtures for roads and other paved areas. BSI, London, 1990, BS 598: Part 109.
23. BRITISH STANDARDS INSTITUTION. *Sampling and examination of bituminous mixtures for roads and other paved areas.* BSI, London, 1990, BS 598: Part 104.
24. HUNTER R. N. and McGUIRE G. R. *Winter surfacing — only the specification is out in the cold.* Institution of Highways and Transportation, London, 1986, **12**, 33.
25. DEPARTMENT OF TRANSPORT AND STANDING COMMITTEE ON HIGHWAY MAINTENANCE. *Preferred method 8 — road rolling.* Cornwall County Council, Truro.

8. Testing

8.1. Introduction

The results of measuring and testing procedures throughout the whole process of flexible road construction are relied on for guidance. Decisions are taken based on the data generated at virtually every stage of the operation, including initial investigation, cost estimation, production control, installation, acceptance and assessment of performance.

Confidence in the accuracy of the data being analysed is, therefore, vital. This can only be achieved by demonstrating that sufficient care has been taken by skilful, knowledgeable individuals in following prescribed methods for proper sampling and testing that yield reliable results. Errors can lead to costly, false assumptions being made about the quality of materials and products.

Intelligent interpretation of the data provided is essential in the light of the precision of the test methods and the reliability of measuring and test equipment. This can be achieved with the aid of appropriate statistical techniques, many of which involve only simple calculations. Good interpretation of reliable data leads to the correct decisions being taken and good roads being built at the right cost.

This chapter considers some of the most common test methods used for examining bituminous mixtures and the factors affecting the test results and their interpretation. Future requirements are also discussed.

8.2. Sampling

Possibly the greatest concern about testing of bituminous mixtures is that laboratory portions are representative of the bulk quantity of the material being examined. Unfortunately, bituminous mixtures are prone to segregation during handling, whereby differently-sized aggregate particles in the mixture separate in either a random or uniform manner. The effect is exacerbated in coarser mixtures which have thin binder films coating the aggregate particles. Thicker binder films resist the relative movement of the particles within the mixture and so aid homogeneity. Where there is a tendency to segregate, strict adherence to standardized sampling practice is of paramount importance. The sampler should be properly trained, competent in implementing prescribed sampling methods and able to work safely without direct supervision.

A number of factors influence segregation of bituminous mixtures. The nature of the specified or designed aggregate grading can markedly affect the tendency of the mixture to segregate. Gap-grading of a type illustrated in Fig. 8.1 can

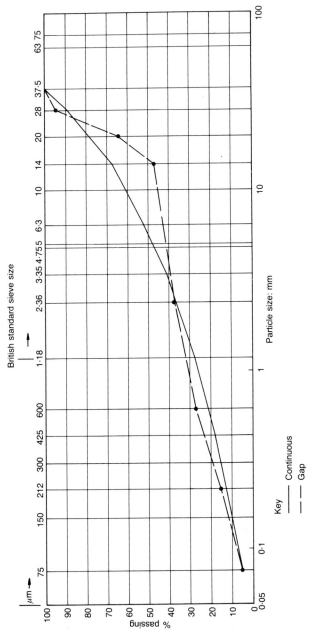

Fig. 8.1. Aggregate grading

be a cause as the aggregate blend comprises high proportions of the larger and smaller particles with a lack of intermediate sizes. Gap-graded mixtures, such as hot-rolled asphalt, have certain advantages and these are covered in chapter 1. Segregation in gap-graded mixtures is hindered if the material has relatively high binder contents and thick binder films.

Continuous gradings, however (Fig. 8.1) are less sensitive to segregation, particularly where the blend comprises a good mix of aggregate particles changing progressively over the full size range with a relatively small maximum size. The difference will not be so evident when, for example, comparing large stone mixtures of DBM and hot-rolled asphalt, due to the uncertainty over the fully continuous nature of the aggregate gradings of dense macadams and the stabilizing effect of the richer sand/bitumen mortar in rolled asphalts.

Segregation of aggregate constituents in stockpiles or feed bins can also persist in the bituminous mixture, depending on the method of manufacture of the mixture. Inspection of aggregate stocks and bins for signs of segregation may determine whether difficulties in sampling a segregated bituminous mixture can be traced back to the methods of handling the aggregate constituents and whether this can then be overcome by taking appropriate measures to improve the handling techniques.

The increasing use of large bins for storage of hot bituminous mixtures brings further potential segregation problems for the sampler. Brock[1] discusses in detail the different methods used to minimize the effects and concludes that both bin charging batchers and rotating chutes (Figs 8.2 and 8.3) are adequate devices to help eliminate segregation.

Ideally, lorries transporting bituminous mixtures should be loaded evenly along their bodies to prevent larger aggregate particles rolling to the base of the conical heaps formed during loading. Brock[2] suggests loading in three different drops in the sequence indicated in Fig. 8.4 to ensure a good mix when discharging from the lorry into the paver hopper. It is therefore important that

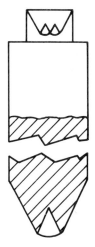
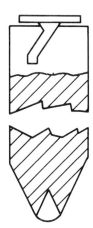

Fig. 8.2. Bin charging batcher[1]   *Fig. 8.3. Bin charging rotating chute[1]*

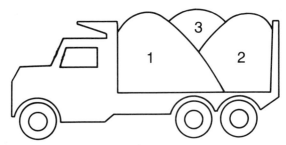

Fig. 8.4. Loading a lorry[2]

the sampler takes due account of the actual loading methods employed to determine the best sampling procedure to be followed at the site of the mixing plant.

BS 598: Part 100[3] describes methods of obtaining bulk samples, both from loaded lorries and during discharge, from the mixing plant. The minimum mass of bulk sample is given in Table 8.1. This is sufficient to provide a laboratory sample for both the supplier and the customer.

Sampling from the loaded lorry using a suitable shovel is frequently preferred by the sampler and this is easily and safely done if a sampling platform has been constructed at the correct height in relation to the height of the sides of the lorry body. Management should ensure that appropriate protective clothing is always worn by samplers and that they know the hazards associated with moving plant and vehicles and are trained in the treatment of burns in the event that skin contact occurs.

A number of increments, depending on the nominal size of the aggregate constituents, are taken from different positions across the top of the load after surface material has been removed from those positions. Alternatively, a sampling pan secured by a hinged bracket to a convenient rigid section of the mixing plant can be operated so that it passes transversely through the centre part of the curtain of material being discharged. Again, a number of appropriate increments are taken to provide the required mass of bulk sample. The risk to the sampler's personal safety using this method is a deterring factor. A distinct advantage of the method, though, is the ability to sample individual batches during production without possible intermingling by other batches. ASTM Standard D979[4] also describes procedures for sampling bituminous mixtures from both belt and skip conveyors.

Segregation can also occur at the laying site. It follows that the sampler should be aware of those locations in the laying process where the taking of either

Table 8.1. Minimum mass of bulk samples (ref. 8.3)

Nominal size of aggregate: mm	Minimum mass of bulk sample: kg
> 20	24
≤ 20	16

coarser or finer samples than is representative of the bulk quantity of material being examined is most likely to occur. Brock[1] makes a number of suggestions on avoiding excessive segregation at the laying site, including

(a)   whenever possible, do not empty the paver hopper between each lorry load
(b)   only operate the paver hopper wings infrequently
(c)   keep the paver hopper as full as possible during loading from the lorry
(d)   do not starve the augers or overfeed them with material
(e)   run the paver and augers continuously whenever possible.

The points made in chapters 6 and 7 about heat loss should be borne in mind in relation to these suggestions.

BS 598: Part 100[3] no longer includes the procedure for sampling from a paver hopper that entailed withdrawing the lorry from the paver after it had discharged approximately half its load before sampling, often causing the cessation of paving operations. This practice is clearly inconsistent with a number of the suggestions. The alternatives are to sample from the material around the augers of the paver and from the material extruded from beneath the paver screed with the sample taken before rolling.

When sampling material from around the augers, two increments should be taken from each side of the paver. The sampler should be aware of a tendency for the augers to throw coarser material to the edges of the paved widths if the auger box is underfilled. Samples should, therefore, only be taken after the box has been properly filled by a continuous slow flow of material from the conveyors.

Generally, there should be less chance of significant segregation actually appearing in the laid material, provided all the necessary precautions are taken upstream of the laying operations. In practice, surface defects caused by segregation and other factors do appear from time to time, frequently only becoming apparent after rolling, and these are discussed in chapter 10. Samples may be taken from the laid material before rolling by placing two sampling trays on the ground just ahead of the paving operation and removing the trays with the aid of wire attachments after the material has been laid. The method is not favoured for sampling wearing course material because of the possibility of leaving surface blemishes at the sampling positions after infilling. It is also inappropriate for relatively thin layers of coarser material because of a real possibility of disturbance as the paver moves over the sampling positions.

For various reasons, samples may need to be taken from the completed layer, e.g. to measure the density of the layer. Samples may be extracted using a core-cutting machine or other equipment suitable for cutting out squares of material. Normally, samples are not taken from a localized spot which is visibly unrepresentative of the whole unless there is a specific reason for examining the material, perhaps to investigate the cause of some apparent difference. Despite this, there is a good argument for devising a plan of random sampling that eliminates intentional (and also minimizes unintentional) bias on the part of the sampler.[5] In this way, there can be no criticism of the sampling regime.

ASTM Standard D3665[6] describes procedures which can be followed to determine the location of samples with the use of a table of random numbers.

CERTIFICATE OF SAMPLING

SAMPLE DESTINATION	SAMPLE IDENTIFICATION NUMBER
MATERIAL DESCRIPTION	TEMPERATURE
PLANT / SUPPLIER	BINDER TYPE
CUSTOMER	BINDER TARGET %
FLUX %	SPECIFIED BINDER %
LOCATION / SOURCE / WAGON / TICKET OF SAMPLE NUMBER NUMBER	REASON FOR SAMPLING
DESTINATION	
DATE OF SAMPLE	TIME OF SAMPLE
METHOD OF B.S. No. Part No. SAMPLING other Method	

Remarks

Signature of Sampler

Name of Sampler (Capitals)

Fig. 8.5. Certificate of sampling

It is recommended that on receipt of the samples at the laboratory, note is made of their condition and any discernable variations to aid interpretation of subsequent test results. A degree of randomness is removed by dividing the whole into smaller lots within which random sampling is carried out.

It may be convenient to reduce the size of the bulk sample before dispatch to the laboratory depending on the planned tests or on whether there is a need to divide the sample between interested parties. This may be done for hot mixtures using a sample divider or riffle box, or by quartering on a clean, hard surface as described in BS 598: Part 101.[7] The laboratory sample obtained should be accompanied by a record of the sampling details. A typical example of a certificate of sampling is shown in Fig. 8.5.

Sampling procedures for constituent aggregates are described in BS 812: Part 102[8] and the subject is discussed in detail by Harris and Sym.[9] Standard methods for sampling bituminous binders are described fully in BS 3195: Part 3[10] and ASTM D140.[11] No method exists which will guarantee that a sample is truly representative of the whole in all respects.[12] There will always be random sampling error and also sampling bias, albeit small.

Large differences between the sample and the total quantity of bituminous mixture are often caused by segregation. Acknowledgement is made in the British Standards for bituminous mixtures[13,14] of the particular difficulties involved in obtaining representative samples of the whole in the provisions made for sampling and testing error by the application of specified tolerances. However, for a uniformly coated mixture, there is a close association between the amount of bituminous binder and the aggregate grading, or more specifically, the aggregate surface area. This fact can be used to determine whether the sample is likely to be unrepresentative. This can be demonstrated by taking a large sample of mixture and intentionally dividing it into test portions having different fine aggregate contents to simulate sampling error due to segregation. Analysis of the constituent parts after the extraction of binder from each portion confirms that the binder content changes in proportion to the change in fine aggregate content. This is because the fine aggregate has a larger surface area per unit volume and so holds a greater quantity of binder coating than the same mass of coarse aggregate particles.

The relationship is illustrated in Fig. 8.6 for a particular bituminous mixture. The percentage of aggregate passing the 3.35 mm test sieve, representing the independent variable, has been plotted against the corresponding binder content. The plot suggests a linear association between the variables which may be described by the best straight line $y = a + bx$.

Regression analysis of the data gave the following information for this material

Intercept (a)	= 2.967
Slope (b)	= 0.047
Correlation coefficient (r)	= 0.996
Percentage fit ($100r^2$)	= 99.2%

The resultant straight line $y = 2.967 + 0.047x$ can then be drawn through the plotted points on the graph. The correlation coefficient gives a measure of the degree of linear association between the variables, whereby values of

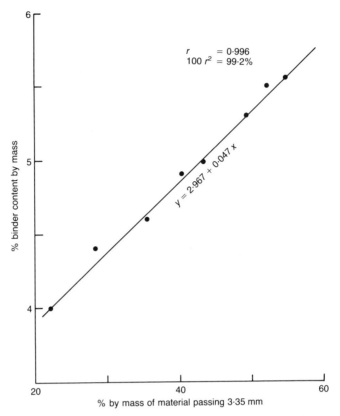

Fig. 8.6. Relationship between binder content and fine aggregate content for a particular uniformly coated bituminous mixture

−1 or +1 indicate perfect association and a value of 0 indicates the absence of a linear association. The percentage fit, indicates the extent to which the variation in binder content has been caused by the variation in material passing the 3.35 mm sieve. The remaining percentage is unaccounted for and may be caused by other independent variables. The equation describing the straight line relationship will depend on the physical and volumetric properties of the constituents of the particular mixture and, especially, the physical differences between the fine and coarse aggregate particles. For example, a steeper slope would indicate a greater capacity of the fine aggregate for holding binder in relation to that of the coarse aggregate particles.

Once the relationship between binder content and coarse/fine aggregate proportions for a particular mixture is established, subsequent plots on the graph during production deviating from the straight line would suggest either poor mixing resulting in lack of uniformity of binder coating or change in constituent proportioning. Points plotted on the graph lying close to the straight line would suggest that segregation is a cause of variation and that the individual test results need not, necessarily, be indicative of an unsatisfactory product. A variation of the analysis based on a summation of each percentage amount passing the fine aggregate sieves was devised by Bryant.[15]

British Standard BS 598: Part 102[16] includes tables for adjusting the binder content found on analysis to correspond with the mid-point of the grading passing the top fine aggregate test sieve specified in the British Standard product specifications, to take account of variations due to sampling and sample reduction procedures. Similar adjustment values are included for aggregate passing the 75 μm test sieve to correspond with the same mid-point on the top fine aggregate test sieve. However, it is debatable whether the mid-point of the specified limits on this sieve should actually be taken as reference, rather than the average percentage passing the sieve from the designed aggregate grading that may be quite different, depending on the specified tolerance, from the mid-point.

Variation in test results due to sampling and testing for binder content of bituminous mixtures has been estimated to be around 65% of the total variation from studies made in the 1960s.[17] Later studies[18] confirmed that this percentage had not changed appreciably. The next section considers the variation that occurs after sampling and that arises during testing of the sample. The variation is described by the precision of the test method.

8.3. Precision

Bituminous mixtures are generally produced to fit the purpose defined by the purchaser. Certain properties of the mixture are influenced by its volumetric composition and the nature of its constituents. Control of these influential factors is important. Knowledge of how certain properties relate to the performance of the product incorporated in the works frequently also leads the specifier to include in his requirements for a number of measurements to be made. Certain desirable features of the constructed pavement may also be measured and these too can be included in the specification.

During the course of the production and installation processes certain changes occur. Test results and other measured data vary due to different factors such as sampling techniques, testing, changes in the quality and nature of constituents and plant variations. Some degree of control is, therefore, required to ensure that the desired properties of the end product are realized. Permitted deviations from the values of the characteristics whose behaviours are to be investigated are specified.

It follows that the tolerances set should allow for errors in sampling and testing, and for plant and material inconsistencies that may be expected to occur when all reasonable care has been taken to avoid them. However, this need not always be the case. It is important, therefore, to be able to estimate the magnitude of these inherent variations so that realistic specification limits can be determined for the variable characteristics measured.

When available, precision data are commonly included in standard documents for test methods. The data may be expressed in different forms but the terms repeatability and reproducibility are commonly used.

There is a distinction between precision and accuracy. In simple terms, precision is concerned with the variability of individual values, irrespective of how close these values are to the true value. Accuracy, on the other hand,

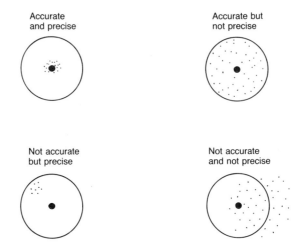

Fig. 8.7. Precision and accuracy

is concerned with how close the measured values are on average to the true value, irrespective of the degree of variability of the individual values. The distinction is illustrated in Fig. 8.7 with the aid of the type of target board that may be found on a firing range. It is clear that an analysis of variable characteristics must include a study of both scatter and location.

The scatter of test results may, of course, be influenced by the inherent variations or precision of the test methods. These testing variations can be determined by experiment and such a procedure is described in BS 5497: Part 1.[19] However, the accuracy of the test method, i.e. how close the overall mean value is to the true value of the characteristic, can be more difficult to determine due to an uncertainty as to what the true value actually is.

When determining the precision of a test method, every effort must be made to minimize the influence of variations from other causes. The experiment concerns both a measure of the variability due to a single operator and a measure of variability between different laboratories, all else being equal.

Testing by a single operator should be carried out each time in circumstances which are as similar as possible and on samples of identical material. Repeat testing on the same test specimen ensures that identical material is used. Unfortunately, most of the tests carried out on bituminous mixtures are destructive and so particular care is needed in the preparation of test portions from laboratory samples which should be derived from a mixture that is considered to be as uniform as possible in terms of both composition and structure. Repeat tests should also be carried out within as short a period of time as possible.

Single operator variation is described by a measure of repeatability determined under 'conditions where mutually independent test results are obtained with the same method on identical test material in the same laboratory by the same operator using the same equipment within short intervals of time'.[19] The variations associated with the inherent variability of constituent aggregates and the difficulties involved in reducing samples of bituminous mixtures to give representative test portions, unavoidably contributes to the measure and so form a real part of it when the same test specimen cannot be used. So, in this case, both material variation and sampling error will tend to increase the magnitude of the reported precision data above that due to testing alone. Depending on the nature of the particular test method under experiment, the error may be minimized by combining measured quantities of the constituents of the mixture to provide the individual test specimens.

The repeatability value r is defined as 'the value below which the absolute difference between two single test results obtained under repeatability conditions may be expected to lie with a probability of 95%'. It is calculated from the following equation.[19]

$$r = 2.8 \; Sr \tag{1}$$

where Sr is an estimate of repeatability standard deviation and 2.8 is a factor derived from the basic statistical model used.

Between laboratory variation is described by a measure of reproducibility determined under 'conditions where test results are obtained with the same method on identical test material in different laboratories with different operators using different equipment'.[19] This is a measure of the maximum variability in results and differs from repeatability due to different operators using different equipment of different calibration status that is used under different environmental conditions.

The reproducibility value R is defined as the value below which the absolute difference between two single test results obtained under reproducibility conditions may be expected to lie with a probability of 95%. It is calculated from the following equation.[19]

$$R = 2.8 \; S_R \tag{2}$$

where S_R is an estimate of reproducibility standard deviation and 2.8 is the same factor as in equation (1).

Harris and Sym[9] in their study of the precision of testing aggregates used a statistical model that included an allowance for variation between laboratory samples. This takes account of an additional variation that arises when it is not possible to obtain test results on the same laboratory sample and contributes to a higher value of R. As similar difficulties are experienced with sampling the two types of product, the model is equally applicable to bituminous mixtures. However, the magnitude of the contribution may be greater with aggregates which are particularly dry and therefore more prone to segregation.

Table 8.2 gives precision data that has been published in the UK for some of the more common test methods.

Precision data published by the American Society for Testing and Materials (ASTM)[21] and AASHTO[22] is given in Table 8.3 for similar tests carried out under conditions described in ASTM Standard Practice C802.[23]

Some of the findings of early research work carried out in the USA[24] on the normal variation of test results of bituminous mixtures from a large number of paving projects are given in Table 8.4. Precision data for the test methods gives a guide to the differences that may be expected to occur when testing

Table 8.2. Precision of British test methods

Test	Standard	Repeatability standard deviation: S_r (Coefficient of variation $= S_r/\bar{x}.100\%$)	Repeatability: r (% of mean)	Reproducibility standard deviation: S_R (Coefficient of variation $= S_R/\bar{x}.100\%$)	Reproducibility: R (% of mean)	Unit
Penetration of bitumens < 50 pen	BS 2000: Part 49: 1983	0.35	1	1.4	4	0.1 mm
Penetration of bitumens > 50 pen	BS 2000: Part 49: 1983	(1.1%)	(3%)	(2.8%)	(8%)	–
Ring and ball softening point of paving grade bitumen	BS 2000: Part 58: 1988		1		2.5	°C
Binder content of bituminous mixtures	BS 598: Part 102: 1989		0.3%		0.5%	–
Adjusted filler content of bituminous mixtures < 75 μ	BS 598: Part 102: 1989		1%		1.5%	–
Percentage refusal density	BS 598: Part 104: 1989		1.2%		1.8%	–
Texture depth by laser texture meter for test result levels 0.8 mm 1.2 mm 1.4 mm	BS 598: Part 105: 1990	0.032 0.047 0.051	0.09 0.13 0.14	0.065 0.078 0.091	0.18 0.22 0.25	mm mm mm
Texture depth by sand patch for test result levels 1.3 mm 1.9 mm 2.3 mm	BS 598: Part 105: 1990	0.062 0.103 0.135	0.17 0.29 0.38	0.073 0.129 0.156	0.20 0.36 0.44	mm mm mm
Laboratory design of hot-rolled asphalt mixtures[20] Density Stability Flow Quotient	BS 598: Part 3: 1985		17 1 0.6 0.7		26 2.2 1.3 1.4	kg/m³ kN mm kN/mm

Table 8.3. Precision of American test methods

Test	Standard ASTM No. (AASHTO No.)	Repeatability standard deviation: Sr (Coefficient of variation $= Sr/\bar{x} \cdot 100\%$)	Repeatability: r (% of mean)	Reproducibility standard deviation: S_R (Coefficient of variation $= S_R/\bar{x} \cdot 100\%$)	Reproducibility: R (% of mean)	Unit
Penetration of bitumen < 50 pen	D5 (T49)	0.35	1	1.4	4	0.1 mm
Penetration of bitumen > 50 pen	D5 (T49)	(1.1%)	(3%)	(2.8%)	(8%)	—
Ring and ball softening point of bitumen	(T53)		2.0		3.0	°C
Bitumen content of bituminous mixtures	D2172 (T164)	0.18%	0.52%	0.29%	0.81%	—
Bulk specific gravity of compacted bituminous mixtures using saturated surface-dry specimens	D2726	0.0124	0.035	0.0269	0.076	—
Theoretical maximum specific gravity of bituminous mixtures	D2041 (T209)	0.004	0.011	0.0064	0.019	—
% air voids in compacted dense and open bituminous mixtures	D3203 (T269)	0.32%	0.91%			—
Surface macrotexture depth using a volumetric technique	E965	(as low as 1%)		(as low as 2%)		
Effect of water on cohesion of compacted bituminous mixtures	D1075 (T165)	6%	18%	18%	50%	—
Viscosity of bitumen residue at 60°C after rolling thin film oven test	D2872 (T240)	(2.3%)	(6.5%)	(4.2%)	(11.9%)	—

Table 8.4. Typical values of standard deviation of test results of bituminous mixtures during paving projects[24]

Test	Average standard deviation
Bitumen content, %	0.17^a–0.42^{b*}
Bulk specific gravity of compacted sample	0.02–0.04
Air voids, %	0.8–1.8
Marshall stability, kg	128
Marshall flow, mm	0.3

* [a] is the value of $\frac{1}{3}$ least variable jobs and [b] is the value of $\frac{1}{3}$ most variable jobs

the same material under the stated conditions and caution should be exercised when comparing this data with the variation of field test results that arises due to sampling, testing and materials variation.

8.4. Test methods

8.4.1. Composition of bituminous mixtures

The composition of a bituminous mixture has a definite influence on its behaviour in service, to an extent depending on the property being measured. The effect of constituents on performance is discussed in detail in chapters 1, 2 and 3.

Richardson[25] described methods used at the beginning of this century for the examination of bituminous mixtures for their bitumen content and mineral aggregate grading. These involved dissolving out the bitumen with an appropriate organic solvent and trapping the mineral matter in a filter or centrifuge apparatus. The recovered mineral aggregate was then passed through a series of sieves and the aggregate graded by the amounts passing the various aperture sizes. The preferred solvent was bisulphide of carbon (carbon disulphide) which was available cheaply and could be readily redistilled for re-use if desired.

The solvent has a boiling point slightly higher than that of methylene chloride (dichloromethane), commonly used today, but is extremely flammable and gives off very poisonous vapour. At the time, pure chloroform (trichloromethane) was found to be an ideal solvent for determining total bitumen content due to a high bitumen solubility. However, it was prohibitively expensive and could not be evaporated or burned off rapidly, as was frequently done when determining residual mineral matter in the extracted binder solution during analysis.

The early analytical methods for fine bituminous mixtures involved determining the bitumen quantity in small samples of the mixtures by subtracting the weight of recovered mineral matter from the initial weight of the sample and expressing this as a percentage of the mixture. To hinder the solvent with dissolved bitumen from carrying over fine particles of mineral matter also, the solution was passed through a filter paper supported in a funnel that was placed in a conical flask to collect the filtered solution. After filtering, the percolate in the flask was allowed to stand overnight to allow any sediment to settle to the bottom. The percolate was decanted and the remnants containing sediment were passed back through the filter paper. The binder solution was evaporated and burned and the bitumen incinerated to determine a correction factor for any mineral matter that passed through with the bitumen, to be added to the weight of mineral aggregates recovered from the filter paper. The used filter paper was also burned to correct for any fine mineral matter remaining trapped in the pores of the paper.

Alternatively, where a more rapid analysis was required, the mineral aggregate was recovered in a centrifuge machine rotating at a speed of 1500 rev/min. This method was also more suitable for analysing mixtures containing relatively large quantities of mineral filler. After centrifuging, the binder solution collected in the centrifuge tubes was decanted leaving the sediment for further centrifuging with new solvent.

When all bitumen was removed from the aggregates, the solvent remaining in the tubes was evaporated and the residue weighed. The binder solution removed from the machine was evaporated and burned and the extracted bitumen incinerated to correct for mineral matter removed with the bitumen in the calculation of their contents from the loss in weight in the centrifuge tube.

Coarser mixtures were analysed using a specially designed centrifuge that could handle larger samples. A filter ring of roofing felt was inserted into the apparatus to catch fine mineral matter thrown out during centrifuging at between 1500 and 1800 rev/min. Facilities were also included to recover solvent for re-use. The separation was considered to be of such efficiency that no correction was made for traces of mineral matter remaining in the extracted bitumen.

Today, the methods of analysis have not greatly changed and the principle of the test remains the same, comprising the following basic operations

- (a)   binder extraction by dissolving in an organic solvent
- (b)   separation of mineral matter from the binder solution
- (c)   determination of binder quantity by difference or binder recovery
- (d)   calculation of soluble binder content
- (e)   grading of mineral aggregates.

This is illustrated in Table 8.5 which summarizes a selection of analytical methods used in Europe, breaking them down into the same basic operations. An additional column is included headed insoluble binder to cater for those products such as tar which are not totally soluble in the solvent used.

A variety of test equipment is used for the various test methods and the main items are illustrated in Figs 8.8–8.16.

Some of the test methods include determination of binder directly by weighing the total recovered binder or a portion of it (Table 8.6).

British Standard 598: Part 102[16] describes a method of direct determination from a portion of the total binder solution collected after extraction by roller bottle. The portion is quickly poured into a number of centrifuge tubes which are then capped securely to prevent evaporation of the solvent and placed in a bucket-type centrifuge. The time centrifuging continues depends on the acceleration developed in the machine to ensure that fine mineral matter separates satisfactorily to the bottom of the tubes. The centrifuged solution

Table 8.5. Analytical test methods for bituminous mixtures

Test method	Binder extraction	Basic operation			Soluble binder	Insoluble binder	Aggregate grading
		Separation of mineral filler	Binder quantity	Ash determination			
(D) DIN 1996: Part 6: 1988 Differential	(i) Hot extraction apparatus or (ii) Cold agitation or Centrifugal filter press using trichloroethylene or toluene	Continuous flow centrifuge	By calculation	For arbitration only DIN 52005	By subtracting mass of total mineral aggregate	By calculation as a function of filler content or by test	DIN 1996: Part 14
Recovery	As above	As above	Total recovery of binder by evaporation and rotary evaporator or distillation apparatus with fractionating column	As above	From mass of recovered binder	As above	As above
(GB) BS 598: Part 102: 1989 Extraction bottle binder by difference	Bottle rolling using methylene chloride or trichloroethylene	Pressure filter	By calculation	—	By subtracting mass of total mineral aggregate	% of mass binder insoluble in solvent pre-determined	BS 812: Part 103
Extraction bottle binder directly determined	Bottle rolling using methylene chloride	Bucket type centrifuge	Recovery of binder from a portion of centrifuged binder solution by evaporation of aliquot portion under reduced pressure	—	From mass of recovered binder residue after evaporation of aliquot portion	As above	Filler directly by pressure filter or Filler by difference after decanting through 75 μ sieve BS 812: Part 103

Method	Apparatus	Filtration	Recovery of binder	Incineration	Filler/aggregate determination		Standard
Sieving extractor	Sieving extractor using methylene chloride	As above	As above	—	As above	As above	Filler by difference after decanting through 75 μ sieve BS 812: Part 103
Hot extractor	Hot extraction apparatus with filter paper lining using trichloroethylene	Filter paper	By calculation	—	By subtracting mass of total mineral aggregate	—	(i) Filler directly by pressure filter or filter paper or (ii) Filler by difference after decanting through 75 μ sieve BS 812: Part 103
(N) Strassentest Automatic binder extraction machine	By recycled solvent spray	Continuous flow centrifuge	By calculation	—	By subtracting mass of total mineral aggregate	—	Vibrating sieving unit of 6 sieves
(NL) Test 65.0 Soxhlet extraction	Modified Soxhlet equipment with pressed filter paper extraction case	Filter paper extraction	Total recovery of binder by evaporation	Incineration of portion of recovered binder	From mass of recovered binder with correction for ash	—	Test 6.0
Decanting-jar centrifuge	Cold agitation and through sieves	Continuous flow centrifuge	By calculation or Recovery of binder from all or portion of centrifuged binder solution by evaporation	Evaporation and incineration of portion of centrifuged binder solution	By subtracting mass of total mineral aggregate or from mass of recovered binder	—	Text 6.0
Automatic decanting-jar centrifuge	Cold agitation and by solvent spray	Continuous flow centrifuge	As above	As above	As above	—	Test 6.0

Table 8.6. Binder determination directly or by difference

Country	Standard	Binder directly	Binder by difference
Germany	DIN 1996: Part 6: 1988	Recovery method*	Differential method
UK	BS 598: Part 102: 1989	Extraction bottle directly determined Sieving extractor	Extraction bottle by difference Hot extractor
Norway	Strassentest		Automatic binder extraction
Netherlands	Test 65.0	Soxhlet extraction*	Decanting-jar centrifuge

* Requires recovery of total binder

Fig. 8.8. DIN hot extractor. Photograph courtesy of Control Testing Equipment Ltd

is then transferred to a burette, care being taken not to disturb the mineral matter that is left in the tubes, and measured into a boiling flask. The solvent is then removed from the solution by boiling in a water bath under vacuum to leave approximately one gram of soluble binder.

Fig. 8.9. Centrifuge extractor. Photograph courtesy of ELE International Ltd

Fig. 8.10. Bottle rolling machine. Photograph courtesy of ELE International Ltd

The flask is dried and allowed to cool in a desiccator to prevent absorption of moisture from the air before weighing. The percentage binder content is then calculated by scaling up in the ratio of the total volume of solvent used to that contained in the measured portion of centrifuged binder solution.

Some methods, as indicated in Table 8.6, rely on the total recovery of the binder in the sample of mixture to determine binder content. The binder may be recovered from solution after extraction from the mixture by solvent and removal of fine mineral matter, using either a rotary film evaporator or a simple distillation apparatus with a fractionating column (Figs 8.17 and 8.18). The simple distillation apparatus may be used for fluxed or cut-back bitumens, as well as penetration grades. The rotary film evaporator is used for penetration grades only.

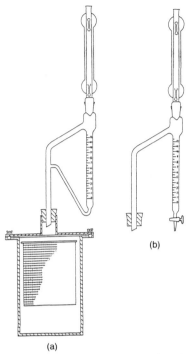

Fig. 8.11. BS hot extractor: (a) solvent density >1 (b) solvent density <1

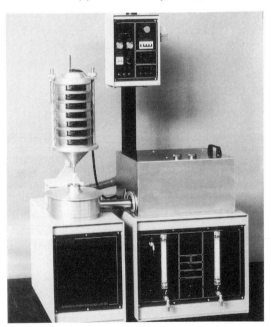

Fig. 8.12. Automatic binder extraction machine. Photograph courtesy of Control Testing Equipment Ltd

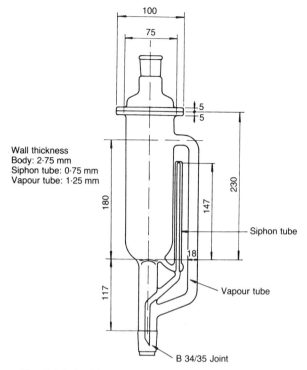

100

75

Wall thickness
Body: 2·75 mm
Siphon tube: 0·75 mm
Vapour tube: 1·25 mm

5
5

180

147

230

117

18

Siphon tube

Vapour tube

B 34/35 Joint

Fig. 8.13. Soxhlet extractor

The rotary film evaporator method[26] is possibly becoming more popular when handling mixtures containing ordinary penetration grade bitumens. The apparatus is operated under reduced pressure and a relatively large sample can be recovered quite quickly. The flow of binder solution into the evaporating flask (Fig. 8.17) is controlled by the induction stopcock so that the rate is approximately equal to that at which the distillate flows into the receiving flask. The evaporating flask is placed in an oil bath at a temperature of 115°C and rotated at 75 rev/min minimizing the possibility of overheating the bitumen.

After all of the binder solution has been transferred to the evaporating flask, the pressure inside the apparatus is increased and the temperature of the oil bath raised to 150°C and above to remove all the remaining solvent. The recovered binder is weighed to give the binder content of the mixture but may also be used for further testing such as bitumen penetration or ring and ball softening point.

The distillation apparatus with fractionating column is shown in Fig. 8.18. In this case,[27] the bulk of the distillation is carried out under atmospheric conditions with the oil bath maintained at a temperature of approximately 100°C while the binder solution is continuously stirred to prevent bumping. The temperature of the oil bath is increased to 175°C or 100°C above the expected softening point of the binder, whichever is higher, as the final portion of binder solution introduced into the distillation flask reduces in volume. When the rate

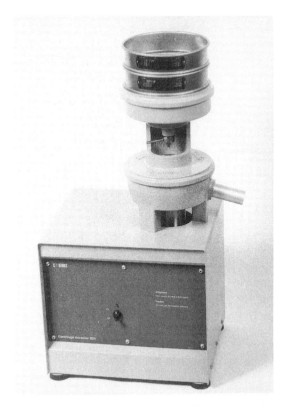

Fig. 8.14. Continuous flow centrifuge. Photograph courtesy of Control Testing Equipment Ltd

Fig. 8.15. Pressure filter. Photograph courtesy of ELE International Ltd

Fig. 8.16. Bucket centrifuge. Photograph courtesy of ELE International Ltd

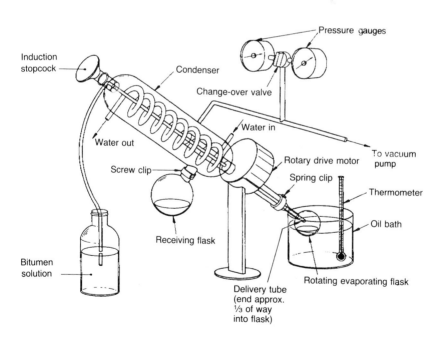

Fig. 8.17. Rotary film evaporator[26]

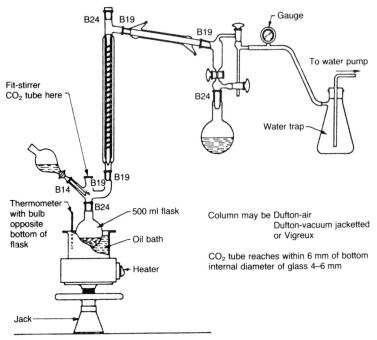

Fig. 8.18. Distillation apparatus with fractionating column[27]

of distillation drops significantly, the pressure in the apparatus is reduced to remove the last traces of solvent. During this last stage, unless very volatile flux is contained in the binder, carbon dioxide gas is passed through the residue in the flask to assist the removal of solvent while hindering the oxidation of the binder. Volatile oils in the binder are recovered from the distilled vapours as they pass through the fractionating column.

The selection of a particular method of determining binder content may be largely a matter of personal preference. Factors influencing choice include cost, exposure to solvent fumes, precision, complexity, rapidity and purpose of test. For quality control testing during production, the time taken to produce a result can be one of the most critical factors. Table 8.7 gives an estimate for those tests detailed in Table 8.5 that can produce a result within two hours and so may be considered for quality control purposes.

Nuclear testing devices are also now available to give a very rapid estimate of the binder content of a sample of mixture. However, although an interesting innovation, for proper assessment of conformance it is still necessary to remove the binder from the mineral aggregate, recover the fine mineral matter from the binder solution and carry out an aggregate grading. Various combinations of test equipment can be adopted to suit a particular need (Fig. 8.19).

Possibly the most common equipment used in Europe to separate out fine mineral matter from binder solution is the continuous flow centrifuge (Fig. 8.14) which can cope with relatively high proportions of filler.

After extracting the binder from the sample of mixture with a suitable solvent

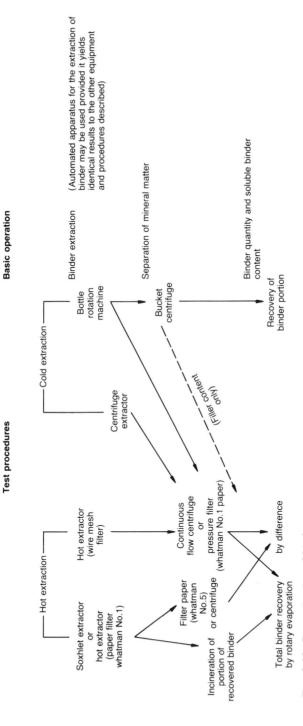

Fig. 8.19. Determination of binder content

Table 8.7. Duration of test for composition

Country	Standard	Up to 2 hours	Over 2 hours
Germany	DIN 1996: Part 6: 1988	Differential method	Recovery method
UK	BS 598: Part 102: 1989	Binder by difference Binder directly determined Sieving extractor method	Hot extractor method
Norway	Strassentest	Automatic binder extraction	
Netherlands	Test 65.0	Decanting-jar centrifuge	Soxhlet extraction

in, e.g. a rolling bottle apparatus, the binder solution containing fine mineral particles is carefully poured through a funnel into a pre-weighed centrifuge cup that rotates inside the machine at 9000 rev/min. The binder solution, due to the centrifugal effect, rises up the sides of the cup and deposits the fine mineral particles on the sides. Pure solvent is passed through the centrifuge until all binder is removed from the revolving cup. The centrifuged solution is discharged and may be collected for recovery of the bitumen or the binder content determined by difference. The fine matter collected in the cup is dried and added to the coarser mineral aggregates recovered from the rolling bottle.

If determining binder content by difference, it is especially important that due allowance is made for any water that may be present in the sample being tested, since this will affect its accuracy. If water is suspected in the sample, its content may be determined by hot extraction of the binder and measurement of the volume of moisture driven off with the solvent in a Dean and Stark apparatus (Fig. 8.11).[16] Different solvents may be used, but trichloroethylene is preferred due to its higher boiling point (87°C) and satisfactory bitumen solubility. The Dean and Stark apparatus illustrated in Fig. 8.11(a) is suitable for use with trichloroethylene which is denser than water.

A different design that allows measurement of water collected at the bottom of the graduated receiver would be required for use with solvents less dense than water (Fig. 8.11(b)). The sample of mixture is placed in the extraction pot with sufficient solvent to permit refluxing in the water-cooled condenser to take place during the distillation. Heat is applied to the pot until the volume of water collected in the receiver remains constant.

8.4.2. Bitumen tests

The Shell Bitumen Handbook[28] discusses the properties of bitumen products and describe the various tests that are used to measure these properties. Currently, the British Standard specification for road bitumens[29] includes test methods which are considered important for classification purposes in the UK.

The penetration test[30] is the most common control test for penetration grade bitumens and the mid-point of the range of penetration values is used to designate the particular grade. It is a measure of the consistency or hardness of the bitumen.

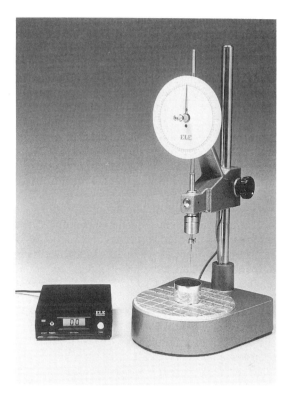

Fig. 8.20. Penetrometer. Photograph courtesy of ELE International Ltd

The penetration value is determined using a penetrometer (Fig. 8.20). This test uses a standard stainless steel needle under a load of 100 g, set at the surface of the material, to penetrate a sample of bitumen which is maintained at a temperature of 25 °C. The penetration value is the distance in tenths of a millimetre that the needle penetrates the bitumen in 5 s. The conditions of loading, time and temperature may be changed to suit special applications.

The test method is convenient and simple to perform in site laboratories as a means of monitoring the quality of supplies. However, penetration has been found not to correlate well with tests used to assess the performance of bituminous mixtures.[31] A much better linear association has been found between bitumen softening point and mixture properties defined by wheel tracking rate,[32] Marshall stability and Marshall quotient.

The physical state of bitumen changes gradually with temperature from solid or semi-solid to a softer more fluid material. A specific melting point cannot be measured. The ring and ball softening point[33] is a temperature that is measured at a specific viscosity of the bitumen at some point during its transition from solid to liquid. The actual viscosity has been found to be around 1300 Pa.s[34] for bitumens with low penetration index and low wax content, equivalent to a consistency as measured by 800 penetration units. This fact enables a viscosity/temperature relationship to be readily established for those bitumens over a wide temperature range.

The softening point temperature is found by placing a 3.5 g steel ball on the surface of a bitumen specimen that has been moulded into a tapered brass ring and allowed to steadily increase in temperature at a rate of 5C° per min until the weight of the ball stretches the bitumen and forces the underside of the specimen to fall 25 mm onto a base plate. The temperature at which the bitumen touches the plate is the softening point. The apparatus used is illustrated in Fig. 8.21. Distilled water is the liquid used in the bath to transfer heat to the specimen for bitumens expected to have softening points up to 80°C. Glycerol is used for more viscous materials. A mechanical stirrer is included to ensure uniform heat distribution throughout the liquid. The equivalent ASTM method D36[35] does not include stirring in the procedure and results in a higher measured value of approximately 1.5C°.[28]

Heukelom[34] developed a bitumen test data chart (BTDC) to describe the viscosity/temperature relationship of bitumens (Fig. 8.22). The chart is designed so that, for bitumens of low penetration index and low wax content designated class S by Heukelom, the relationship can be described by a single straight

Fig. 8.21. Ring and ball softening point apparatus. Photograph courtesy of ELE International Ltd

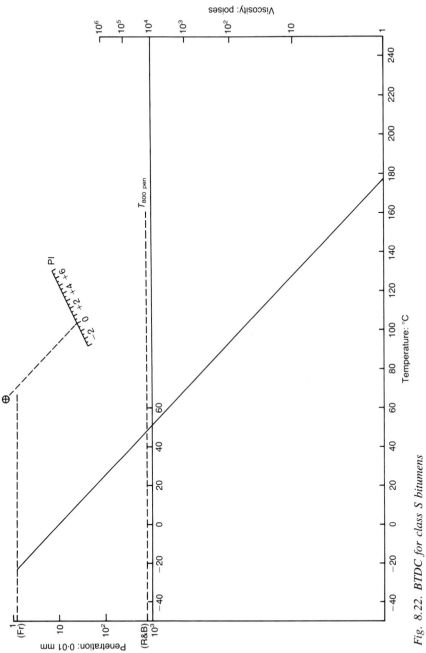

Fig. 8.22. BTDC for class S bitumens

line constructed by joining the plotted points for penetration and ASTM softening point. The line can then be extrapolated to give an estimate of Fraas breaking point (*Fr*), the temperature at which the bitumen cracks under controlled test conditions, and viscosities at higher temperatures. Alternatively, the line for a bitumen can be drawn using penetration values measured at two different temperatures.

Waxy bitumens give differently shaped curves and require further test data for their construction. The curves characteristically comprise two parallel misaligned straight lines joined by an ill-defined transition zone over which the waxy constituents change state from crystalline to molten. The behaviour at lower temperatures may be similar to that of equivalent bitumens but for low wax contents. At higher temperatures when the wax is molten, the bitumen has correspondingly lower viscosities.

Blown bitumens are produced by blowing air through a low viscosity grade of bitumen or heavy oil at high temperatures which results in a product of low temperature-susceptibility suitable mainly for certain industrial applications.

Bitumens which have been blown also give differently shaped curves. The viscosity/temperature relationship is represented by two intersecting straight lines, where the line at higher temperatures is characteristic of unblown bitumens of the same origin and that at lower temperatures has a less steep slope, indicative of lower temperature susceptibility over that range.

The Penetration Index (PI) of a bitumen is a measure of its temperature susceptibility varying from about -3 for highly temperature susceptible bitumens to about $+7$ for bitumens of low susceptibility. The chart in Fig. 8.22 also includes a means of determining the PI. The indices are found by drawing lines through the PI reference point parallel to the lines constructed from the bitumen test data and reading the values at their intersection with the PI scale. In its applications, bitumen is used to coat aggregate particles in films, some of which are so thin as to be transparent. Regardless of the film thickness, bitumen makes an excellent binder because it is adhesive, cohesive, self-healing, resistant to abrasion, and waterproof.

With the passage of time the bitumen will suffer a gradual loss of these desirable properties. This lessening of effectiveness is caused by hardening which is the result of continued exposure to heat, light, air, and moisture, which are ever present in the environment. Although a degree of hardening may, on the contrary, impart certain beneficial properties to the bituminous mixture in particular circumstances. Sparlin[36] also found that ultraviolet energy is capable of producing measurable increases in the hardness of bitumen films, characterized by their viscosity, when they are sealed from the atmosphere and held at room temperature.

The hardening of bitumen binder is usually discussed in terms of the separate reactions that occur in service and result in a progressive loss of desirable properties. The reactions are

- (*a*)   oxidation (a process involving the loss of electrons, frequently with the gain of oxygen or loss of hydrogen)
- (*b*)   volatilization (the loss of lighter constituents)
- (*c*)   thixotropy (an isothermal gel-sol transformation brought about by

shaking or other mechanical means e.g. a gel on shaking may form a sol which rapidly sets again when allowed to stand)
(d)   syneresis (the separation of liquid from a gel or jelly-like substance on standing)
(d)   polymerization (the formation of large molecules, comprising repeated structural units).

In general, it is agreed that the hardening of these binders in service is due in large part to oxidation and volatilization, and to a lesser degree to thixotropy, syneresis and polymerization.

During manufacture of bituminous mixtures, the rate of oxidation can be particularly high in the mixer of a conventional batch mix plant[37] where thin films of bitumen are exposed to a large volume of oxygen in the air under high temperature conditions. If mixing is unduly prolonged, significant hardening may occur with consequential loss of the desirable bitumen properties.

Oxidation of bitumen is the combination of the hydrocarbon compounds with oxygen and results in a direct union, with the elimination of a portion of the hydrogen and carbon.

Abraham[38] characterized these reactions as follows

$$C_xH_y + O \rightarrow C_xH_yO \qquad (1)$$

$$C_xH_y + O \rightarrow C_xH_{y-2} + H_2O \qquad (2)$$

$$C_xH_y + 2O \rightarrow C_{x-1}H_y + CO_2 \qquad (3)$$

These reactions require progressively greater amounts of activating energy (heat and light) to promote them. An oxidation reaction of type (1) occurs mainly on the exposed surface of bitumen. It forms a protective skin on the outer surface, and if this protective skin is not abraded or washed off, the rate of oxygen absorption slows, profoundly modifying further reaction with the oxygen and evaporation of volatile oils from the binder and hardening is retarded.

However, reaction in the absence of light must be considered as the main long-term cause of the deleterious hardening of bituminous binders which have access to oxygen through, say, an interconnecting network of air voids in a compacted asphalt layer.

The susceptibility of different bitumens to hardening is, therefore, important. To assess the resistance of a bitumen to hardening it is usually subjected to the thin-film oven test[39] or the rolling thin-film oven test.[40] In the former test method, 50 g samples of bitumen are prepared in weighed containers to give a thickness of approximately 3 mm and placed in a ventilated oven maintained at 163 °C onto a shelf that rotates at 5−6 rev/min. The sample containers are removed from the oven after 5 h and weighed to determine loss due to evaporation. The same samples may be melted carefully at a low temperature and re-mixed to determine the loss in penetration value or other test property after heating. The loss in weight and penetration is normally expressed as a percentage of the original value before heating.

The results of viscosity measurements made on bitumen subjected to the

Fig 8.23. Rolling thin film oven. Photograph courtesy of Control Testing Equipment Ltd

rolling thin-film oven test method have been shown to correlate well with the changes produced during mixing and laying of dense-graded bituminous mixtures under normal conditions of temperature control.[41] In this case, 35 g samples are poured into glass bottles which are secured horizontally into openings in a vertical, circular carriage fitted inside the oven (Fig. 8.23).

The carriage is rotated at a rate of 15 rev/min and heated air blown by an air jet into each bottle at its lowest point of travel for 85 min. The test temperature of 163°C should be reached within 10 min, otherwise the test is discontinued. The method of rotation during the test ensures that a fresh surface of bitumen is continuously being exposed to air. At the end of the test, the percentage loss in weight and penetration may be found as before.

Bitumen is soluble in carbon disulphide and this solvent can be used to detect the presence of coke or mineral matter in the bitumen. The degree of solubility is also important information in the extraction of bitumen for analysis of composition. For health and safety reasons, trichloroethylene is normally used as an equivalent solvent to determine bitumen solubility and the test method is described in BS 2000: Part 47.[42] This involves dissolving the sample of bitumen in the solvent and filtering through a layer of powdered glass in a sintered crucible. The insoluble matter is then washed, dried and weighed.

Direct or indirect measurement of flow behaviour or viscosity of bitumens may be of interest at particular temperatures e.g. at 60°C or 135°C roughly corresponding to maximum in-service and layer compaction temperature respectively. Viscosity at particular temperatures may be determined by measuring the time for a fixed volume of bitumen to flow through a capillary or standard orifice of various different types of viscometers.[43] The viscosity may be reported as simply efflux time or converted into fundamental units by multiplying by the calibration constant of the particular viscometer.

An absolute measure of viscosity is the ratio between the applied shear stress and rate of shear. The cgs unit of measurement is called the poise (1 g/cm.s) and the increasingly more common SI unit is the Pascal second (Pa.s) (equivalent to 10 poises). When this ratio is independent of the rate of shear, the substance is said to be Newtonian. The viscosity of most bitumens changes with the rate of shear and so they are not strictly Newtonian, although they are sometimes treated as such by researchers.[44] However, at higher temperatures, they may be regarded as essentially Newtonian. More complex flow behaviour is termed non-Newtonian.

The ratio of viscosity to the density of bitumen gives a measure of its resistance to flow under gravity and is called kinematic viscosity. It is expressed in cgs units of cm^2/s or the stoke. The SI unit is m^2/s and 1 centistoke $= 1 \times 10^{-6} m^2/s = 1 mm^2/s$.

The sliding plate micro-viscometer[45] permits the direct determination of the absolute viscosity of liquids and visco-elastic substances in the range of $10-10$ billion Pa.s. The operation of the viscometer is based on the principle of shearing a sample between two parallel flat plates of dimensions $2 \times 3 \times 0.6$ cm under the action of a constant shearing stress. The sample is spread and pressed between the two plates to a thickness of between 50 and 100 μm. Thicknesses out of this range are influenced by plate effects.

To ensure an accurate measurement of viscosity, it is necessary to have a uniformly thick film. The thickness may be determined by weighing and assuming a specific gravity of 1 for the bitumen without introducing any significant error. The plates are then clamped in the micro-viscometer, allowing a few minutes for the sample to reach the bath test temperature before the test begins. A load is applied and the movement of the glass plate is recorded on a millivolt recorder. Measurements are made at different rates of shear by varying the applied load and the viscosity determined by interpolation to the particular shear rate desired.

Once a suitable grade of bitumen has been selected, the ideal bitumen content of the bituminous mixture may be estimated by following certain mixture design procedures.

8.4.3. Mixture design

In the UK, a great deal of the common types of bituminous mixtures are manufactured to recipe formulations, based on observed satisfactory performance over a period of time in the field.[13,14] However, for certain types, a need was identified for a mix design procedure that could be followed to optimize the use of available materials and to provide the necessary data that would enable the selection of an appropriate binder content for adequate durability of the mixture without sacrificing stability. Increasing traffic volumes on major roads also demanded a means of improving the resistance to deformation of surface courses.

The Marshall method[46] for the design and control of bituminous mixtures was developed in the USA in the 1940s and was later adopted by the UK Air Ministry in the 1950s as a means of designing bituminous surfacings for airfield pavement construction that would offer high stability, good rideability, weatherproofing, durability, and a smooth, dense surface.[47] Marshall asphalt using continuously-graded aggregates was generally favoured.

The Marshall method involved an enhanced level of supervision and testing effort that gave rise to a more highly controlled and consistent material. In 1973 a revision of BS 594[48] included for the first time a laboratory design procedure based on the Marshall method for the composition of the bitumen/sand/filler component of hot-rolled asphalt wearing course for roads and other paved areas. The standard included a table of adjustments of bitumen and filler contents for different percentages of coarse aggregate required in the desired mixtures. However, the current edition of the standard[13] now specifies a laboratory design procedure for the complete mixture.

British Standard 598: Part 107[49] describes a method of test for the determination of the composition of designed rolled asphalt wearing course, the properties of which are discussed in chapters 1, 2 and 3. The test procedure generally takes at least three consecutive days to complete when followed properly. A quantity of dried and blended mineral aggregate constituents is prepared for each batch of mixture that is sufficient for a compacted cylindrical specimen of 101.6 mm in diameter and approximately 63.5 mm high.

The aggregates are heated for at least four hours in an oven maintained at a temperature of 110C° above the softening point of the specified binder.

The binder is poured into a number of small containers so it can be individually heated up to the required temperature to minimize undue binder hardening. Three separate batches are mixed at each binder content for one minute at a temperature of not less than that required for compaction. Each batch is transferred to a steel cylindrical mould, the inside surface of which has been smeared with silicone grease and the base covered with a disc of paper to prevent sticking, and the mixture spaded with a spatula to prevent any bridging of aggregate. After covering the top of the mixture with another paper disc, the steel mould is secured on top of a compaction pedestal made to specified dimensions and of particular materials, including a $200 \times 200 \times 450$ mm laminated hardwood block, and the mixture is compacted on each side at a temperature of 92C° above the softening point of the binder by 50 blows of a 7850 g flat-footed steel hammer with a falling height of 457 mm. Before removing the compacted specimens from the moulds, they are cooled under water to prevent distortion. Each specimen is then extruded from the mould and allowed to dry. The procedure is repeated for mixtures of at least nine binder contents differing consecutively by 0.5% of total mix. The specimens are tested for relative density, Marshall stability and flow.

The relative density is determined by weighing the specimen in air and in water and calculating the volume displaced from the difference of the two weights i.e. assuming a relative density of 1.0 for water, and then dividing the result into the mass in air. The compacted aggregate density is also calculated by multiplying the relative density of the specimen by the percentage by mass of the total aggregate in the mixture.

The specimens are then immersed in water at 60°C for at least 45 minutes before being individually placed in a testing head comprising two segments of specified dimensions shaped to take the curved sides of the moulded specimen. The testing head is immediately placed centrally on a compression testing machine and a load applied to the specimen at a constant rate of deformation of 50 mm/min. Both the maximum load applied and the vertical deformation of the specimen at the point of maximum resistance are recorded and reported as stability and flow, respectively.

Mean values of each property are calculated for each set of three specimens of the same binder content. In the case of stability, the individual measured values are firstly corrected for variations in specimen volume before averaging. These values are then plotted against percentage binder content, as shown in the example in Fig. 8.24. The mean of each of the binder contents at the maximum value from the curves of mix density, stability and compacted aggregate density is then calculated.

A factor which depends on the coarse aggregate content of the mixture is added to the mean to give the design binder content. This factor has been derived from the correction necessary in the correlation between design binder content using complete mixture and that using the binder/sand/filler component only to ensure equivalent performance in terms of stability, durability and workability. The different factors are shown in Table 8.8.

The Marshall method is also used for the design and control of asphaltic concretes produced extensively in other countries for road surfacing. These mixtures are quite similar to Marshall asphalts used as surfacings on airfields in the UK. Asphaltic concrete has now become a generic term for any bituminous mixture with continuously-graded aggregate constituents.

For asphaltic concrete mixtures used as surfacing courses, the aggregate constituents are carefully graded down through all sizes to produce a relatively fine but dense material with a controlled void content. This type of aggregate structure can be quite sensitive to changes in the percentage of added binder but as long as due account is taken of traffic and climatic conditions in the selection of the binder content, the mixture will provide a very durable and stable surfacing product.

In designing these mixtures, the inter-relationship between aggregate packing, binder content, and air voids in the total mix is critical. The binder content

Table 8.8. Factors for the calculation of design binder contents of hot-rolled asphalt wearing course [49]

Coarse aggregate content, %	Addition factor
0	0
30	0.7
40	0.7
55	0

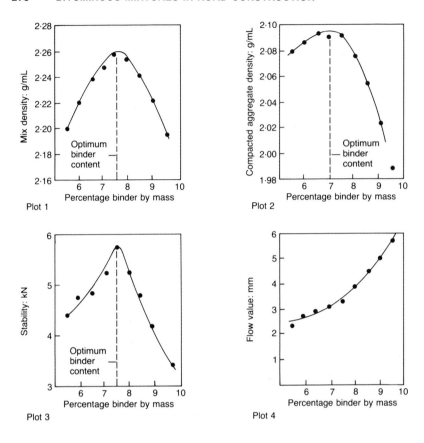

The plotted points through which the curves are drawn are averages of three specimens for each binder percentage.
Plot 4 is not used in the calculation of design binder content
Example calculation
The optimum binder content for mix density from plot 1 = 7·5%.
The optimum binder content for compacted aggregate density from plot 2 = 7·1%.
The optimum binder content for stability from plot 3 = 7·5%.

The mean of the three optima $\dfrac{7·5 + 7·1 + 7·5}{3}$ = 7·4% to which is added the appropriate factor from Table 8·8 to arrive at the design binder content

Fig. 8.24. Example of evaluation of design binder content[49]

should be sufficiently high for durability, without overfilling the voids in the mineral aggregate to the extent that the mixture will not develop strength from the applied loads of traffic, due to the volume of binder preventing close aggregate packing.[50]

Measurement of voids and the capacity for further densification by traffic after construction are therefore essential elements of the design procedure for these mixtures. In some cases, the specified air voids content may have to reflect an inability of the laboratory compactive effort to closely simulate the predicted final density in the field, especially where the aggregate constituents have a tendency to degrade under the impact of an increased number of blows of the Marshall hammer.

The *Asphalt Institute Manual Series No. 2* (MS-2)[51] describes such procedures. As well as Marshall, the Hveem method of mixture design, developed in California, is included. An interesting feature of the Hveem procedure is the preliminary determination of an approximate binder content by the Centrifuge Kerosene Equivalent (CKE) test. Dry fine aggregate is placed in a centrifuge cup fitted with a screen and disc of filter paper. The aggregate is saturated in kerosene and then centrifuged in a machine capable of achieving an acceleration of 400 times gravity for a period of two minutes. The weight of retained kerosene is the CKE.

A surface capacity test is also carried out on the coarse aggregate. In this case, dry aggregate of 9.5−4.75 mm is placed in a metal funnel and immersed in lubricating oil. The oil is then allowed to drain from the sample for 15 minutes and the weight of retained oil determined. The surface area of the aggregates is then calculated from the grading by multiplying each cumulative aggregate fraction by a corresponding surface area factor and summing. These data are then used to determine the approximate binder content from a series of charts given in the Manual. A number of compacted specimens are prepared at different binder contents including those on either side of the approximate content. Tests are carried out on the specimens to determine their resistance to both deformation and to the action of water.

An analysis of density and voids provides the final information necessary to determine an optimum binder content that satisfies the specified design criteria.

Both Hveem and Marshall methods are used in the USA. However, it is the Marshall method which is adopted most frequently outside the USA for the design and control of asphaltic concrete. Using the Marshall compaction apparatus, the density of the compacted specimen is increased up to a point by increasing the number of blows of the compaction hammer. MS-2[51] recommends three levels of compaction for light, medium and heavy traffic conditions, corresponding to 35, 50 and 75 blows respectively on each side of the specimens. As noted earlier, an important consideration is the prediction of the amount of densification that will occur in the road layer due to trafficking after construction.

The total amount of densification is also a function of temperature and, to a large extent, the degree of initial compaction achieved during construction. Investigations[50] have shown that, for a properly designed mixture, final density should be achieved typically within a period of trafficking over three summers. To maximize use of the mixture design data, the density of the laboratory compacted specimen should equal the final layer density. It is not possible, owing to different variable factors, to be too precise but the laboratory density may be regarded as a best estimate.

Once acceptable laboratory compaction is established that simulates as closely as possible the conditions in the field, a voids analysis can be made for each of the compacted specimens. The volumetric proportions of interest are air voids, voids in the mineral aggregate (VMA), and voids filled with binder

(VFB). Note that mineral aggregate may be defined as any hard, inert, mineral material such as crushed rock, gravel, slag or sand that has been produced in specified sizes for further processing in the manufacture of bituminous mixtures or other products. Air voids are defined as the total volume of air between the coated aggregate particles of the compacted mixture. This does not include air in the aggregate pores trapped beneath the binder films. VMA is defined as the volume occupied by the air voids and the amount of binder not absorbed into the pores of the aggregate. Both air voids and VMA are expressed as percentages of the total volume of the compacted specimen. VFB is defined as the percentage of the VMA filled with binder.

Percentage of air voids is probably the most important criterion for the evaluation of the performance of asphaltic concretes and selection of the appropriate binder content. The relevance of air voids content is discussed in chapter 1. Minimum percentage contents of VMA are recommended by the Asphalt Institute to ensure that the mixture is neither deficient in binder nor air voids. There should be enough room in the aggregate structure to take sufficient binder for durability of the mixture but still leaving a sufficient volume of air voids to avoid problems with plastic deformation. Limits placed on VFB control the balance between the effective binder content i.e. excluding binder absorbed into the pores of the aggregate, and air voids content.

The air voids content is calculated from the following equation

$$V_t = \frac{S_t - S_b}{S_t} . 100$$

where V_t = % air voids
S_t = theoretical maximum density of loose mixture
S_b = bulk density of compacted specimen.

The theoretical maximum density of the mixture may be determined from the individual densities of its constituents. However, the calculation of air voids content is quite sensitive to differences in aggregate particle density (specific gravity) and so the basis of the determination of aggregate density is important. Selection may be made from aggregate particle densities on an oven-dried basis (bulk), saturated surface dry basis, or on an apparent basis, their values increasing in magnitude in the same order. A change in aggregate density of 0.02 corresponds to a change in air voids content of around 0.6%. The determination of oven-dried particle density is based on a volume that includes the water permeable pores in the aggregate whereas that of the apparent particle density is based on a volume that excludes those water permeable pores. The volume of water is invariably higher than the volume of binder that is absorbed by the aggregate. Use of either of those aggregate densities alone does not necessarily offer an accurate basis for the analysis of the volumetric proportions of a bituminous mixture. Generally, an approximation is made to the effective particle density, (i.e. the density based on the volume of water-permeable material but excluding voids accessible to binder, expressed as a ratio of this value to the density of water), by simply using one of the measured aggregate densities or an average of two.

Such a procedure may be satisfactory for many applications, provided the effect on calculated air voids is understood. The Asphalt Institute recommends a direct determination of theoretical maximum density of the bituminous mixture. Such a procedure is described in ASTM D2041.[52]

A weighed sample of loose bituminous mixture is placed inside a vacuum container and covered with water at 25°C. Air is removed from between the coated particles by gradually increasing vacuum to give a residual pressure of 30 mm Hg or less. The vacuum is maintained for 5 to 15 minutes and the container shaken to assist the expulsion of air. The theoretical maximum density is calculated from one of the following equations, depending on whether the volume of loose mixture is determined by weighing the container in water or air.

$$\text{Weight in water} \quad S_t = \frac{A}{A-C}$$

where A = dry mass of sample
$\quad C$ = mass of sample in water

$$\text{Weight in air} \quad S_t = \frac{A}{A+D-E}$$

where D = mass of container filled with water
$\quad E$ = mass of container filled with sample and water.

The test is best carried out on mixtures having a binder content at or near their optimum, to avoid breaking of thin binder films under vacuum or coagulation of rich mixtures trapping air which would otherwise be removed. Recent work[53] also suggests that the test method should take account of the time-dependent nature of binder absorption, by conditioning the sample for, say, four hours in an oven at 143°C before test. Other methods used in some European countries replace water with hydrocarbon solvent. In this case, a vacuum is not necessary to expel entrapped air but because the solvent readily penetrates the binder films, a measure of binder absorption cannot be derived and higher theoretical maximum densities are determined.

ASTM D2726[54] describes a method of determining the bulk density of compacted specimens of dense mixtures. The specimen is weighed in water after immersion for three to five minutes at a temperature of 25°C. The specimen is then removed from the water and blotted with a damp towel before weighing in air. The difference in weights gives the mass of an equal volume of water, corresponding to the volume of the specimen for a density of water of 997 kg/m³ The bulk relative density is then calculated from the following equation

$$S_b = \frac{A}{B-C}$$

where A = dry mass of specimen in air
$\quad B$ = mass of saturated surface dry specimen in air
$\quad C$ = mass of specimen in water

This method of test assumes a volume for all specimens that includes both internal and surface voids. The procedure described by BS 598: Part 107,[49] however, for measuring the relative density of compacted specimens of hot-rolled asphalt assumes a volume that excludes any surface voids and any internal voids that may be accessible to water so that the volume determined may be slightly lower, thus giving rise to a slightly higher density.

The choice of methods is probably not so critical for carefully prepared laboratory specimens of dense mixtures. However, careful thought should be given to the choice for the measurement for the density of extracted cores which are unavoidably subject to greater surface voids due to coring and surface effects at the top and base of the core. The choice should be based on a consideration of what can be realistically achieved.

The density of compacted specimens may also be determined by sealing any surface voids with paraffin wax. The procedure is described by ASTM D1188[55] and is frequently the preferred method for extracted cores of coarser materials.

Since VMA is defined as the volume occupied by air voids and the amount of binder not absorbed into the pores of the aggregate, its calculation must be based on the aggregate particle density determined on an oven-dried basis (bulk specific gravity)[56,57] as follows

$$\text{VMA} = 100 - \frac{S_b \cdot P_a}{G_{ab}}$$

where S_b = bulk density of compacted specimen
P_a = % aggregate by mass of total mixture
G_{ab} = aggregate particle density on an oven-dried basis.

VMA determined on any other basis may be acceptable, providing its effect on calculated voids is fully understood.

The optimum binder content of asphaltic concretes may then be determined by averaging the individual binder contents corresponding to maximum compacted mixture density, maximum stability and the specified air voids content. Providing the mean meets all other design criteria, full scale production trials at this binder content may begin. In some cases, selection of a binder content without averaging but still meeting all the specified design criteria may be considered.

More recently, a need has been identified to develop a mixture design procedure that is based on fundamental engineering properties and that more accurately predicts the in-service behaviour of bituminous mixtures than do procedures such as Marshall that are empirically-based and include measurements of relatively poor precision.

A research programme began in the USA on a new Asphalt-Aggregate Mixture Analysis System (AAMAS), originally funded through the National Cooperative Highway Research Program and then part of SHRP.[58]

An initial mixture design is carried out, based on volumetric analysis of compacted specimens of bituminous mixtures prepared by using Marshall apparatus or some other preferred method to determine a suitable composition of mixture for further testing. Specimens of this composition are then prepared for measurement of its fundamental properties. At this stage, the selection of the method of laboratory compaction is important to simulate the conditions in the field as closely as possible.

Different compactors tend to produce specimens that, although of the same composition, have quite different engineering properties. Sousa et al.[59] have reported an investigation into the effect of different methods of specimen preparation on permanent deformation characteristics of bituminous mixtures. Three methods were selected for investigation that subjected the mixtures to shearing motions similar to those induced by site compactors. The compactors used were the Texas gyratory, kneading, and rolling wheel apparatus. Rolling wheel compaction was recommended to simulate field compaction most closely. However, gyratory compaction may be considered more convenient for the preparation of small specimens. Specimens are also prepared for conditioning for moisture damage and age hardening before testing.

The critical mixture properties examined are resistance to fatigue cracking, permanent deformation, and thermal cracking. The tests used to evaluate these properties include indirect tensile strength, resilient modulus, and creep. The measured properties are then used to predict performance and for comparison with the requirements of the structural pavement design.

8.4.4. Adhesion

Moisture damage of asphalt layers is an important distress mode and better methods of assessing the water sensitivity of bituminous mixtures were researched as part of the SHRP work. Methods that are currently used are discussed.

Sustainable adhesion of binder to stone in a bituminous mixture is necessary for a long and serviceable life in flexible pavement. Failure of the pavement can sometimes be attributed to loss of adhesion or stripping. Most aggregates have a greater affinity for water than for bitumen due to bitumen's lower surface tension and thus, inferior wetting power. It is therefore difficult for bitumen to displace water which has already formed a film over the aggregate surface, except for certain special processes. In some cases, an excessive dust coating on the aggregate will also prevent an adequate bond with bitumen and hasten the occurrence of de-bonding.

Adhesion between the aggregate surface and bitumen may be broken by water forcing its way under the bitumen film and stripping it free from the surface. The risk of stripping is usually greater with softer bitumens and with acidic or high silica aggregates. Low silica aggregates such as limestone are thought to have a greater chemical attraction for the polar carboxylic acid components of bitumen. However, although properly manufactured and placed bituminous mixtures generally perform satisfactorily in the presence of water, all have a potential for damage by moisture and a number of tests have been developed in an attempt to measure their susceptibility to it.

Static immersion tests are carried out, whereby clean, dry, single-sized aggregate is coated with a specified percentage of binder and, after being allowed to stand at room temperature for a specified period, the sample is immersed in previously boiled, distilled water. The temperature of the water and the period of immersion vary, but typically may be 20°C for 48 hours or 40°C for three hours. At the end of the immersion period, the aggregate particles are visually examined for signs of stripping and an estimate made to the nearest 5% of the residual coated area. Alternatively, the number of particles showing evidence of stripping may be counted and expressed as a percentage of the total number of particles.

Mechanical immersion tests may also be carried out involving measurement of the change in the mechanical properties of compacted specimens of bituminous mixtures after extended periods of immersion in water. ASTM D1075[60] describes such a procedure using statically compacted cylindrical specimens 4 in high and 4 in in diameter, suitable for the measurement of compressive strength. However, Marshall specimens are frequently used as an alternative. In this case, six Marshall specimens are prepared at the desired binder content and their bulk densities determined in order that they can be divided into two groups of three which have similar values of mean bulk density because the test results are sensitive to the voids content of the specimen which controls the amount of water absorbed. The first group is tested for Marshall stability after immersion in water at 60°C for 30 min and the second group tested after immersion at the same temperature but for 24 h.

An immersion index of retained strength is then calculated from the following formula

$$\text{Index} = \frac{S_2}{S_1} \times 100$$

where S_1 = average Marshall stability of group 1 (30 min immersion)
      S_2 = average Marshall stability of group 2 (24 h immersion).

A minimum index of 75% is often specified for satisfactory resistance to damage by moisture.

The immersion wheel-tracking test[61] is a method that takes account of the role played by traffic in the process of stripping. The apparatus consists of three solid tyred wheels each approximately 5 cm wide which pass over three specimens of compacted bituminous mixture to simulate the action of traffic (Fig. 8.25). The wheels travel with a reciprocating motion at a frequency of 25 cycles/min and a stroke of about 28 cm. Each wheel is loaded to give a total weight of 18 kg bearing on the specimen contained in a mould approximately 3 cm deep, 30.5 cm long and 10 cm wide, maintained horizontally in a water bath so that the level is just above the top of the specimens. The water is maintained at a temperature of 40°C during the test.

Resistance to stripping is assessed according to the time necessary to produce failure, defined as the point at which the rate of penetration of the wheel into the specimen sharply increases. Typically, failure within four hours of tracking may correspond to very low resistance to stripping and failure after 20 h to high resistance to stripping. The test is normally stopped within 24 h.

8.4.5. Percentage refusal density

The specifying authority generally prefers that a certain level of compaction of bituminous materials is achieved at the time of construction. This will ensure that air voids in those mixtures that depend on their density to provide satisfactory performance, will not be so high as to significantly expose the binder to the potentially damaging effects of air and water. Early distress such as ravelling at the surface will also be avoided. Because it is known that improved compaction of the asphalt layers leads to reduction in stresses

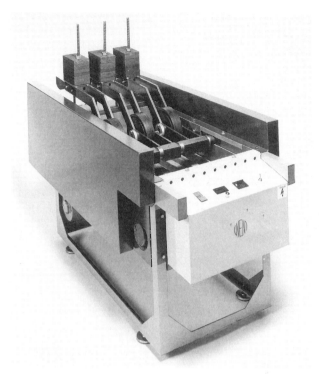

Fig. 8.25. TRL immersion wheel tracking machine. Photograph courtesy of Wessex Engineering and Metalcraft

developed in the lower pavement by traffic loading, the purchaser will wish to ensure that such a benefit is realized early on.

The percentage refusal density (PRD) is a measure of the relative state of compaction of cores extracted from the pavement layers. Currently, it is only applicable to dense coated macadam roadbases and basecourses.

It is the ratio, expressed as a percentage, of the bulk density of the core to its density after reheating and re-compacting to refusal by a particular prescribed procedure. A description of the procedure is given in BS 598: Part 104.[62] A minimum value of PRD of 93% after completion of site compaction is generally specified. Although not necessarily a specification requirement, the corresponding air voids content for a typical coated macadam roadbase may be around 9%.

A core of 150 mm diameter is dried in an oven to constant mass. After cooling, the core is weighed in air and then its surface is coated with molten wax. Once the wax has hardened, the core is again weighed in air and then weighed in water at 20°C. The bulk density G of the core is calculated using the following equation

$$G = \frac{A}{(B-C) - \left(\dfrac{B-A}{D}\right)}$$

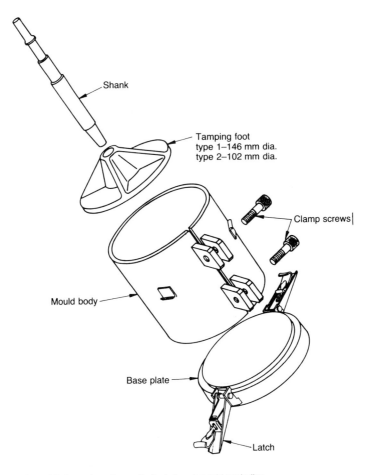

Shank

Tamping foot
type 1–146 mm dia.
type 2–102 mm dia.

Clamp screws

Mould body

Base plate

Latch

Essential dimensions (inspection) of closed mould excluding
base plate
   Internal diameter 152·45 ± 0·5 mm
   Length   170 ± 0·5 mm

Fig. 8.26. Compaction apparatus for PRD test[62]

where A = the mass of the dry core
     B = the mass of the waxed core
     C = the mass of the waxed core in water
     D = the density of the wax.

The wax is then removed from the surface of the core, and the core is placed into a split mould with base plate (Fig. 8.26).

The inside surfaces of the mould are coated with silicone grease and the baseplate covered with a disc of filter paper to prevent sticking. The mould with the core is placed in an oven until the temperature required before compaction is reached. The mould is removed and the core is covered with another disc of filter paper. The core is then compacted with a vibrating hammer using a 102 mm diameter tamping foot around the internal circumference of the mould in a prescribed sequence for a period of two minutes. Surface irregularities are then removed by vibrating with a 146 mm diameter tamping foot. The core is then inverted and compacted on its opposite side in the same fashion as before. After cooling, the core is then weighed in air and in water and its density calculated using the following equation

$$\text{Refusal density } H = \frac{1}{1 - \dfrac{F}{E}}$$

where F = the mass of the compacted core in water
E = the mass of the compacted core in air.

The PRD is then determined as follows

$$\text{PRD} = \frac{100\, G}{H}$$

The PRD test introduces a very useful measurement in the analysis of bituminous mixtures for prediction of performance. In an earlier section on mixture design, the importance of the laboratory reference density used in the Marshall method was discussed and it was noted this should simulate as closely as possible the final density in the road. However, there is no guarantee that this will actually be the case. Measurement of refusal density and calculation of the corresponding air voids content may offer some additional insurance against the possibility of plastic deformation occurring, should the air voids content at refusal be sufficiently greater than zero.

The main drawdacks of the PRD test for control purposes are the time it takes to achieve a result and the intensity of labour required. The adverse effect this also has on client-contractor relationships has been reported by various engineers since the introduction of the test for acceptance purposes.[63] Nevertheless, it seems likely that this test, or a variation of it, will be used for the foreseeable future.

8.4.6. Texture depth

The need for good texture on a road surface as an aid to resistance to skidding, especially at higher speeds, is discussed in chapters 2 and 3.

The UK *Specification for highway works*[64] gives requirements for the average texture depth to be achieved over 1000 m sections of carriageway, using two different test methods. Texture depth may be measured by the sand patch test or using the TRL laser texture meter. It is a measure of the combination of microtexture and macrotexture of the surface.

The minimum specified values of average texture depth are different for each test method (see Table 8.9).

The specified values for use of the laser texture meter are derived from work carried out by the UK TRL on newly-laid chipped rolled asphalt.[65] Regression analysis of the results of a correlation study between sensor measured texture depth (SMTD) and texture depth determined by sand patch SP gave the

Table 8.9. Minimum texture depths for different test methods

Test method	Section tested	Minimum value of average texture depth, mm
Sand patch	1000 m	1.5
	Each set of ten individual measurements	1.2
Laser texture meter	100 m	1.03
	Each 50 m length	0.9

following formulae

$$SMTD = 0.41 \ SP + 0.41$$

$$SP \quad = 2.42 \ SMTD - 0.98$$

Using these formulae, the SMTDs that are equivalent to sand patch results of 1.2 mm, 1.5 mm and 2.0 mm are 0.90 mm, 1.03 mm and 1.23 mm, respectively. The formulae are applicable only to newly-laid chipped rolled asphalt with sand patch texture depths in the range 1.0–2.0 mm. A subsequent precision experiment carried out in 1990 gave the following correlation.[66]

$$SP = 1.591 \ SMTD$$

However, it must be stressed that this relationship was applicable to particular laser texture meters used on particular materials at a particular site. Ideally, a calibration experiment should be carried out at each site where such a correlation is desired. The same experiment gave the precision data shown in Table 8.10, expressed as a percentage of the measured value.

On this occasion, the repeatability of the SMTD was better than that of the SP, but the reproducibility or between-operator precision was worse. It is for this reason that, at the present time, SP results generally take precedence over SMTD results in contractual situations.

British Standard 598: Part 105[67] describes the two methods of test for the determination of texture depth. The sand patch method may be used on any type of surface, but the use of the laser texture meter remains restricted to newly laid chipped rolled asphalt wearing course.

The advantages of measurement by the laser texture meter are ease and speed of use. The principle of the meter is that infra-red light from a rapidly pulsed laser is projected onto the road surface and the reflected light from the spot

Table 8.10. Relationship between laser texture meter and sand patch results

Test method	Measured value, mm	Repeatability, %	Reproducibility, %
Laser texture meter	1.03	10.7	19.9
Sand patch	1.5	14.7	17.6

Low speed surface texture measuring equipment

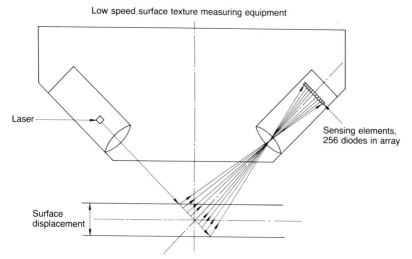

Fig. 8.27. Principle of laser displacement transducer[67]

so formed is focused by a receiving lens onto any array of photo sensitive diodes (Fig. 8.27). The position of the diode receiving most light gives a measure of the distance to the road surface at that instant and the depth of the texture is computed from a series of such measurements as the meter is propelled along the road surface. The meter is controlled by a microcomputer system and is wheeled and operated by hand. It prints the average SMTD for each completed 10 m measured, together with an overall average for each 50 m length.

The meter is operated at a speed of between 3–6 km/h on a diagonal line across the carriageway lane width left to right in the direction of traffic flow. The meter must be calibrated by the manufacturer on a regular basis strictly in accordance with a prescribed procedure and the sensitivity of the meter must be checked before use. The sensitivity is checked by running the meter over a check mat and comparing the drop-out percentage, which is the proportion of the series of height measurements from the surface which the receiver fails to detect, with that quoted by the manufacturer. The difference should not be greater than three.

Conditions to avoid when the meter is being operated include low angle sunlight illuminating the road surface, condensation on the lens, build-up of material on the tyres, and foreign matter or moisture on the road surface.

Texture depth by sand patch is determined by taking ten individual measurements at approximately 5 m spacing along a diagonal line across the carriageway lane width, avoiding positions within 300 mm of the edge. The average value of these measurements gives the texture depth for the corresponding 50 m length of lane.

Before making a measurement the surface is dried, if necessary, and any foreign substances removed. A 50 ml measuring cylinder with 30 mm maximum internal diameter is then filled with rounded silica sand predominantly all passing a 300 μm test sieve which has been previously washed and dried. Care should be taken not to compact the sand by any vibration and it is finally struck off level with the top of the cylinder. The volume of sand is then poured into a heap on the road surface and spread in a rotary motion with the aid of a hard rubber disc so that the surface depressions are filled with sand to the level of the peaks. The operator should avoid being over-exuberant with movement of the disk to avoid undue scatter of sand across the surface. The diameter of the sand patch is then measured with a steel rule to the nearest 1 mm at four diameters approximately 45° apart. Knee pads for the operator are recommended.

The texture depth is calculated in mm using the following formula

$$TD = \frac{63,660}{D^2}$$

where D = the mean diameter of the sand patch in mm.

The specified test method is not suitable for wet, moist or sticky surfaces. However, a test method was developed by Wimpey Laboratories Ltd in 1980[68] which permits measurements to be made under such conditions. In this case, the sand patch test is carried out with the surface being tested submerged in water. The test was developed at a time when it was considered important for the contractor to rapidly monitor the texture depth of hot-rolled asphalt with pre-coated chippings to control the chipping process which markedly affects the level of texture achieved. As this test is unaffected by wet or hot surfaces, measurements could be made immediately after rolling.

The principle of the test is that if only sand and water are present with the exclusion of air, there can be no surface tension forces acting on the grains of sand which would otherwise cause them to cohere. A 'wet' sand patch test was therefore developed whereby the road surface to be tested is submerged in water. A water-retaining ring is placed on the surface. It is then sealed to the surface by applying a plaster of Paris/sand grout mixture to its outside rim. Once the grout has set, usually after about five minutes, the ring is filled with water.

The rest of the procedure is followed in accordance with the specified standard, except that extra care may be needed in spreading the sand into a circular patch to avoid disturbing it by turbulence in the water. Results obtained may be regarded as being the same as those given by the standard test.

The need for a test that can begin rapidly after surfacing operations are completed has probably diminished due to the contractors' increasing confidence in achieving the required texture by following established procedures that closely control rate of spread of chippings, laying and compaction temperatures, and rolling technique, and by the increased use of vibrating rollers.

The introduction of requirements for relatively high levels of texture depth led to localized problems with the loss of pre-coated chippings from rolled asphalt wearing course due to their bunching and reduced contact with the asphalt mortar. This occurred when applying heavier rates of spread to give the cover necessary to meet the texture requirements. This and other problems experienced with hot-rolled asphalt wearing course are discussed in chapters 7 and 10.

8.4.7. Wheel tracking rates

The principal function of the wheel tracking test is to determine the resistance of bituminous mixtures to plastic deformation. The UK Transport Research Laboratory (TRL) has done studies which led to the following relationship being derived between the laboratory wheel tracking rate and trafficking of hot-rolled asphalt wearing course, for deformation to be less than 0.5 mm per annum.

$$d < \frac{14000}{N + 100}$$

where d = the wheel tracking rate in mm/h at 45°C
N = the number of commercial vehicles per lane per day.

So, for example, a road carrying 6000 commercial vehicles/lane/day should be surfaced with a wearing course having a maximum wheel tracking rate of 2.3 mm/h at 45°C. A maximum rate of 2 mm/h has since been recommended.[69]

Further work has been carried out to compare the wheel tracking rates of rolled asphalts with their Marshall stabilities and Marshall quotients (stability divided by flow) which are readily determined as part of the routine laboratory mixture design procedure. Choyce et al.[70] established the following relationship for mixtures excluding those having binder contents below that corresponding to the maximum Marshall stability

wheel tracking rate (in mm/h at 45°C) = 116 (stability in kN at 60°C)$^{-2}$

A correlation coefficient of 0.88 was quoted. This suggests that a wheel tracking rate of no greater than 2 mm/h at 45°C may be achieved with a laboratory mixture having a Marshall stability at 60°C of at least 7.6 kN, or thereabouts. British Standard 594: Part 1,[13] in turn, gives recommendations on the test criteria that may be specified for the Marshall stability of laboratory designed rolled asphalt wearing courses that are suitable for different traffic categories expressed in terms of commercial vehicles/lane/day. The acceptable stabilities are from 2−10 kN depending on the traffic category.

A draft for development (DD 184)[32] was published by the British Standards Institution in 1990 describing a method for the determination of the wheel tracking rate of cores of bituminous wearing course. This is the first step in standardizing the test equipment used and the test procedure followed and will allow direct comparison of test results to be made between laboratories. The draft, as yet, is not applicable to laboratory-prepared and compacted specimens.

Six 200 mm diameter cores that have been marked to indicate direction of traffic flow are used to obtain a test result. Each core is bedded in a holding medium of plaster of Paris onto a flat steel base plate and held in place by plywood clamping blocks which are bolted to the baseplate to provide rigid support. A mounting table is used during assembly to ensure that the surface of the core is properly level for testing. Cores of chipped rolled asphalt which cannot provide a smooth flat surface are inverted and their undersides tracked after cutting with a circular saw to form the flat surface.

The core, held in its clamping assembly, is placed in the tracking machine and maintained at a temperature of 45°C. The tracking apparatus consists of a wheel fitted with a 50 mm wide, solid rubber, treadless tyre and loaded to apply a force of 520 N to the surface of the core held in a table which moves backwards and forwards in a fixed horizontal plane at a frequency of 21 cycles/min over a total distance of 230 mm. The machine is set in motion and regular readings taken from a dial gauge, measuring the vertical displacement of the wheel, at no more than 5 min intervals. The wheel tracking path should correspond to the direction of traffic flow initially marked on the core. The test is continued until a displacement of 15 mm is recorded or for a period not exceeding 45 min.

The vertical displacements are then plotted against time and the rate of increase of track depth determined from the mean gradient of the last third of the graph. The tracking rates of each core specimen are then averaged to give the test result for the particular material in mm/h.

8.5. Analysis of test results

Test results alone give only part of the story. Interpretation of the results and analysis of trends are needed to provide the full information. No single test result can be a reliable estimate of a true value because of the variations that occur due to sampling and testing. It is, therefore, preferable that any decision based on a single result is avoided whenever possible. Practically, a balance must be struck between cost and the risk associated with inferior quality of data output for satisfactory analysis. Those responsible for determining the means by which items are examined should also be continually searching for quicker and cheaper methods that produce the same or more reliable results.

The assessment of quality by an inspector or quality controller based on an analysis of test results can be made easier with the use of statistical methods. Fortunately, most of the results of tests on bituminous mixtures have a normal distribution[71] i.e. their frequency over the range of results produced can be illustrated by the area under a bell-shaped curve. This fact means that simple statistical calculations can be made to describe their variability from a relatively small sampled proportion of the whole quantity under examination.

Those critical variables such as binder content or mixture densities determined by testing can be collated during sufficiently long production runs when steady state conditions prevail and a programme of monitoring, analysis and control can then be established.

The application of statistical techniques is limited when production runs are short due to the scarcity of data gathered and the lower probability that steady state conditions are achieved. The benefit of the techniques is therefore fully realised for major projects involving large tonnages of few different types of products.

Much of what follows is derived from a presentation by C.A.R. Morris on continuous Quality Improvement.

All measuring and test apparatus should be calibrated and properly adjusted before operation to ensure that variation in results is not largely due to inaccurate or misused equipment. This is discussed further in a later section on laboratory management. All the rest results obtained relating to the variable characteristic of interest should be recorded on a data sheet that includes details of the nature of the data, the purpose of the test, the dates of sampling and testing, the test method, the sampling plan and the name of the person who did the test.

The test results can then be systematically recorded on a chart such as a tally sheet or histogram where the range of values is sufficiently extensive to accommodate all expected results. The construction of the tally sheet or histogram is checked for the symmetrical bell shape that is characteristic of a normal distribution. Should a normal distribution of data be uncertain, the data can be plotted on normal probability paper to confirm a straight line representation of the normal distribution.[72]

The boundaries of capability of a particular process should be known by management in order to realistically assess the risk associated with the need to comply with the customer's specified requirements. The process capability can be determined by an analysis of normally distributed data taken from a process that is under statistical control i.e. one which is not subject to any assignable or special causes of variation. All processes have inherent variations present due to chance causes that influence measured data or test results in the same way. Variations due to common causes are therefore predictable where variations due to special causes are unpredictable.

Special causes influence some or all measurements in different ways and produce intermittent, unusual variations that indicate specific underlying changes to the process itself. A sudden change in constituent materials or in the operation of proportioning devices on an asphalt plant affecting mixture compositions or temperature would be an example of a special cause of variation. Detection of special causes is discussed later.

The mean and standard deviation of the collection of test results are calculated usually by the use of a statistical calculator or a suitable computer program.

The standard deviation may alternatively be calculated from

$$s = \sqrt{\left(\frac{\Sigma(x-\bar{x})^2}{n-1}\right)}$$

where x is the test result, $\bar{x}$ is the mean value and n is the number of test results.

The standard deviation is a measure of the spread of results or their deviation from the mean. It defines the width of a normal distribution (Fig. 8.28). Another measure of spread that is commonly used is the range, which is the difference between the smallest and largest value in the group of test results. The range is convenient when looking at only a few results. Either measure of spread can be used to assess the capability of a process. The standard deviation can be estimated from the mean range of a number of sub-groups of values by dividing by a factor that depends on the number of values in each sub-group.[73]

It can be seen from Fig. 8.28 that 99.7% of values lie within three standard deviations on either side of the mean. It follows that in order to achieve substantial compliance with the specified requirements, the tolerance must be greater than six standard deviations. This is assuming that the mean equals

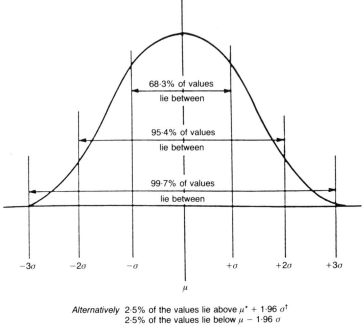

Alternatively 2·5% of the values lie above μ^{*} + 1·96 $\sigma^{\dagger}$
2·5% of the values lie below μ − 1·96 σ

Similarly   0·2% of the values lie outside μ ± 3·09 σ

$^{*}\mu$ = Population mean
$^{\dagger}\sigma$ = Population standard deviation

Fig. 8.28. Normal distribution[72]

the specification target. Measures are, therefore, required to indicate both spread and position of the mean in relation to the specification limits.

The following process capability indices are calculated to provide those measures of capability

$$C_{p} = \frac{\text{USL} - \text{LSL}}{6s}$$

$$C_{pk} = \frac{\text{USL} - \bar{x}}{3s} \text{ or } \frac{\bar{x} - \text{LSL}}{3s}$$

where USL and LSL = the upper and lower specification limits, respectively
s = the standard deviation
$\bar{x}$ = the mean value

For a process to be capable of meeting the specification requirements, both C_{p} and C_{pk} should be greater than one.

When only a small number of values are available, the specification target and limits may be directly compared with the mean and the smallest and largest values from the process to give an indication of capability.

When the analysis shows that the process is not capable of meeting the specification, an estimation can be made of the percentage outside specification using the following equation and Table 8.11

$$Z = \frac{(y - \bar{x})}{s}$$

where Z = standardized normal variate
$\quad\quad\quad y$ = the upper or lower specification limit
$\quad\quad\quad \bar{x}$ = the mean value
$\quad\quad\quad s$ = the standard deviation

Z is, in effect, the number of standard deviations between the mean and the upper or lower specification limit.

Example — use of Z, the standard normal variate.

$\quad\quad\quad y$ = lower specification limit = 38
$\quad\quad\quad \bar{x}$ = 41
$\quad\quad\quad s$ = 2.1

So $Z = \dfrac{38 - 41}{2.1} = -1.43$

From Table 8.11, substitution of Z equal to 1.43 gives a function of 0.0764. This corresponds to 7.64% of the area under the normal distribution curve that is below $\bar{x}$ − 1.43 standard deviations. It can therefore be said that about 8% of product will fall below the lower specification limit.

So it can be seen how a collection of test results can give a picture of how consistent a particular process is. All too often, in other areas too, the odd test result is used as a basis of judgment and then filed and forgotten but it is clearly necessary to analyse groups of data to enable the best assessment to be made.

Interpretation of test results obtained during the production process is also aided by the use of control charts. In many cases, simple mean and range charts or individuals and moving range charts are suitable. However, where the detection of trends and slight changes in data is required, the cumulative sum chart (CUSUM) may be used to highlight the small but persistent changes. The control chart is used to distinguish between unusual patterns of variability (special causes) and natural or random patterns inherent in the process (common causes). When detected, variations due to special causes are marked on the control chart and are investigated by those responsible for the operation and control of the process and the necessary corrective action taken to eliminate the cause. Reduction in the variations due to common causes, however, would probably require a significant change in the way that the process is operated or a complete change in the process itself.

An example of the type of chart that may be used is shown in Fig. 8.29. For mean/range charting, it is usual to plot the averages of, say, 2−5 consecutive results forming a sub-group and the corresponding ranges of each sub-group. Averaging reduces the variation, effectively smoothing out the plot and making it easier to detect changes in the process. The mean gives an indication of the general location of the variable in relation to the target.

Table 8.11. Proportions under the normal distribution curve

$Z=\dfrac{(x-\mu)}{\sigma}$	0.00	0.01	0.02	0.03	0.04	0.05	0.06	0.07	0.08	0.09
0.0	0.5000	0.4960	0.4920	0.4880	0.4840	0.4801	0.4761	0.4721	0.4681	0.4641
0.1	0.4602	0.4562	0.4522	0.4483	0.4443	0.4404	0.4364	0.4325	0.4268	0.4247
0.2	0.4207	0.4168	0.4129	0.4090	0.4052	0.4013	0.3974	0.3936	0.3897	0.3859
0.3	0.3821	0.3783	0.3745	0.3707	0.3669	0.3632	0.3594	0.3557	0.3520	0.3483
0.4	0.3446	0.3409	0.3372	0.3336	0.3300	0.3264	0.3238	0.3192	0.3156	0.3121
0.5	0.3085	0.3050	0.3015	0.2981	0.2946	0.2912	0.2877	0.2843	0.2810	0.2776
0.6	0.2743	0.2709	0.2676	0.2643	0.2611	0.2578	0.2546	0.2514	0.2483	0.2451
0.7	0.2420	0.2389	0.2358	0.2327	0.2296	0.2266	0.2236	0.2206	0.2177	0.2148
0.8	0.2119	0.2090	0.2061	0.2033	0.2005	0.1977	0.1949	0.1922	0.1894	0.1867
0.9	0.1841	0.1814	0.1788	0.1762	0.1736	0.1711	0.1685	0.1660	0.1635	0.1611
1.0	0.1587	0.1562	0.1539	0.1515	0.1492	0.1469	0.1446	0.1423	0.1401	0.1379
1.1	0.1357	0.1335	0.1314	0.1292	0.1271	0.1251	0.1230	0.1210	0.1190	0.1170
1.2	0.1151	0.1131	0.1112	0.1093	0.1075	0.1056	0.1038	0.1020	0.1003	0.0985
1.3	0.0968	0.0951	0.0934	0.0918	0.0901	0.0885	0.0869	0.0853	0.0838	0.0823
1.4	0.0808	0.0793	0.0778	0.0764	0.0749	0.0735	0.0721	0.0708	0.0694	0.0681
1.5	0.0668	0.0655	0.0643	0.0630	0.0618	0.0606	0.0594	0.0582	0.0571	0.0559
1.6	0.0548	0.0537	0.0526	0.0516	0.0505	0.0495	0.0485	0.0475	0.0465	0.0455
1.7	0.0446	0.0436	0.0427	0.0418	0.0409	0.0401	0.0392	0.0384	0.0375	0.0367
1.8	0.0359	0.0351	0.0344	0.0336	0.0329	0.0322	0.0314	0.0307	0.0301	0.0294
1.9	0.0287	0.0281	0.0274	0.0268	0.0262	0.0256	0.0250	0.0244	0.0239	0.0233
2.0	0.0228	0.0222	0.0216	0.0211	0.0206	0.0201	0.0197	0.0192	0.0187	0.0183
2.1	0.0179	0.0174	0.0170	0.0165	0.0161	0.0157	0.0153	0.0150	0.0146	0.0142
2.2	0.0139	0.0135	0.0132	0.0128	0.0125	0.0122	0.0119	0.0116	0.0113	0.0110
2.3	0.0107	0.0104	0.0101	0.0099	0.0096	0.0093	0.0091	0.0088	0.0086	0.0084
2.4	0.0082	0.0079	0.0077	0.0075	0.0073	0.0071	0.0069	0.0067	0.0065	0.0063
2.5	0.0062	0.0060	0.0058	0.0057	0.0055	0.0053	0.0052	0.0050	0.0049	0.0048
2.6	0.0046	0.0045	0.0044	0.0042	0.0041	0.0040	0.0039	0.0037	0.0036	0.0035
2.7	0.0034	0.0033	0.0032	0.0031	0.0030	0.0029	0.0028	0.0028	0.0027	0.0026
2.8	0.0025	0.0024	0.0024	0.0023	0.0022	0.0021	0.0021	0.0020	0.0019	0.0019
2.9	0.0018	0.0018	0.0017	0.0016	0.0016	0.0015	0.0015	0.0014	0.0014	0.0013
3.0	0.0013									
3.1	0.0009									
3.2	0.0006									
3.3	0.0004									
3.4	0.0003									
3.5	0.00025									
3.6	0.00015									
3.7	0.00010									
3.8	0.00007									
3.9	0.00005									
4.0	0.00003									

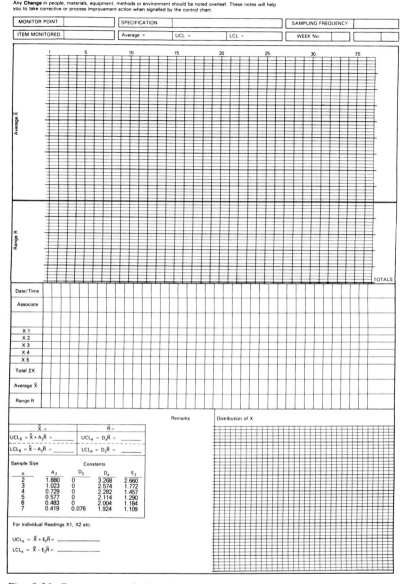

Fig. 8.29. Process control chart for variables data

Monitoring the range will indicate changes in spread or scatter of results.

Figure 8.29 gives equations for determining appropriate control limits that can be superimposed on the chart. Points outside the control limits or unusual patterns within the limits would indicate that the process is out of control. Patterns to look out for include

(a)   7 points in a row on one side of the mean
(b)   7 points in a row that is ascending or descending
(c)   less than 2/3 of points within middle 1/3 region of the chart.

In many cases, it will be more convenient to plot individual results e.g. when the frequency of testing is relatively low. Individuals/moving range charts may then be appropriate. The same control chart shown in Fig. 8.29 may be used. The individual values are plotted in place of the averages of the sub-groups. The moving range is determined from each successive sub-group of, say, three results on a moving basis. An example of the individuals/moving range chart for bitumen content is shown in Fig. 8.30.

The control limits for individual results may alternatively be calculated from the following

$$\text{UCL} = \bar{x} + 3\,s \tag{1}$$

$$\text{LCL} = \bar{x} - 3\,s \tag{2}$$

where ULC and LCL = the upper and lower control limits respectively
$\bar{x}$ = the mean result
s = the standard deviation

In the example shown, this would give limits of 3.52−4.54 for a standard deviation of 0.17. If the standard deviation were to be estimated from the range, the following value would result

$$s = \bar{R}/d_2$$
$$= 0.3/1.693$$
$$= 0.177$$

where s = the standard deviation
$\bar{R}$ = the mean range
d_2 = Hartley's Constant (for a sample size of 3, $d_2 = 1.693$).

Substitution of the estimated value in equations (1) and (2) would then give the same control limits as those shown in Fig. 8.30. The same set of test results can be used to construct a CUSUM chart which is a highly informative means of graphical presentation of data. The CUSUM plot indicates trends not necessarily revealed by other control charts. The changes in slope on the CUSUM indicate changes in mean value in a clearly defined manner. The essential principle is that a target value is subtracted from each observation or result and the cumulative sum of the deviations from target is calculated and plotted against its sample number. British Standard 5703 offers guidance in the use of CUSUM techniques for data analysis and quality control purposes.[74]
It is necessary to collect sufficient data to determine a standard error of the process for the particular variable concerned. A standard error is the standard deviation of an estimated statistic and is used in this case to fix the scaling factor in the graph. An accepted convention in scaling the CUSUM chart is that a distance on the y-axis (CUSUM) equal to one test interval on the x-axis should represent approximately two standard error units. The standard error is determined from the following

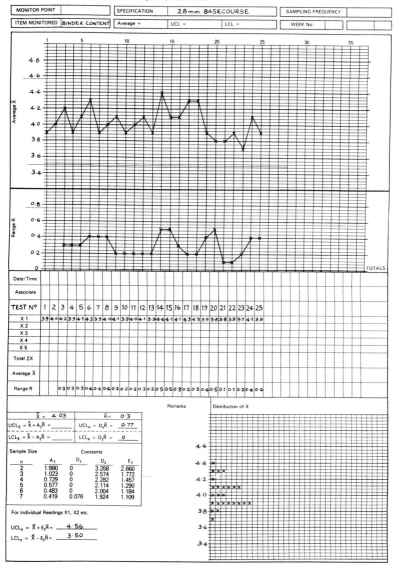

Fig. 8.30. Individuals/moving range chart for binder content

$$\sigma_e = \frac{8}{9(m-1)} \Sigma(Y_j - Y_{j+1}) \tag{3}$$

where m = the number of results taken for the estimate
the function $\Sigma(Y_j - Y_{j+1})$ = the sum of successive differences (Table 8.12, column 6)
Y_j, Y_{j+1} = the successive values

Table 8.12 illustrates the various stages of calculation for CUSUM charting. The deviation of each value from the target is given in column 4 and the cumulative sum of these deviations from tests 1 to 25 is given in column 5. Using equation (3) yields a standard error of 0.15. Therefore, the interval of the plotted variable is 0.3. Should the calculated interval not be convenient for the graph paper used, it is acceptable to round down to the nearest convenient unit.

$$\sigma_e = \frac{8}{9} \times \frac{1}{24} \times \frac{4}{1} = 0.15$$

$$2 \times 0.15 = 0.3$$

In Fig. 8.31, an individuals chart is constructed above a CUSUM plot for the same test results.

Although open to individual interpretation, it can be argued that the CUSUM plot clearly defines 6 segments where the path is either generally parallel to the x-axis or at a specific angle to it. Where the path is horizontal, the mean value can be considered to be at or near to the target value. When sloping upwards, the path indicates a mean value higher than the target and lower when sloping downwards.

Comparison of the two charts shows that the CUSUM plot gives a much clearer display of changes in the process. Take, for example, test results 14 to 16 on the individuals chart. An initial high value may prompt immediate investigation to determine the cause but is it an isolated result or an indication of plant malfunction or perhaps a change in the quality of constituent materials?

Subsequent test results show a downward trend and so may indicate that, whatever was wrong, the process is correcting itself. Reference to the CUSUM plot, though, shows a sustained upward slope, indicating that a high level is being maintained and that action is required.

A simple method of assessing approximate averages is to incorporate a slope guide or protractor on the chart. Any inclination of the CUSUM path may then be compared with the protractor as shown on the figure. The construction of the protractor is described in BS 5703: Part 1.[74]

It is often necessary to compare two different sets of results to try and determine whether it is likely that they belong to the same population, that is, have the same population mean and variation. A technique is available to determine whether the difference between their means is statistically significant.

A 't' test can be performed to decide if the observed difference between small samples is due to chance only or to some real cause. The conclusion to the test has a specified probability of being correct, normally 0.95.

Take, for example, two sets of core densities, as shown in Table 8.13.

Significance level = 0.05 i.e. 1 in 20 risk

$$\text{Standard Error } (\sigma_e) = S_p \sqrt{\left(\frac{1}{n_1} + \frac{1}{n_2}\right)}$$

where

$$S_p = \sqrt{\left[\frac{(n_1-1)S_1^2 + (n_2-1)S_2^2}{n_1+n_2-2}\right]}$$

Table 8.12. Example of CUSUM charting of test results for 28 mm basecourse (figures in percentages)

Binder content target value = 4.0%

1 Date	2 Test number	3 Calculated value	4 Deviation from target	5 Cumulative sum of deviation	6 Successive differences
Day 1	BC 1	3.94	−0.06	−0.06	+0.06
	2	4.00	0	−0.06	+0.06
Day 2	3	4.16	+0.16	+0.10	−0.29
	4	3.87	−0.13	−0.03	+0.27
Day 3	5	4.14	+0.14	+0.11	+0.12
	6	4.26	+2.26	+0.37	−0.36
Day 4	7	3.90	−0.10	+0.27	+0.10
	8	4.00	0	+0.27	+0.10
Day 5	9	4.10	+0.10	+0.37	−0.20
	10	3.90	−0.10	+0.27	+0.08
	11	3.98	−0.02	+0.25	+0.10
	12	4.08	+0.08	+0.33	−0.15
Day 6	13	3.93	−0.07	+0.26	+0.47
	14	4.40	+0.40	+0.66	−0.27
	15	4.13	+0.13	+0.79	−0.02
Day 7	16	4.11	+0.11	+0.90	+0.16
	17	4.27	+0.27	+1.17	+0.02
	18	4.29	+0.29	+1.46	−0.36
Day 8	19	3.93	−0.07	+1.39	−0.09
	20	3.84	−0.16	+1.23	+0.01
	21	3.85	−0.15	+1.08	+0.01
	22	3.86	−0.14	+0.94	−0.14
Day 9	23	3.72	−0.28	+0.66	+0.34
	24	4.06	+0.06	+0.72	−0.12
	25	3.94	−0.06	+0.66	

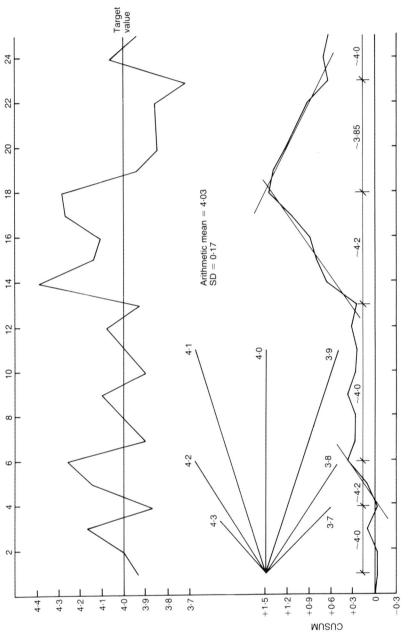

Fig. 8.31. CUSUM chart for binder content

Table 8.13. Test to compare the means of two sets of core densities (small samples)

	Core density	
	Set 1	Set 2
	2.328	2.351
	2.341	2.383
	2.350	2.378
	2.382	2.396
	2.399	2.354
	2.365	2.374
Mean	2.361 x_1	2.373 x_2
Standard deviation	0.026, S_1	0.017, S_2
Number of results	6, n_1	6, n_2

where $\quad S_p$ = combined standard deviation for the 2 independent samples
n_1 = number of results in Set 1
n_2 = number of results in Set 2
S_1 = standard deviation of Set 1
S_2 = standard deviation of Set 2
So $\quad\quad\quad S_p = 0.022$
So $\quad\quad\quad \sigma_e = 0.013$
Degrees of freedom $= (n_1 - 1) + (n_2 - 1)$
$= 10$

Critical value from Table 8.14 (two-sided test)

$$= \pm 2.23$$

$$\text{Test value} = \frac{\text{observed difference}}{\sigma_e}$$

$$= \frac{2.361 - 2.373}{0.013}$$

$$= -0.92$$

The test value is less than the critical value. Therefore, there is no significant difference between the sets of results.

The calculation of the standard error SE is specific to the test for the difference between two means when the samples sizes are small. Since, in this case, $n_1 = n_2$, the equation for the pooled standard deviations S_p may be reduced to

$$S_p = \sqrt{\left(\frac{S_1^2 + S_2^2}{2}\right)}$$

Table 8.14. Critical values for the 't' test

Degrees of freedom	Significance level					
	Two-sided test			One-sided test		
	10% (0.10)	5% (0.05)	1% (0.01)	10% (0.10)	5% (0.05)	1% (0.01)
1	6.31	12.71	63.66	3.08	6.31	31.82
2	2.92	4.30	9.92	1.89	2.92	6.97
3	2.35	3.18	5.84	1.64	2.35	4.54
4	2.13	2.78	4.60	1.53	2.13	3.75
5	2.02	2.57	4.03	1.48	2.02	3.36
6	1.94	2.45	3.71	1.44	1.94	3.14
7	1.89	2.36	3.50	1.42	1.89	3.00
8	1.86	2.31	3.36	1.40	1.86	2.90
9	1.83	2.26	3.25	1.38	1.83	2.82
10	1.81	2.23	3.17	1.37	1.81	2.76
11	1.80	2.20	3.11	1.36	1.80	2.72
12	1.78	2.18	3.06	1.36	1.78	2.68
13	1.77	2.16	3.01	1.35	1.77	2.65
14	1.76	2.15	2.98	1.35	1.76	2.62
15	1.75	2.13	2.95	1.34	1.75	2.60
16	1.75	2.12	2.92	1.34	1.75	2.58
17	1.74	2.11	2.90	1.33	1.74	2.57
18	1.73	2.10	2.88	1.33	1.73	2.55
19	1.73	2.09	2.86	1.33	1.73	2.54
20	1.72	2.08	2.85	1.32	1.72	2.53
25	1.71	2.06	2.78	1.32	1.71	2.49
30	1.70	2.04	2.75	1.31	1.70	2.46
40	1.68	2.02	2.70	1.30	1.68	2.42
60	1.67	2.00	2.66	1.30	1.67	2.39
120	1.66	1.98	2.62	1.29	1.66	2.36
Infinite	1.64	1.96	2.58	1.28	1.64	2.33

The degrees of freedom is the number of independent observations or values under a given constraint. When calculating standard deviation, the constraint is that all the deviations must add up to 0. So, for a group of 6 results, the first 5 deviations calculated can take any values but the sixth value must be such that the sum of deviations is 0. Therefore, there are 5 independent values and so 5 degrees of freedom i.e. $n-1$.

Table 8.14 gives critical values for both two-sided and one-sided tests. A one-sided test is used only if there is interest in or there are changes anticipated in one direction. This decision and the significance level chosen defines the critical region under the normal distribution curve (Fig. 8.32).

In the example shown in Table 8.13, the test value does not lie in the critical region and so no significant difference between the means has been detected at the 5% level.

The above test assumed that the two sets of results had the same standard deviation or, rather, that the sets had deviations belonging to the same population of standard deviations. It should be checked that the two standard deviations do not differ significantly as it is possible that both conditions of significance may not be satisfied. The F test is used to determine whether two

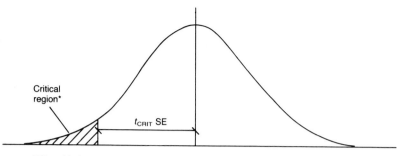

* The critical region lies beyond the number of
standard errors from the mean that is equal to
the critical value t_{CRIT}

Fig. 8.32. Critical region for 't' test

standard deviations are significantly different. The F value is calculated from
the following formula

$$F = \frac{S_1^2}{S_2^2}$$

where S_1 is the larger standard deviation, S_2 is the smaller standard deviation.

The same example gives an F value of 2.34. The critical value taken from
Table 8.15 is 7.15 which is greater than the test value and so, confirms that
there is no significant difference between the variability of the two sets of data.

There is a multitude of applications of statistical techniques for analysing
test results and other data. Here, only a small selection has been considered
that may be found useful in interpreting results of test on bituminous mixtures
and their regular application should contribute, in particular, to gaining a much
better understanding of processes so that decisions based on the results of testing
are more likely to be correct in all the circumstances. In addition to the other
references given on statistical analysis at the end of this chapter, BS 2846:
Parts 1−7[5] gives guidance on the statistical interpretation of data.

8.6. Laboratory management

The product of the laboratory is the test report. The report may be prepared
for a purchaser who is interested in some property of a material or is seeking
assurance that the material meets certain requirements. The purchaser may
also be looking for advice that would require interpretation of test results and
recommendations for further action. The report may be required for internal
use only as a source of management information. The nature and frequency
of reporting will vary greatly depending upon its purpose. For quality control
purposes, brief daily reports are appropriate to facilitate rapid evaluation of
the ongoing quality of manufactured products.

A programme of testing may be required as part of an investigation into
an identified problem or of a product development project. In such a case,
a report would only be prepared once sufficient data is collected and properly
analysed and would be structured in a more formal way to give the reader

Table 8.15. Critical values for the F test. Two-sided at 5% significance level

Degrees of freedom for smaller variance	Degrees of freedom for larger variance														
	1	2	3	4	5	6	7	8	9	10	12	15	20	60	Infinity
1	647.80	799.50	864.20	899.60	921.80	937.10	948.20	956.70	963.30	968.60	976.70	984.90	993.10	1010.00	1018.00
2	38.51	39.00	39.17	39.25	39.30	39.33	39.36	39.37	39.39	39.40	39.41	39.43	39.45	39.48	39.50
3	17.44	16.04	15.44	15.10	14.88	14.73	14.62	14.54	14.47	14.42	14.34	14.25	14.17	13.99	13.90
4	12.22	10.65	9.98	9.60	9.36	9.20	9.07	8.98	8.90	8.84	8.75	8.66	8.56	8.36	8.26
5	10.01	8.43	7.76	7.39	7.15	6.98	6.85	6.76	6.68	6.62	6.52	6.43	6.33	6.12	6.02
6	8.81	7.26	6.60	6.23	5.99	5.82	5.70	5.60	5.52	5.46	5.37	5.27	5.17	4.96	4.85
7	8.07	6.54	5.89	5.52	5.29	5.12	4.99	4.90	4.82	4.76	4.67	4.57	4.47	4.25	4.17
8	7.57	6.06	5.42	5.05	4.82	4.65	4.53	4.43	4.36	4.30	4.20	4.10	4.00	3.78	3.67
9	7.21	5.71	5.08	4.72	4.48	4.32	4.20	4.10	4.03	3.96	3.87	3.77	3.67	3.45	3.33
10	6.94	5.46	4.83	4.47	4.24	4.07	3.95	3.85	3.78	3.72	3.62	3.52	3.42	3.20	3.08
12	6.55	5.10	4.47	4.12	3.89	3.73	3.61	3.51	3.44	3.37	3.28	3.18	3.07	2.85	2.72
15	6.20	4.77	4.15	3.80	3.58	3.41	3.29	3.20	3.12	3.06	2.96	2.86	2.76	2.52	2.40
20	5.87	4.46	3.86	3.51	3.29	3.13	3.01	2.91	2.84	2.77	2.68	2.57	2.46	2.22	2.09
60	5.29	3.93	3.34	3.01	2.79	2.63	2.51	2.41	2.33	2.27	2.17	2.06	1.94	1.67	1.48
Infinity	5.02	3.69	3.12	2.79	2.57	2.41	2.29	2.19	2.11	2.05	1.94	1.83	1.71	1.39	1.00

a clear and full explanation of the work carried out. The contents of the report may also be specifically described by the purchaser as a contractual requirement.

Whatever the purpose of the test report, it is the responsibililty of the laboratory manager to establish a system that will satisfy the requirements of the customer which means that all activities associated with the laboratory process upstream of the issue of the final report are carried out both efficiently and effectively. In other words, the manager has to be sure that everything is done correctly and, from an overall viewpoint, that the system is addressing the relevant issues to achieve its aim.

The laboratory should also be prepared to offer a follow-up service after issue of the report and not simply regard reporting as the last stage in the process because, as stated at the beginning of this chapter, test results invariably form the basis of important decision-making. European Standard EN 45001[76] specifies the general criteria for the operation of testing laboratories and offers useful guidance for assuring technical competence.

The scope of testing in the laboratory should be clearly defined to establish the overall requirements in respect of facilities, personnel, working methods, materials and any special safety measures. All laboratory personnel should be made aware of their individual responsibilities and undergo proper induction procedures.

Induction of new members of laboratory staff should include a brief introduction to the aims, objectives and organizational structure of the laboratory, explanation of rules and disciplinary procedures, health, safety and welfare policies and responsibilities. A description of the roles and duties of the individual must be given and the importance of an impartial approach stressed so that the outcome of tests may not be influenced by, for example, deliberate bias in sampling. At the same time, the policy of the organization on security of information should be made quite clear. Where necessary, an individual may be placed in the care of an experienced member of staff until he is able to carry out the tests competently without direct supervision.

Frequent reviews of the individual's progress should be made by the laboratory manager over the first few weeks. Periodical reviews of employee skills and knowledge should be made to assess recruitment and training and development needs. Once the needs have been identified, a training and development plan would be prepared which would address objectives such as updating in new techniques or technology, improving performance, and preparation for promotion. Personal training files may be held by individuals indicating the degree of competence achieved over a period of time in carrying out particular tests and other practices.

The working environment of the laboratory should be made as conducive as possible to the satisfactory output of reliable data. This requires that the laboratory is maintained at the required temperature and humidity conditions, devoid of excessive vibration, noise, dust and other nuisance, and is furnished to minimize the risks to the health and safety of employees, particularly with respect to the use of organic solvents to which the Control of Substances Hazardous to Health Regulations (COSHH)[77] apply.

The laboratory equipment should conform to the appropriate standards and be sufficient to complete all the tests included under the scope of testing. It should be properly maintained and instructions issued on adjustment and operation. The laboratory manager should maintain a comprehensive inventory of all equipment with details of any malfunctions, repairs and calibration status.

To ensure that test equipment is provided and maintained to the specified standard, a programme of regular servicing and calibration has to be established. Wherever possible, the schedule of calibration procedures for measuring and test equipment and for reference standards of measurement e.g. standard weights, thermometers, etc. should comply with national standards or with the manufacturers' written instructions. Table 8.16 lists some British Standards relating to testing of bituminous mixtures that specify calibration requirements for test equipment used in a variety of laboratory test procedures.

Table 8.16. Relevant British Standards for calibration of measuring and test equipment

BS	Part	Date	Title
410		1986	Specification for test sieves
593		1989	Specification for laboratory thermometers
598	102	1989	Sampling and examination of bituminous mixtures Analytical test methods
812	100	1990	Testing aggregates General requirements for apparatus and calibration
846		1985	Specification for burettes
1377	1	1990	Soils for civil engineering purposes General requirements and sample preparation
1610	1 and 2	1985	Materials testing machines and force verification equipment
1780		1985	Specification for Bourdon tube presssure and vacuum gauges
1792		1982(1988)	Specification for one-mark volumetric flasks
1797		1987	Schedule for tables for use in the calibration of volumetric glassware
2000	0 Addendum 1	1987	Methods of test for petroleum and its products Standard reagents and thermometers
2648		1955	Performance requirements for electrically-heated laboratory drying ovens
5781	1	1992	Quality Assurance requirements for measuring equipment Metrological confirmation system for measuring equipment

Examples of the minimum frequency of routine calibration are given in Table 8.17. The schedule shown is not exhaustive and will vary according to the specific advice of manufacturers, purchasers' instructions, degree of exceptional use of equipment and the requirements of the particular laboratory.

Calibration reporting forms should identify the equipment and include details of the type of test, specification, test method, test results, date and place of test, and the name and signature of the calibrator. Calibration of reference standards of measurement by a qualifying calibration facility that can provide traceability to a national or international standard of measurement should not be overlooked. Calibration intervals should be appropriate to the nature and usage of the equipment.

The laboratory manager is responsible for ensuring that the relevant standards for sampling and test methods are made readily available to all technicians. In some cases, it may be necessary, for whatever reason, to follow non-standard procedures, in which case the specific requirements need to be fully documented.

Many laboratories now operate quality management systems[78] as a means of improving performance. The laboratory system may be part of an overall company quality system. Irrespective of how it is managed, the essential elements remain the same

(a)   documentation
(b)   implementation
(c)   internal audit
(d)   review.

Broadly, management determines the best practical means of ensuring that it understands and satisfies the needs of its customers and documents the methods to achieve these aims. To ensure that the quality policy of the organization is being implemented, the relevant activities are audited against the requirements of the documented procedures and any deficiencies are corrected. Reviews are then made to assess whether the system is working effectively and economically and to consider ways of improving it.

Once a quality management system has been established, some organizations will apply to the National Measurement Accreditation Service (NAMAS) for accreditation of their laboratories whereby verification is made by independent assessors that specified criteria for the satisfactory operation of testing facilities are being met. This seal of approval firmly establishes the credibility of the laboratory. Indeed, some purchasers now demand it for certain test requirements.

It is necessary to follow procedures that will ensure that samples of materials received at the laboratory are correctly identified and registered and their integrity preserved until ready for testing. Each sample is given a unique reference number and details of its condition recorded together with the essential details from the certificate of sampling where used (Fig. 8.5). The sample quantity should also be checked and recorded and compared with the specified requirements of the particular sampling and test method. Sample details can be conveniently kept in a register for easy reference.

The importance of good records that ensure traceability from sampling point

Table 8.17. Calibration guide for measuring and test equipment

Item	Method and/or reference	Maximum calibration interval
Laboratory weigh scales	By NAMAS certificated weights	3 months and at least once/year by specialist firm
Thermometers (mercury in glass)	By NAMAS certificated thermometer BS 593: 1989	12 months
Thermometers (bimetallic thermocouples, thermistors, electronic)	By NAMAS certificated thermometer. Also check in boiling water if within scale range.	6 months Frequent, depending on use e.g. daily, weekly
Laboratory ovens	BS2648: 1955 Temperature at mid-point of usable space of empty oven by calibrated thermometer	12 months
Volumetric glassware	Dependent on whether class A or B glassware. Refer to BS relevant to the particular glassware e.g. BS 846: Part 1985 and BS 1792: 1982. Also by weight of water, certificated weights and BS 1797: 1987	12 months
Test sieves	BS 410: 1986 Visual check per set. Grading check per set using either reference aggregate or reference set of sieves.	1 week 3 months
Centrifuge	BS 598: Part 102: 1989. Use NAMAS certificated strobe meter, for example.	6 months
Compression testing machines	BS 1610: Parts 1 and 2.	12 months
Proving rings	By NAMAS certificated test house	12 months
Marshall test equipment	BS 598: Part 107: 1990. Mass and dimensional check. Rate of blows of automatic compactor.	12 months 6 months
Vacuum gauges	BS 1780: 1985. NAMAS certificated gauge or mercury manometer corrected for air pressure.	6 months
Stopclocks/watches	British Telecom speaking clock.	6 months
Bottle rotation machines	BS 598: Part 102: 1989. Visual calibration of speed of rotation.	12 months
Bitumen penetrometer	BS 2000: Part 49: 1983. NAMAS certificated needles. Timer and loader application to be checked.	At least once/year by specialist firm.
Mastic asphalt hardness machine	BS 5284: 1976. Indentor, timing and force application to be checked.	6 months
TRL Mini-texture meter	(i) Certification of calibration factor by the manufacturer. (ii) Sensitivity on the daily check mat by the user. Corrected for air pressure.	12 months Daily when in use.
Rolling straight edge	TRL SR 290/1977. (i) Certification by the manufacturer. (ii) Calibration check by the user.	6 months or 30 km Daily when in use.
Nuclear density gauge	Manufacturer's instructions	

or source to final test report cannot be exaggerated. Certain records also serve as the objective evidence of the operation of a quality management system and are required to be kept to prove the effectiveness of its implementation. Every effort should also be made to reduce the amount of paper and so a procedure for the control of records that specifies the required period of retention and disposal arrangements should be established. Nowadays, increasing use is being made of data storage on magnetic tape, computer file and microfilm. Records should be stored in an orderly manner, maintained in a clean state, protected from damage and, where necessary, indexed for ready retrieval. The test report takes many forms, but particular care should be taken in the presentation of the data and their ease of comprehension by the recipient. For the individual organization, corporate identity can be important in establishing how it is perceived by the external customer and, so, the opportunity can be taken to portray the desired image through inclusion on the report sheet of the company logo, colour, type-style, etc.

Every laboratory test report should include at least the following information

(a)   name and address of testing laboratory
(b)   unique identification number of the report
(c)   name of customer and project
(d)   description and identification of the sample
(e)   date of receipt of the sample
(f)   designation of test method
(g)   deviation from the designated test method
(h)   date of testing
(i)   signature and title of person accepting technical responsibility for the test report
(j)   date of issue.

Other information detailed in reports of site testing should include

(a)   location of the site
(b)   date of installation of materials
(c)   weather conditions
(d)   location of sampling point.

Should a series of test results form the basis of a larger investigation or other project, guidance may be required on an appropriate structure for interim or final reporting. It is recommended that the report consists of the following sections

(a)   acknowledgements
(b)   table of contents
(c)   summary
(d)   introduction
(e)   apparatus
(f)   procedure
(g)   results
(h)   discussion
(i)   tables

(*j*) figures
(*k*) conclusions
(*l*) references
(*m*) appendices.

Finally, communication and teamwork are two of the most important elements of successful management and this applies right across the departmental boundaries of any testing organization. The manager needs to communicate, co-ordinate and facilitate.

References

1. BROCK J. D. *Segregation — causes and cures*. Astec Industries Inc. Chatanooga, Tennessee, 1986, Tech. Bulletin No. T-115.
2. BROCK J. D. *Segregation — causes and cures*. Astec Industries Inc. Chatanooga, Tennessee, 1979, Tech. Bulletin No. T-101.
3. BRITISH STANDARDS INSTITUTION. *Sampling and examination of bituminous mixtures for roads and other paved areas: methods for sampling for analysis*. BSI, London, 1987, BS 598: Part 100.
4. AMERICAN SOCIETY FOR TESTING AND MATERIALS. Standard practice for sampling bituminous paving mixtures. Construction: road and paving materials; pavement management technologies. *Annual book of ASTM standards*. ASTM, Philadelphia, D979−89.
5. NATIONAL ASPHALT PAVEMENT ASSOCIATION. *Quality control for hot mix asphalt manufacturing facilities and paving operations*. NAPA, Riverdale, Maryland, 1987, Quality improvement series 97/87.
6. AMERICAN SOCIETY FOR TESTING AND MATERIALS. Standing practice for random sampling of construction materials. *Annual book of ASTM standards*. ASTM, Philadelphia, 1990, **04.03**, D3665−82.
7. BRITISH STANDARDS INSTITUTION. *Methods for the preparatory treatment of laboratory samples*. BSI, London, 1987, BS 598: Part 101.
8. BRITISH STANDARDS INSTITUTION. *Methods for sampling*. BSI, London, 1989, BS 812: Part 102.
9. HARRIS P. M. and SYM R. Sampling of aggregates and precision tests. *Standards for aggregates*. Pike D.C. (ed.), Ellis Horwood Ltd, Chichester, West Sussex, 1990, Chapter 2.
10. BRITISH STANDARDS INSTITUTION. *Sampling petroleum products: methods for sampling bituminous binders*. BSI, London, 1987, BS 3195: Part 3.
11. AMERICAN SOCIETY FOR TESTING AND MATERIALS. Standard practice for sampling bituminous materials. *Annual book of ASTM standards*. ASTM, Philadelphia, **04.03**, D140−88.
12. SMITH R. and JAMES G. V. *The sampling of bulk materials*. Royal Society of Chemistry, London, 1981.
13. BRITISH STANDARDS INSTITUTION. *Hot-rolled asphalt for roads and other paved areas. Specification for constituent materials and asphalt mixtures*. BSI, London, 1992, BS 594: Part 1.
14. BRITISH STANDARDS INSTITUTION. *Coated macadam for roads and other paved areas. Specification for constituent materials and for mixtures*. BSI, London, 1993, BS 4987: Part 1.
15. BRYANT L. J. Effect of segregation of an asphaltic concrete mixture on extracted asphalt percentage. *Proc. Ass. Asphalt Paving Technologists*. 1967, **36**, 206.
16. BRITISH STANDARDS INSTITUTION. *Sampling and examintion of bituminous mixtures for roads and other paved areas: analytical test methods*. BSI, London, 1989, BS 598: Part 102.

17. GRANTLY E. C. Quality assurance in highway construction. Part 4 — Variations of bituminous construction. *Public roads*, 1969, **35**, No. 9, 201.
18. FROMM H. J. *Proc. Ass. Asphalt Paving Technologists*. 1978, **47**, 372.
19. BRITISH STANDARDS INSTITUTION. *Precision of test methods: guide for the determination of repeatability and reproducibility for a standard test method by inter-laboratory tests*. BSI, London, 1987, BS 5497: Part 1.
20. NICHOLS J. C. *Precision of tests used in the design of rolled asphalt*. Transport Research Laboratory, Crowthorne, 1991, RR 281.
21. AMERICAN SOCIETY FOR TESTING AND MATERIALS. *Annual Book of ASTM standards*, ASTM, Philadelphia, 1990, **04.03.**
22. AMERICAN ASSOCIATION OF STATE HIGHWAY AND TRANSPORTATION OFFICIALS. *Standard specification for transportation materials and methods of sampling and testing, Part II. Methods of sampling and testing*. AASHTO, Washington DC, 1986.
23. AMERICAN SOCIETY FOR TESTING AND MATERIALS. Standard practice for conducting an inter-laboratory test program to determine the precision of test methods for construction materials. *Annual book of ASTM standards*. ASTM, Philadelphia, 1990, **04.03**, C802−87.
24. NATIONAL ASPHALT PAVING ASSOCIATION. *Statistical methods for quality control at hot mix plants*. NAPA, Maryland, 1973 (1983).
25. RICHARDSON C. *The modern asphalt pavement*. John Wiley & Sons, New York, 1908, 2nd edn.
26. BRITISH STANDARDS INSTITUTION. *Recovery of bitumen binders by dichloromethane extraction (rotary film evaporation method)*. BSI, London, 1991, DD 193.
27. BRITISH STANDARDS INSTITUTION. *Petroleum and its products: recovery of bituminous binders by dichloromethane extraction*. BSI, London, 1991, BS 2000: Part 105.
28. SHELL BITUMEN UK. *The Shell bitumen handbook*. Shell Bitumen UK, 1990.
29. BRITISH STANDARDS INSTITUTION. *Bitumen for building and civil engineering. Specification for bitumens for road purposes*. BSI, London, 1989, BS 3690: Part 1.
30. BRITISH STANDARDS INSTITUTION. *Penetration of bituminous materials*. BSI, London, 1983, BS 2000: Part 49.
31. JACOBS F.A. *Hot-rolled asphalt: effect of binder properties on resistance to deformation*. Transport and Research Laboratory, Crowthorne, 1981, LR 1003.
32. BRITISH STANDARDS INSTITUTION. *Method of determination of the wheel tracking rate of cores of bituminous wearing courses*. BSI, London, 1990, Draft for Development DD 184.
33. BRITISH STANDARDS INSTITUTION. *Petroleum and its products: softening point of bitumen (ring and ball)*. BSI, London, 1988, BS 2000: Part 58.
34. HEUKELOM W. *An improved method of characterising asphaltic bitumens with the aid of their mechanical properties*. Shell International Co Ltd, London, 1974.
35. AMERICAN SOCIETY FOR TESTING AND MATERIALS. Construction: roofing and waterproofing materials. *Annual book of ASTM standards*. ASTM, Philadelphia, 1987, **04.04**, D36.
36. SPARLIN R. F. The effect of ultraviolet light on viscosity of thin films of asphalt cements. *Proc. ASTM*, 1958, **58**, 1316.
37. BROCK J. D. *Oxidation of asphalt*. ASTEC Industries Inc., Tennessee, 1986, Tech Bulletin No T-103.
38. ABRAHAM H. *Asphalts and allied substances*. D Van Nostrand, New York, 1945, 5th Edn, 1464.
39. BRITISH STANDARDS INSTITUTION. *Petroleum and its products: loss on heating of bitumen and flux oil*. BSI, London, 1982, BS 2000: Part 45.

40. AMERICAN SOCIETY FOR TESTING AND MATERIALS. Standard test method fo effect of heat and air on a moving film of asphalt (Rolling thin-film oven test). *Annual book of ASTM standards*. ASTM, Philadelphia, 1990, **04.03**, D2872−88.

41. DICKINSON E. J. *Bituminous roads in Australia*. Australian Road Research Board, Vermont South, Australia, 1984.

42. BRITISH STANDARDS INSTITUTION. *Petroleum and its products: solubility of bituminous binders*. BSI, London, 1983, BS 2000: Part 47.

43. AMERICAN SOCIETY FOR TESTING AND MATERIALS. Standard test method for kinematic viscosity of asphalts (bitumens). *Annual book of ASTM Standards*. ASTM, Philadelphia, 1990, **04-03**, D2170−85.

44. DAVIS R. L. *Relationship between the rheological properties of asphalt and the rheological properties of mixtures and pavements*. ASTM, 1985, STP 941.

45. GRIFFEN R. L. *et al*. A curing rate test for cutback asphalts using the sliding plate micro-viscometer. *Proc. Ass. of Asphalt Paving Technologists*, Seattle, Washington, 1957, **26**, 437.

46. *The Marshall method for the design and control of bituminous paving mixtures*. Marshall consulting & testing laboratory, Mississippi, 1949.

47. PROPERTY SERVICES AGENCY. *A guide to airfield pavement design and evaluation*. PSA, Croydon, 1989.

48. BRITISH STANDARDS INSTITUTION. *Hot-rolled asphalt for roads and other paved areas*. BSI, London, 1973, BS 594.

49. BRITISH STANDARDS INSTITUTION. *Sampling and examination of bituminous mixtures for roads and other paved areas. Method of test for the determination of the composition of design wearing coarse rolled asphalt*. BSI, London, 1990, BS 598: Part 107.

50. FOSTER C. R. *Development of Marshall procedures for designing asphalt paving mixtures*. NAPA, Maryland, 1982, Information Series 84.

51. ASPHALT INSTITUTE. *Mix design methods for asphalt concrete and other hot-mix types*. Asphalt Institute, Maryland, USA, 1988, Manual Series No. 2.

52. AMERICAN SOCIETY FOR TESTING AND MATERIALS. Standard test method for theoretical maximum specific gravity of bituminous paving mixtures. *Annual book of ASTM standards*, ASTM, Philadelphia, 1990, **04.03**, D2041−90.

53. KANDHAL P. S. *et al*. Evaluation of asphalt absorption by mineral aggregates. *J. Ass. Asphalt Paving Technologists*, Seattle, Washington, 1991, **60**, 207−229.

54. AMERICAN SOCIETY FOR TESTING AND MATERIALS. Standard test method for bulk specific gravity and density of compacted bituminous mixtures using saturated surface-dry specimens. *Annual book of ASTM standards*, ASTM, 1990, **04.03**, D2726−90.

55. AMERICAN SOCIETY FOR TESTING AND MATERIALS. Standard test method for bulk specific gravity and density of compacted bituminous mixtures using paraffin-coated specimens. *Annual book of ASTM standards*, ASTM, Philadelphia, 1990 **04.03**, ASTM D1188−89.

56. AMERICAN SOCIETY FOR TESTING AND MATERIALS. Standard test method for specific gravity and absorption of coarse aggregates. *Annual book of ASTM standards*, ASTM, Philadelphia, 1990, **04.03**, ASTM C127−88.

57. AMERICAN SOCIETY FOR TESTING AND MATERIALS. Standard test method for specific gravity and absorption of fine aggregate. *Annual book of ASTM standards*, ASTM, Philadelphia, 1990, **04.03**, ASTM C128−88.

58. VON QUINTUS H. L. *et al*. *Asphalt-aggregate mixtures analysis system. Philosophy of the concepts. Asphalt concrete mix design — Development of more rational approaches*. ASTM, 1989, STP 1041, 15−38.

59. SOUSA J. B. *et al.* Effect of laboratory compaction method on permanent deformation characteristics of asphalt–aggregate mixtures. *J. Ass. Asphalt Paving Technologists*, Seattle, Washington, 1991, **60**, 533–585.

60. AMERICAN SOCIETY FOR TESTING AND MATERIALS. Standard test method for effect of water on cohesion of compacted bituminous mixtures. *Annual book of ASTM standards*, ASTM, Philadelphia, 1990, **04.03**, D1075–88.

61. MATHEWS D. H. *et al.* The immersion wheel tracking test. *J. App. Chem.*, 1962, 505–509.

62. BRITISH STANDARDS INSTITUTION. *Sampling and examination of bituminous mixtures for roads and other paves area. Methods of test for the determination of density and compaction.* BSI, London, 1989, BS 598: Part 104.

63. HOWE J. H. The Percentage Refusal Density test. *J. Inst. Highways & Transportation*, London, Mar. 1986, **33**, No. 3, 15–18.

64. DEPARTMENT OF TRANSPORT. *Specification for highway works.* HMSO, London, 1992, **1**.

65. HOSKING J. R. *et al. Measurement of the macro-texture of roads. Part 2: a study of the TRRL mini-texture meter.* Transport Research Laboratory, Crowthorne, 1987, RR120.

66. SYM R. *Methods of Test for the Determination of Texture Depth.* BSI, London. Unpublished report to BSI Committee RDB/36/4 on BS 598: Part 105: Statistician's Report, 1990.

67. BRITISH STANDARDS INSTITUTION. *Sampling and examination of bituminous mixtures for roads and other paved areas. Methods of tests for the determination of texture depth.* BSI, London, 1990, BS 598: Part 105.

68. HILLS J. F. *et al.* The measurement of texture depth under adverse conditions. *Highways and Public Works.* London, Mar/Apr 1981, **49**, No.s 1852/1853, 8–12.

69. SZATKOWSKI W.S. Rolled asphalt wearing courses with high resistance to deformation. The performance of rolled asphalt surfacings. *Proc. Inst. Civ. Engrs Conf.*, London, 1980, 107–122.

70. CHOYCE P. W. *et al.* Resistance to deformation of hot rolled asphalt. *J. Inst. Highways and Transportation.* London, Jan 1984, **31**, No. 1, 28–32.

71. ROBERTS F. L. *et al. Hot mix asphalt materials, mixture design, and construction.* NAPA Education Foundation, Maryland, 1991.

72. OAKLAND, J. S. *Statistical process control — a practical guide.* Heinemann, London, 1986.

73. NATIONAL ASPHALT PAVING ASSOCIATION. *Statistical methods for quality control at hot-mix plants.* NAPA, Maryland, USA, 1983, QIP95.

74. BRITISH STANDARDS INSTITUTION. *Guide to data analysis and quality control using CUSUM techniques.* BSI, London, 1980, BS 5703: Parts 1–4.

75. BRITISH STANDARDS INSTITUTION. *Guide to statistical interpretation of data.* BSI, London, 1985, BS 2846: Parts 1–7.

76. BRITISH STANDARDS INSTITUTION. *General criteria for the operation of testing laboratories.* BSI, London, 1989, EN 45001.

77. THE ROYAL SOCIETY OF CHEMISTRY. *COSHH in laboratories.* R. S. Chem., London, 1989.

78. BRITISH STANDARDS INSTITUTION. *Quality systems.* BSI, London, 1987, BS 5750: Parts 1–3.

9. Surface dressing and other surface treatments

9.1. Introduction

In chapters 1 and 2 the various characteristics of the surface layer of a road are described. Surface texture and its relationship with skid resistance are vital to ensure that the running surface of the road provides the best possible interaction with the vehicle tyres, and safety of the road user.

In a new road the requisite surface texture is designed into the running surface in accordance with the specification requirements for aggregate properties and texture depth for both bituminous and cement concrete road. During its service life the surface becomes polished under the action of traffic and the skid resistance will fall below the minimum specified value. Clear guidance on appropriate intervention values is given in Department of Transport standards and advice notes discussed later in this chapter.

9.2. Surface dressing

For a structurally-sound road, one of the most cost-effective ways to restore its skid resistance is the application of surface dressing. All classes of road and all categories of site can be surface dressed providing the work is correctly specified, appropriate materials selected, work is properly executed and necessary aftercare is exercised.

The reason to surface dress will usually be one or more of the following

(a)   to provide or restore a skid resistant surface
(b)   to arrest the disintegration and loss of aggregate from the road surface
(c)   to seal the surface of the road and thus protect its structure from damage by water.

Surface dressing consists of spraying a thin film of binder onto the road surface, followed by the application of one or two layers of stone chippings (Fig. 9.1).

The chippings are then rolled to promote contact between the chippings and the binder and to initiate the formation of an interlocking mosaic. The factors which influence the performance and the service life of a surface dressing are traffic, existing road surface, size and type of chipping, binder, rate of spread of binder and environmental conditions.

Fig. 9.1. Surface dressing. Photograph courtesy of Colas Ltd

Traffic

The volume of commercial vehicles/lane/day governs the rate of embedment of chippings into the old road surface; a commercial vehicle is defined as a vehicle of unladen weight greater than 1.5 t. Vehicles of lower axle weight have no significant effect on embedment.

Existing road surface

The degree to which chippings will penetrate into the old road surface is related to the hardness of the old road surface and to the number of commercial vehicles in the traffic lane under consideration. Where an old surface is porous, allowance must be made for the effective reduction in binder film thickness caused by penetration of binder into the underlying surface. Severe cracks in the old road surface will ultimately reflect through the surface dressing.

Size and type of chipping

Chippings should not be so small that they are rapidly embedded into the underlying surface nor so large that they may be dislodged by traffic. They should have strength characteristics and resistance to polishing appropriate to the road being surface dressed.

Binder

The function of the binder is to seal cracks and bind the chippings to the underlying surface. The viscosity of the binder must be such that it can wet the chippings adequately at the time of application, prevent dislodgement of chippings when the road is opened to traffic and not become brittle during periods of prolonged low temperature and function effectively for its design life.

Rate of spread of binder

The rate of spread of binder has to be such that there is sufficient binder to retain chippings during the early life of a dressing, without filling a proportion of voids in the layer of chippings in excess of that required to ensure adequate surface texture depth during the life of the dressing.

Environmental conditions

These may be related to location, climate or exceptional traffic conditions. The rate of cure of binders can be influenced by whether the surface dressing is laid in an exposed or sheltered area e.g. under trees or bridges, the time of year in which surface dressing is carried out and the weather conditions during the early life of a surface dressing may have an important influence on performance. Traffic may impose exceptional stresses due to braking, acceleration or turning movements at junctions or roundabouts which increase the risk of chippings being dislodged, or by additional loading on long inclines which increases embedment of chippings.

These factors are logically addressed in the latest edition of TRL Road Note 39 *Design guide for road surface dressing*.[1]

Further advice on all the practical aspects of surface dressing are contained in the Road Surface Dressing Association publication *Code of Practice for Surface Dressing*.[2]

Surface dressing is a technical process which requires the highest degree of management, supervision and control, for which responsibility should be placed in the hands of competent and experienced engineers. The extracts from Road Note 39[1] give an indication of a proper design and selection process for surface dressing. However, there is no substitute for a comprehensive study of this document to ensure that a durable and cost effective treatment is applied.

Figure 9.2 shows a flow diagram for the logical selection of the correct treatment of a chosen site. The number in the bottom right hand corner of each box refers to the relevant section of Road Note 39.[1]

The quantity, type and speed of traffic carried by a road is of primary importance in the design process and Fig. 9.3 shows the method used to assign roads to the various traffic categories used in Road Note 39.

9.2.1. Road hardness

The hardness of the road surface affects the extent to which the applied chippings become embedded into the road surface during the life of the dressing. Choice of chipping size is directly related to road hardness. The use of chippings which are too small will result in early embedment of the chippings into the surface leading to a rapid loss of texture depth and in the worst case fatting up of binder to cover the entire surface of the road. The use of chippings which are too large may result in immediate failure to the treatment due to the stripping of the aggregate under the applied stresses of the traffic and can also result in excessive surface texture and consequent noise. The hardness of the road surface also influences the rate of spread of binder required for a given size of chipping; the rate of spread must be decreased where the road surface is soft to compensate for the greater penetration of the chippings into the surface under the action of traffic.

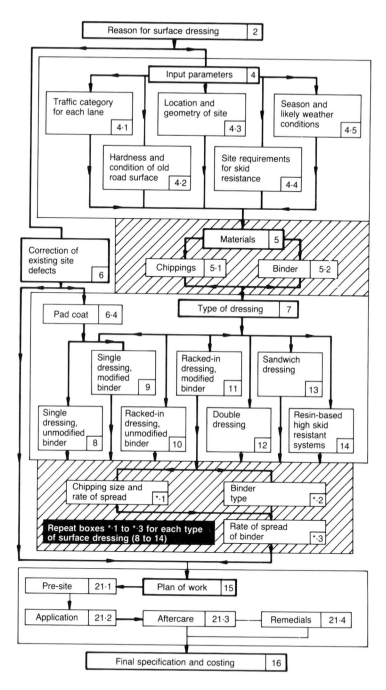

Fig. 9.2. Flow diagram for surface dressing[1]

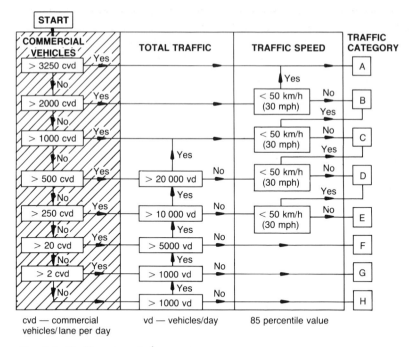

cvd — commercial vehicles/lane per day   vd — vehicles/day   85 percentile value

Fig. 9.3. Traffic categories[1]

The hardness can be measured using a special loaded probe produced by Tar Industry Services of Chesterfield. Since virtually all road surfaces except concrete are temperature susceptible, the penetration readings obtained have to be related to the road temperature. Table 9.1 gives a descriptive list of the various hardness categories.

Other factors that influence the embedment of chippings are road surface

Table 9.1. Road surface hardness categories[1]

Hardness	Description of surface
Very hard	Surfaces such as concrete or exceptionally lean bituminous mixtures with dry stony surfaces into which there will be negligible penetration of chippings under heavy traffic
Hard	Surfaces containing some hard bituminous mortar into which chippings will penetrate only slightly under heavy traffic
Normal	Surfaces into which chippings will penetrate moderately under heavy and medium traffic
Soft	Surfaces into which chippings will penetrate considerably under heavy and medium traffic
Very soft	Surfaces into which even the largest chippings will be submerged under heavy traffic, usually rich in binder

temperature and vehicle speed. If a section of road is colder throughout the year less chipping embedment will occur. This is the reason why more binder should be applied in areas of shadow under trees and bridges. In these locations, typically 10−20% more binder is required. Where vehicle speeds are generally lower, the chippings are loaded for a longer period and more embedment occurs. To avoid fatting up less binder should be applied on steep inclines especially those that are in full sunlight. Typically the application rate at these locations is reduced by 10%.

9.2.2. Skid resistance

Skid resistance is defined by the mean summer SCRIM coefficient (MSSC) at either 50 km/h or 130 km/h. SCRIM values are measured using the sideway-force coefficient routine investigation machine. In 1973, the TRL[3] introduced the concept of risk rating which is a measure of the relative skidding potential at an individual site determined by its geometry and location. Within each category, there is a range of target sideways force coefficient (SFC) values, the target value for each site depending on its risk rating. A road will need to be surface dressed if the MSSC value is inadequate for the risk rating on the site. The investigatory levels of skid resistance for motorways and trunk road are given in Department of Transport standard HD 15/87[4] and Advice Note HA 36/87.[5] Advice for non-trunk roads is given in *Highway Main-tenance, A Code of Good Practice.*[6] In these documents, the investigatory levels are those at which the measured skid resistance of the road needs to be reviewed. In the review process it may be decided to restore skid resistance by surface dressing or the risk rating of the site may warrant change.

The skid resistance of a road is achieved by a combination of the macrotexture or overall surface roughness (typically peak to trough height 0.2−3 mm) and the microtexture or surface rugosity of the chippings (typically peak to trough height up to 0.2 mm). Fig. 9.4 shows both microtexture and macrotexture.

Microtexture has some influence on skidding resistance at all speeds by allowing the tyres to grip the road surface but is predominant at lower vehicle speeds (up to about 50 km/h). At higher speeds, macrotexture becomes more important since it provides the series of minute channels that allow rainwater to flow to the side of the road and enable the vehicle tyres to contact the road surfacing i.e. prevent the tendency to aquaplane.

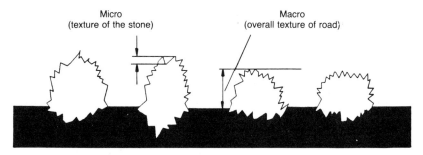

Fig. 9.4. Micro and macrotexture[1]

The petrographic characteristics of the aggregate largely influence its resistance to polishing and this is measured using the polished stone value (PSV) test.[7] This test subjects the aggregate to simulated trafficking in an accelerated polishing machine. A high PSV indicates good resistance to polishing. The majority of road aggregates have a good microtexture prior to trafficking so most road surfaces have a high skid resistance when new. Under traffic, the exposed aggregate surfaces become polished and within a year the skid resistance falls to an equilibrium level.

There are many sources of high PSV aggregate in the UK, mostly gritstones. Unfortunately, high PSV tends to be associated with poor abrasion resistance or poor resistance to weathering and as a result only a few of the high PSV aggregates are suitable for surface dressing. Thus, blanket specifications, imposed by some local authorities for all sites, are a waste of valuable resources and more thought should be given to the type of site to enable more abundant lower PSV aggregates to be used effectively. Catt of Warwickshire County Council has developed a rational approach to the selection of suitable aggregate types.[8] The method takes into account all the factors relating to skid resistance and can demonstrate the suitability of aggregates which have previously been considered unsuitable.

The macrotexture is determined by the nominal size of aggregate used. However, resistance to abrasion is also important. Aggregates with poor abrasion resistance, determined by the aggregate abrasion value test (AAV), will quickly be worn away, with a consequent loss of macrotexture. Maximum AAVs for various traffic categories are specified in Department of Transport Technical Memorandum H16/76.[9] The macrotexture of the road surface is measured using the sand patch test.[10] In this test a known volume of sand is spread onto the road surface to form a circular patch so that the valleys are filled to the level of the peaks. The texture depth of the road is determined by dividing the volume of the sand by the cross-sectional area of the patch.

The rate at which skid resistance falls off with increasing vehicle speed is influenced by the macrotexture. The TRL has shown[3] that there is a correlation between macrotexture, expressed in terms of average texture depth, and skid resistance, measured by percentage decrease in braking force coefficient (BFC) between 50–130 km/h, as shown in Table 9.2.

Mature surface dressings in good condition typically have texture depths between 1.5 mm and 2 mm. Thus the high-speed skid resistance of the surface dressing is maintained throughout the life of the dressing.

Table 9.2. Relationship between texture depth and reduction in BFC[3]

Texture depth: mm	Reduction in BFC: %
0.5	30
1.5	10
2.0	negligible

9.2.3. Materials
Chippings

All chippings should comply with the general requirements for size, shape and strength included in BS 63: Part 2.[11] They can either be clean and uncoated or coated with a thin layer of bitumen depending on the type of binder selected and the design of surface dressing.

Uncoated chippings need to be clean to ensure good adhesion to the binder film. The presence of dust can delay or even prevent adhesion and is acute at low temperatures and with the smaller sizes of chipping.

Coated chippings have a thin film of binder applied at a coating plant. This binder film eliminates surface dust and helps to ensure rapid adhesion to the binder. The binder film should be as thin as possible and need not be continuous as a small number of uncoated pinholes will not affect performance. If the binder film is too thick, chippings will tend to stick together in warm weather and will not flow evenly through the chipping spreader. Coated chippings are used to improve adhesion with cutback bitumens, particularly in the cooler conditions at the extremes of the season. They should not be used with emulsions as the shielding effect of the binder will delay the break of the emulsion.

The guiding design principles of Road Note 39[1] are to select the size and type of chipping taking into account the following factors.

(a) The size of chipping has to offset the gradual embedment into road surfaces of different degrees of hardness caused by the action of traffic.

(b) Microtexture and macrotexture have to be maintained during the life of the surface dressing by selecting a chipping of appropriate size, PSV and AAV. Additional guidance on the selection of suitable aggregate to provide appropriate levels of skidding resistance for various traffic categories is given in TRL report LR 510.[3]

(c) The lane traffic category and the road surface hardness category has to be related to the size of chipping required. Figs 9.6 and 9.7 illustrate this relationship for both single and racked-in types of surface dressing.

(d) The rate of spread of chippings has to be sufficient to cover the binder film after rolling without excess chippings lying free on the surface of the dressing.

(e) Binder type and viscosity has to be appropriate for the time of year, traffic category and degree of stress likely to be encountered; the application rate appropriate to the traffic category, road hardness and nominal size of chipping.

Types of binders used in surface dressing

The essential function of the binder is to seal the road surface and provide the initial bond between the chippings and the road surface. The subsequent role of the binder depends on the traffic stresses applied to the surface dressing from low stressed, low traffic country roads, to highly stressed sites carrying large volumes of traffic. Clearly a range of binders is necessary if optimal performance is to be achieved under these widely differing circumstances. Three broad ranges of bitumen-based surface dressing binders are available

(a)   traditional cutback and emulsified bitumens
(b)   epoxy-resin thermoset binders
(c)   proprietary polymer modified bitumens (hot applied and emulsions).

Tar and tar-bitumen blends are not included since their use has dramatically declined over the last few years and in a number of European countries they are no longer specified because of health reasons.

Bitumen emulsions. The principal surface dressing binder is K1-70 bitumen emulsion. This is a cationic bitumen emulsion containing approximately 70% of 200 or 300 pen bitumen modified by the addition of approximately 1% kerosene. The relatively low viscosity of bitumen emulsions enables them to be sprayed at 75−85°C through swirl jets. Emulsions are able to satisfactorily coat most aggregates, even through water films. Dusty aggregates should be avoided because the dust can prevent the coalescing globules of bitumen from penetrating and wetting the aggregate.

The relatively low viscosity of K1-70 allows the binder to flow on the road surface so it is important that the chippings are applied to the sprayed emulsion film as soon as possible. Once the chippings have been applied to the emulsion film, the time taken for the emulsion to break is subject to several variables: ambient temperature; humidity; aggregate water absorption; wind speed; the physio-chemical properties of the emulsion and the aggregate; the mechanical forces applied to the surface dressing by rolling and traffic. Pre-coated chippings must not be used with emulsions.

Cutback bitumen. This is usually 100 pen or 200 pen bitumen which is diluted with kerosene to comply with an STV viscosity specification usually 50 s, 100 s or 200 s. The suffix X on cutback bitumens e.g. 100X, indicates that they have been doped with a specially-formulated thermally-stable passive adhesion agent. This additive assists wetting of the aggregate and the road surface during application and resists stripping of the binder from the aggregate in the presence of water. The effectiveness of this additive can be demonstrated by a laboratory test; the immersion tray test.[1] If surface dressing is being carried out in marginal weather conditions or if the aggregate is wet, it is recommended that an active wetting agent is added to cutback bitumens immediately before spraying. Cutback bitumens are sprayed through swirl or slot jets at temperatures between 140−160°C. During spraying, 10−15% of the kerosene evaporates and a further 50% evolves from the surface dressing in the first few years after application.

Selection of binder. The choice between these two conventional binders is a balance of cost, initial adhesion, performance and ease of spraying. Excellent performance is obtained with both binders if they are applied correctly.

Generally, bitumen emulsions are more vulnerable during the early life of the dressing. They are more tolerant to damp chippings and so are often preferred at the beginning and end of the surface dressing season i.e. during April/May and August/September when the weather is variable and less suitable for cutback bitumens. Demand for emulsions during these two periods is less predictable so emulsion suppliers may increase emulsifier contents to improve storage stability. This, together with dusty chippings and cool weather conditions, may result in a semi-cured emulsion state often known as the cheesy state. Here, the bitumen droplets have agglomerated but not coalesced and the dressing is extremely vulnerable. It is essential then that the emulsion is completely broken before uncontrolled traffic is allowed on the dressing. This can be a serious disadvantage on heavily-trafficked roads.

Cutback bitumens have benefits in terms of immediate cohesive grip of the chippings but generally the residual binder properties are more variable because of the differences in evolution of the diluent.

Epoxy-resin thermoset binders. These are bitumen-extended epoxy-resin binders used with high PSV aggregates such as calcined bauxite. Epoxy-resin based binders are used in high performance systems such as Shell Grip and Spray Grip that are fully capable of resisting the stresses imposed by traffic on the most difficult sites e.g. roundabouts or approaches to traffic lights and designed accident black spots. These binders are classified as thermosetting as the epoxy-resin components cause the binder to cure by chemical action and harden and it is not subsequently softened by high ambient temperatures nor by spillage of fuel. The dressing acts as an effective seal against the ingress of oil and fuel, which is particularly important on roundabouts where regular fuel spillages occur. With this type of binder very little embedment of the chippings into the road surface takes place and the integrity of the surface dressings is largely a function of the cohesive strength of the binder.

The extended life of this binder justifies the use of an exceptionally high PSV and durable aggregate. The initial cost of this surface dressing system is high compared with conventional binders, but its exceptional wear resistant properties and the ability to maintain the highest levels of skid resistance throughout its life make it a cost-effective solution for difficult sites by significantly reducing the number of skidding accidents. Recent statistics show the cost to the community of a serious accident to be £26 500 and a fatal accident to be £760 000.[12]

Proprietary polymer modified binders. Fig. 9.5 shows a relationship between binder type and performance requirement.[13] Precise categorization is not possible since there is a significant variation in severity along the length of every class of road which is not solely dependent on traffic numbers and vehicle weight.

A high proportion of the road network can be classified as easy or average in terms of surface dressing. These roads can largely be surface dressed using bitumen emulsion or cutback bitumen although it may be necessary on certain highly-stressed sections to consider a more enhanced treatment i.e. at sharp bends, road junctions and in shaded areas under trees and bridges. In these areas it is probable that the same binder can be used but the specification will be changed using one of the following

- (a) rate of application of binder
- (b) chipping size and rate of application
- (c) double chip application
- (d) double binder, double chip application.

These latter techniques are widely practised in France on highly-trafficked roads, and with correct procedures their use can significantly reduce the risk of flying chippings and consequent windscreen breakage. Conventional

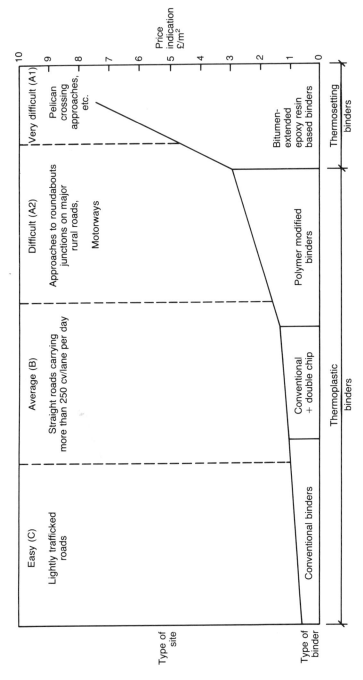

Fig. 9.5. Relationship between binder type and performance requirement[13]

emulsions and cutbacks have the lowest cost and with careful workmanship, proper compliance with specification and correction aftercare, good results can be obtained.

As a further aid to consistent performance under the variable weather in the UK, a number of proprietary adhesion agents are available which act in two basic ways. Some types (active) modify surface tension at the stone/binder interface to improve the preferential wetting of chippings by the binder in the presence of water. Other types (passive) improve the adhesion characteristics of the binder and give an improved resistance to the subsequent detachment of the binder film in the presence of water throughout the early life of the dressing. The effectiveness of the former type of adhesion agent deteriorates after a few hours at elevated temperature and so it is generally mixed with the binder a few hours before spraying. The latter type of agent is more temperature stable and is generally added to the binder by the supplier.

At the top end of the performance scale, bitumen extended epoxy-resin binders have been developed that are fully capable of resisting the very high stresses imposed by traffic at the most difficult sites. Additional thermosetting binders are also available with a reduced resin content which are less costly. The curing time to achieve full performance is increased, but its use can be considered where the stresses are slightly less severe.

The middle ground between easy and most severely stressed sites is difficult to specify with certainty, and the types of binder to use in this area have taken time to develop. End users have been reticent to pay more for binders to use on site where under the best conditions a standard 'might just have worked'. It is necessary to remember that the price of failures is high in both monetary and potential accident terms.

When considering the specification for these sites two distinct objectives should be borne in mind

(a) to provide a treatment which is more foolproof or tolerant of difficult conditions so that it will always give results at least equal to the best obtainable with conventional binders
(b) to provide a treatment which will enable surface dressings to be used on sites where conventional binders would normally fail.

The first of these objectives is frequently overlooked and in this area the addition of a thermoplastic polymer is less likely to satisfy the requirement and provide the commercial justification for its use.

Types of additive. With these performance requirements in mind, it is possible to examine the ways in which the various polymers can be used with bitumen to modify its performance. The properties of bitumen that would be improved by the addition of a modifier are as follows.

(a) Temperature susceptibility, which can loosely be defined as the extent to which a binder softens over a given temperature range. It is clearly desirable for a binder to exhibit a minimum variation in viscosity over a wide range of ambient temperatures.
(b) Cohesive strength, which enables the binder to hold chippings in place when they are subject to stress by traffic. This property can also be

coupled with the elastic recovery properties of the binder which enables a surface dressing to maintain its integrity even when subject to high strain levels.

(c)   Adhesive power or 'tack' is a difficult property to define, but it is essential, especially in the early life of a surface dressing before any mechanical interlock or embedment of the chippings has taken place.

It is also desirable for the modified binder to

(a)   maintain its premium properties for long periods in storage and then during its service life

(b)   be physically and chemically stable at storage, spraying and service temperatures

(c)   be capable of use with conventional or easily modified spraying equipment

(d)   be cost effective.

Many polymers have been used over a number of years with varying degrees of success.

Rubbers. Polybutadiene, polyisoprene, natural rubber, butyl rubber, chloroprene, random styrene/butadiene rubber and EPDM have all been used with bitumen and their main effect is to increase viscosity with only a small effect on elasticity. In some instances, the rubbers have been used in a vulcanized (cross-linked) state, but this is difficult to disperse in bitumen requiring high temperatures, long digestion times and can result in a heterogeneous binder with the rubber acting mainly as a flexible filler.

Thermoplastics. Polyethylene, polyvinyl chloride, polystyrene, and ethyl vinyl acetate are the principal thermoplastic polymers that have been tried as bitumen modifiers. Their behaviour is similar to that of bitumen i.e. they soften when heated and harden when cooled. They can be blended with bitumen at elevated temperature using comparatively low shear mixers and mainly increase the viscosity of the binder and stiffness at ambient temperature. They do not enhance the elastic properties of the binder and tend to separate in storage, which can lead to uneven dispersion of the polymer in the finished work.

Thermoplastic rubbers. Thermoplastic rubbers (TR) combine both elastic and thermoplastic properties. They consist of links of block co-polymers of styrene and butadiene coupled in pairs to give di-blocks or star configuration. These block co-polymers are more commonly known as styrene-butadiene-styrene (SBS) polymers and have been used for many years to modify bitumen to provide high performance coatings for roofing membranes, and more recently for a range of road applications including surface dressing.

The unique properties of thermoplastic rubbers are thermoplastic behaviour at elevated temperatures coupled with a rubbery type behaviour at ambient temperature. The strength of TR is provided by the hard polystyrene end blocks which associate in domains to form a three dimensional network. On heating to above 100°C the polystyrene softens, the domains disassociate and the polymer flows. The process is completely reversible and the domains reform on cooling. The viscoelastic nature of normal bitumen and its susceptibility to temperature and loading time are well known. This susceptibility is reduced by TR addition and the extent of the change is closely related to polymer concentration.

Providing the correct blending bitumen is used, noticeable changes in viscosity and temperature susceptibility can be obtained with about 3% polymer in the blend. It is necessary to raise the polymer content to about 6% before a full three-dimensional network is established and maximum improvements in elasticity are displayed. Improved initial tack can be obtained by careful selection of polymers generally of the styrene-isoprene-styrene (SIS) type which are used in combination with SBS polymers in a proprietary binder Shelphalt SX.

The choice of either an emulsion or cutback formulation for modified binders is not an easy one. Generally speaking, emulsions are more vulnerable during the early life of the dressing before the emulsion has fully broken. The rate of break is influenced by a number of factors including emulsion formulation, weather and mechanical agitation during rolling and trafficking. Under certain conditions, the emulsion goes through a cheesy state during cure where the bitumen droplets have agglomerated but not coalesced and the dressing is extremely vulnerable. The time of this condition is variable and is generally worsened by the presence of high polymer contents which can cause skinning and retard breaking. It is essential that the emulsion is completely broken before uncontrolled traffic is allowed on the dressing, which can be a serious disadvantage on heavily trafficked roads. Development work on chemical after-treatments (sprayed on) to promote breaking is showing success.

Cutback bitumens have benefits in terms of immediate cohesive grip of the road and chippings, but generally the residual binder properties are more variable due to the differences in the evolution of the diluent and its absorption into the road surface. Nevertheless, cutback formulations are the preferred choice for high speed and heavily trafficked roads due to their predictable early life performance.

The development of polymer modified surface dressing binders to complement conventional and epoxy-resin systems offers the highway engineer a range of binders suitable for all categories of site. Regardless of the improved properties of a particular binder, it is the performance of the aggregate binder system and application mode which dictate the success of the dressing.

9.2.4. Types of dressing

The original concept of a normal single layer surface dressing has been developed over the years and there are now a number of techniques available which vary according to the number of layers of chippings and applications of binder. Each of these techniques has particular advantages and associated cost, and it is quite feasible that along the length of a road a variety of techniques would be used dependent on the stresses at any particular feature. Fig. 9.6 shows the types of technique available.

Single surface dressing

The single surface dressing system has the least number of operations, uses the least amount of material and is sufficiently robust for most easy sites and a number of average sites (Fig. 9.7). There is a limit to the stresses this system will withstand before more robust techniques are required.

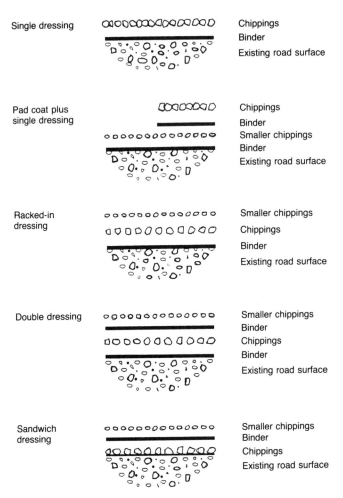

Fig. 9.6. Types of surface dressing[1]

Pad coats

A pad coat consists of a single dressing with small sized chippings which is applied to a road which has uneven surface hardness, possibly due to extensive patching by the utilities. The pad coat is used to provide a more uniform surfacing which can subsequently be surface dressed. The chipping size for a pad coat is traditionally 6 mm with a slight excess of chippings. Pad coats can also be used on very hard road surfaces such as concrete or heavily chipped asphalts to reduce the effective hardness of the surface, but using a racked-in system with a polymer modified binder is now the preferred option.

After laying and compaction by traffic, excess chippings should be removed before the road is opened to unrestricted traffic. Pad coats may be left for a short time before the main dressing, either a single dressing or a racked-in system, is applied. Either system will embed into the pad coat and derive

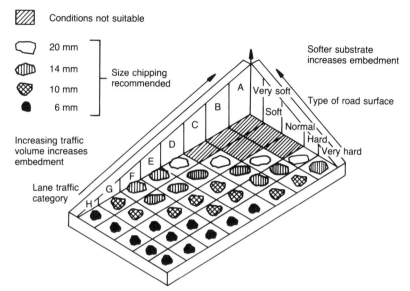

Fig. 9.7. Use of different size chippings in single surface dressings[1]

immediate mechanical strength, lessening the demand on the cohesive strength of the binder and reducing the risk of failure. All loose material should be swept from the surface of the road before the final dressing is applied.

Double surface dressing

Traditionally, surface dressing has been carried out by a single application of binder followed by chippings. For high-speed and heavily-trafficked roads, double chipping applications are now being successfully used, with small chippings employed as a secondary aggregate to lock the larger chippings in place.

There are two main types of double surface dressing systems: the single spray double chip or racked-in method and the double spray double chip method. Fig. 9.8 shows how different chipping sizes are used for varying hardness characteristics in racked-in surface dressings.

A schematic representation shows the sequence of spray and chip for single spray double chip or racked-in method (Fig. 9.9(a)) and the double spray double chip methods (Fig. 9.9(b)).

Single spray double chip or racked-in method

This type of dressing uses a single application of binder. The 14 or 10 mm principal chippings are spread at approximately 80–90% cover which leaves a window of binder between the chippings. This window is filled with small chippings 6 mm or 3 mm respectively to achieve high mechanical interlock of the principal chippings. The rate of spread of binder on this type of surface dressing is usually slightly higher than that required on a single size dressing.

The initial surface texture and mechanical strength of the dressing are high. The configuration of the principal chippings is different from that achieved

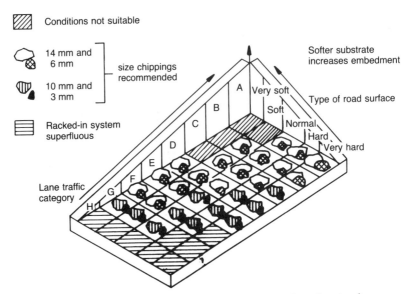

Fig. 9.8. Use of different size chippings in racked-in surface dressings[1]

with a single dressing. The mosaic associated with the latter cannot be formed because the principal chippings are locked in place. In time, small chippings not in contact with binder are lost to the system without damaging vehicles or windscreens, resulting in increasing surface texture despite some coincident embedment of the larger chipping.

On a racked-in dressing, vehicle tyre contact is mainly on the large aggregate with little contact on the small aggregate. Thus the microtexture of the small aggregate is less critical and consideration should be given to the use of cheaper lower PSV aggregates such as limestone for the small chippings.

The advantages of the racked-in system are

(a)   virtual elimination of flying 14 mm chippings
(b)   early stability of the dressing through good mechanical interlock
(c)   good adhesion of larger size chippings
(d)   a rugous surface texture with an initial texture depth of over 3 mm.

Since 1987, there has been a marked increase in the use of racked-in dressing on main and heavily-trafficked roads. The use of this technique has dramatically reduced the incidence of broken car windscreens.

Double spray double chip method

This dressing uses two applications of binder and chippings. A typical sequence would be to apply the binder plus 14 mm clean chippings. This dressing is lightly rolled and followed immediately by a second application of binder and clean 6 mm chippings. This technique provides an excellent dressing with a texture depth over 3 mm. This type of dressing is most suitable on roads which can be closed and lanes coned-off, as the initial stability can be rather low.

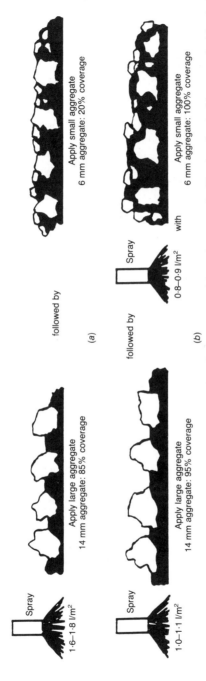

Fig. 9.9. Different surface dressing systems with typical application rates: (a) single spray double chip (b) double spray double chip[14]

The advantages of double surface dressing systems are summarized below.

(a)   Their initial strength is greater, with less reliance on the strength of the binder than single surface dressings.

(b)   Because they are physically locked in, larger chippings can be selected than would normally be the case for a given road hardness. This prolongs the surface texture of the new dressing and reduces the chance of failure resulting from the onset of cold weather before chipping embedment has taken place.

(c)   The number of non-adhering large chippings lying on the road surface is significantly reduced. Because they are locked in place by the smaller chippings, the chance of windscreen breakage is reduced.

(d)   The initial and long-term surface texture of the single spray double chip system tends to be higher than for the single surface dressing system.

Modified binders are especially effective with the racked-in or double surface dressing systems to maximize the improvement in mechanical interlock and adhesive cohesive stability.

Sandwich surface dressing

Sandwich surface dressing is used in situations where the road surface condition is binder rich. The first layer of chippings is spread without a layer of binder and then a conventional dressing is applied. Sandwich dressings can be considered as double dressings in which the first binder film has already been applied. The degree of binder richness at the surface has to be sufficient to hold the first layer of chippings in place while the rest of the operation proceeds. The chipping sizes need to reflect the extent of the excess binder on the surface, and with the relevant rate of spread of the second binder application. This technique has been widely used in France but there is little experience with it in the UK at the moment.

Selection of type of dressing

The choice of an appropriate surface dressing system depends on a number of factors. Fig. 9.10 shows a flow diagram to aid selection. The sets of criteria are arranged so that the harshest conditions dictate the system to be used, minimizing the risk of failure. The decision process enables a good 'percentage' decision to be made with the known input criteria.

There may be situations where the basic design needs to be amended to take into account unusual or one off criteria. Possible reasons might include dressing out of season due to temporary roadworks, occasions when a road has limited structural life, when the traffic intensity is expected to change significantly in the planned life of the road, or when the road has strategic importance for reasons other than traffic flow.

9.2.5. Surface dressing equipment

Improved design of surface dressings and advances in binder technology and aggregate quality have been major contributory factors towards improving the performance of surface dressings. Advances in the design of application

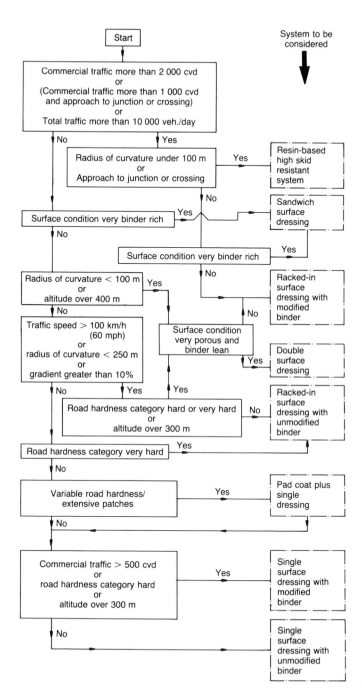

Fig. 9.10. Selection of type of surface dressing¹

machinery in harmony with the above has enabled the whole process to be co-ordinated with respect to speed of operation, ability to comply with specification and continued improvement to standards of safety.

Sprayers

Figure 9.11 shows a modern surface dressing sprayer.

The uniformity of transverse distribution has dramatically improved over the last ten years. It is probable that further improvements can be made, especially at the ends of the bar and at the centre of an expanding bar, where respectively heavy and light application problems can occur. The current distribution requirement for $\pm 15\%$ variation is not precise enough for major roads. Most contractors can now provide bars with $\pm 10\%$ variation or even better. The use of hydraulically expanding spray bars enables the joint to be maintained at the centre of the road while varying the width at the road edge, thus eliminating unsightly and sometimes dangerous joints from the wheel tracks. Although most conventional binders are still sprayed from distribution bars fitted with swirl jets, the higher initial viscosities of the thermosetting binders and some polymer modified binders necessitate the use of slot jets.

Advanced designs of slot jets are now available which give similar standards of transverse distribution to those obtained with swirl jets, and they will probably be increasingly used in the UK market for conventional binders. By using this type of jet it is possible to reduce the quantity of diluent in a binder with subsequent savings in cost and reduced application temperatures resulting in a reduction of fire risk.

Programmable bitumen distributors are now available which are extremely sophisticated spray machines using two bars to achieve differential rates of spread across the width of the machine. Thus application rates can be changed for areas of patching, wheel tracks, fatty areas and hungry areas. The rate

Fig. 9.11. Modern surface dressing sprayer. Photograph courtesy of Colas Ltd

of spread is automatically reduced by up to 30% in the wheel tracks and can be increased between wheel tracks or in shaded areas. Both longitudinal and transverse binder distributions can be pre-programmed. Vehicle speed, binder temperature, spray bar width and application rates are controlled by a micro-processor.

Chipping spreaders

Self-propelled chipping spreaders (Fig. 9.12) are now in regular use and these machines not only allow work to proceed more rapidly, but also spread the chippings more evenly and accurately at the required rate per square metre. They also have the advantage that the chippings are applied much closer to the road surface. As a result, the likelihood of chippings bouncing off the road surface and exposing binder or bouncing to another part of the dressing with the binder uppermost is reduced. These chippers are also intrinsically safer because they are driven forward rather than reversing as is the case with lorry mounted spreaders.

The latest development in self-propelled chipping spreaders include four-wheel drive for pulling heavy tippers up steep inclines without a judder, a mechanism for breaking lumps of lightly coated chippings within the hopper, methods for spraying additive onto the binder film and incorporation of a pneumatic-tyred roller to ensure that the chippings are pressed into the binder film at the earliest possible moment while it is still hot.

Rollers

Rolling techniques have been substantially improved by increasing use of pneumatic-tyred rollers (Fig. 9.13) which secure the initial bond of the chipping

Fig. 9.12. Self-propelled chipping spreader. Photograph courtesy of Colas Ltd

Fig. 9.13. Pneumatic-tyred roller. Photograph courtesy of Temple Asphalt Ltd

to the binder film without crushing the carefully selected aggregate and thus preventing the attendant problems of fatting up later in the life of the surface dressing. The use of vibrating rubber drum-type rollers improves the initial mosaic formation, although it must be recognized that any rolling technique is largely a preliminary process undertaken before the main stabilization of the dressing by subsequent slow speed trafficking.

Sweepers

High-speed roads require thorough sweeping to remove all excess chippings from surface dressing before allowing unrestricted traffic onto the dressing. For this, a full-width suction sweeper should be used with a power broom collecting chippings thrown off the dressing. However, 14 mm chippings pose a problem as current full-width machines are not always able to collect this large stone. Therefore, conventional single suction-with-sweep machines can be used when the dressing is stable and will withstand disruption by the brush. Properly organized and implemented sweeping will do much to alleviate the damage which can be caused by surface dressings in their early life by chipping loss.

Traffic control and aftercare

Traffic control and aftercare are vital to the successful performance of the surface dressing. Ideally, when traffic is first allowed onto the new dressing, it should be behind a control vehicle, which should travel at a speed of about 15 mph. If there are gaps in the traffic, it may be necessary to introduce additional control vehicles. The objective should be to ensure that vehicles do not travel at more that 15 mph on the new dressing, and that sharp braking and acceleration of these vehicles does not occur.

The length of time involved will depend very much on the type of binder used and the prevailing weather conditions. It is likely to be longest where emulsions are being used and the weather conditions are humid. The use of wet or dusty chippings or the early onset of rain delays adhesion and necessitates a longer period of traffic control. Traffic passing slowly over a dressing immediately after completion creates a strong interlocking mosaic of chippings, a result only otherwise obtained by long periods of rolling.

9.2.6. Surface dressing failures

The majority of surface dressing failures fall into one of the following categories

(*a*) loss of chippings immediately after laying or during the first few days
(*b*) loss of chippings during the first winter
(*c*) loss of chippings in later years
(*d*) bleeding during the first hot summer
(*e*) fatting up in subsequent years.

The type and rate of application of the binder and size of chippings has, of course, a large influence on performance; incorrect application rates can result in failures of any of these types and is one of the most frequent causes of failure. Every care should be taken to ensure that the selected application rates are maintained throughout the surface dressing operation.

Very early loss of chippings may be due to slow break in the case of emulsions, or due to poor wetting in the case of cutback bitumens. The use of high binder content cationic rapid setting emulsions (more than 70% binder) largely overcomes the former problem and the use of adhesion agents and/or precoated chippings, the latter.

Many surface dressing failures are first seen at the onset of the first prolonged frosts when the binder is very stiff and brittle. These can normally be attributed to a combination of inadequate binder application rate, the use of too large a chipping or inadequate embedment of the chippings into the old road due to the surface being too hard and/or there being insufficient time between applying the dressing and the onset of cold weather. Inadequate binder application and/or the use of excessively large chippings will exacerbate this problem. Loss of chippings in the long term usually results from a combination of low durability binders and, again, poor embedment of chippings.

Bleeding may occur within a few weeks on dressings laid early in the season, or the following summer on dressings laid late the previous year. It results from the use of binders of too low viscosity for the high ambient temperatures prevailing, or from high temperature susceptibility binders such as road tar or some tar-bitumen binders.

Fatting up is complex in nature and may result from any or a combination of the following factors.

(*a*) Excessive application rate.
(*b*) Gradual embedment of chippings into the road surface causing the

relative rise of binder between the chippings to a point where the chippings disappear beneath the surface. This is largely dependent on the intensity of traffic, particularly on the proportion of heavy commercial vehicles in the total traffic. The composition of the binder also affects the process; high solvent content cutbacks can soften the old road surface and accelerate embedment.

(c) Crushing of chippings. Most chippings will eventually split or crush under heavy traffic with the loss from the dressing of many small fragments. The binder-to-chipping ratio therefore tends to increase, thus adding to the fatting up process. A dilemma here is that many of the best aggregates for skid resistance are often the worst for crushing resistance.

(d) Absorption of dust by the binder. Bitumen binders of high durability tend to absorb dust that falls upon them. Soft binders can absorb large quantities of dust, thereby increasing the effective volume of the binder. This effect, coupled with chippings embedment, leads to the eventual loss of surface texture.

9.2.7. Safety

The following highlights some potential hazards associated with the handling of surface dressing binders, but should not be used as a substitute for health and safety information from individual suppliers or the Institute of Petroleum *Code of safe practice*[15] and the Road Surface Dressing Association *Code of practice for surface dressing*.[2]

Handling temperatures

As a result of their viscosity, cut-back grades of bitumen have to be heated so they can be sprayed through jets. The recommended handling temperatures for cut-back grades of bitumen are detailed in Table 9.3. The temperatures have been calculated on the basis of viscosity measurements and are supported by operating experience.

Cutback bitumens are commonly handled at temperatures well in excess of their flash point i.e. the temperature at which the vapour given off will burn in the presence of air and an ignition source. Thus it is essential to exclude sources of ignition in the proximity of cut-back handling operations by the use of suitable safety notices.

To minimize the risks from burns, fumes, flammable atmosphere and fire,

Table 9.3. Spraying temperatures for swirl and slot jets[1]

Grade of cutback binder (STV seconds)	Spraying temperature: °C	
	Swirl jets	Slotted jets
50	140 ± 10	115 ± 10
100	150 ± 10	125 ± 10
200	160 ± 10	135 ± 10

Table 9.4. Recommended handling temperatures for cutback binders[15]

Grade of cutback binder (STV seconds)	Temperature: °C		
	Minimum pumping*	Spraying†	Maximum safe handling‡
50	65	150	160
100	70	160	170
200	80	170	180

* Based on a viscosity of 2000 centistokes

† Based on a viscosity of 30 centistokes, conforms with the maximum spraying temperatures recommended in Road Note 39[1]

‡ Based on generally satisfactory experience of the storage and handling of cutback grades in contact with air, subject to the avoidance of sources of ignition in the vicinity of tank vents and open air operations

every operation should be carried out at as low a temperature as possible, compatible with efficient working, and always below the maxima given in Table 9.4. Some bitumen emulsions are applied at ambient temperatures, but the majority of high bitumen content cationic emulsions are applied at temperatures up to 85°C.

Operation of surface dressing sprayers using cut-back bitumens

The sprayer should be located on level ground before being heated. The flame tube heater can be used only if a minimum of 150 mm of bitumen covers the heater tubes, and must not be used during product transfer or spraying. The heat should be extinguished and left to cool for at least 30 min before spraying. During heating the operator should remain in attendance to ensure that the bitumen is not overheated and there are no operations in the vicinity which might release cut-back vapours before the flame tube is cool.

To minimize fire risks and avoid excessive fuming, the temperature of the cutback in the sprayer should be kept as near as possible to the optimum for the grade and should never exceed the maxima given in Table 9.4. During spraying, clear warning signs against smoking and naked flames should be displayed at appropriate points on site, visible to the general public.

Operation of surface dressing sprayers using bitumen emulsions

The same general handling precautions apply as to cutback bitumens. The two major differences are that bitumen emulsions are normally heated to 56–85°C. A maximum temperature of 90°C is recommended because overheating to temperatures approaching 100°C can produce 'boil-over' due to the presence of water. Flammable atmospheres are not generated by bitumen emulsions at normal working temperatures. It is therefore less important that flame tube heaters are thoroughly cooled before product transfer or spraying.

Precautions, protective clothing and hygiene for personnel

When spraying either emulsion or cutbacks all operators should wear protective outer clothing i.e. eye protection, clean overalls and impervious shoes and gloves, to protect against splashes and avoid skin contact. Operators working near the spraybar should wear orinasal fume masks. For work alongside open highways, operators should wear clean high visibility jackets and reflective patches.

Although bitumen emulsions are applied at relatively low temperatures protection is essential because emulsifiers are complex chemical compounds and prolonged contact may result in allergic reactions or other skin conditions. Barrier creams, applied to the skin before spraying, assist in subsequent cleaning should accidental contact occur. However, barrier creams are not adequate substitutes for gloves or other impermeable clothing. If any bitumen comes into contact with the skin, operatives should wash thoroughly as soon as practicable and always before going to the toilet, eating or drinking.

9.2.8. The future for surface dressing

The importance of skid resistance in reducing accident potential was publicly acknowledged in January 1988 by the Minister for Roads and Traffic, when he announced that the agents who maintain motorways and trunk roads would be continually monitoring levels of skid resistance, to ensure that mean summer SFCs on these high-speed roads do not fall below minimum investigatory levels.

The need to maintain high skid resistance and the intense pressure on local authorities to obtain value for money, provide increased technical and economic justification for increasing the use of surface dressings on motorways. The advances and innovations in surface dressing technology which have been achieved in the last ten years will need to continue if surface dressing on these high profile sites is to become commonplace. The importance of the design of surface dressings, the development of polymer modified binders and new chipping techniques, improved training of operatives and greater attention to aftercare are vital if the industry is to break down the resistance in some quarters to the use of surface dressing on high-speed roads and make the process more acceptable to motorists. Specifying authorites for their part must understand the need for higher initial costs, to implement improved standards of supervision and application techniques and to encourage further innovation of materials and plant, so improving long-term performance and effectively reducing whole-life costs.

9.3. Slurry sealing

Slurry sealing was first introduced into the UK in the mid 1950s to stop runway deterioration on military airfields. It was not until the introduction of quick setting cationic emulsions that its use became widespread on highways. Two grades of slurry seal are included in BS 434[16] and the Department of Transport Specification for highway works.[17] One grade is laid at a nominal 1.5 mm thickness and the other is laid at a nominal 3 mm thickness.

The main purpose of slurry sealing is to seal the road surface from the ingress of air and water and arrest surface disintegration. The material also fills minor surface depressions providing a more even riding quality and improving the appearance of the road surface. The texture depth achieved with slurry seals is relatively low and therefore the material is not suitable for roads carrying high-speed traffic. Slurry sealing only provides a thin veneer treatment to an existing paved area, thus it does not add any significant strength to the road structure nor will it be durable if laid on an inadequate substrate. The process has mainly been restricted to lighter trafficked situations such as housing estate roads, footways, light airfield runways, car parks. Specially designed slurry seals providing a higher texture depth have been used on motorway hard shoulders.

The mixed slurry comprises aggregate (crushed igneous rock, limestone or slag) an additive and a bitumen emulsion. The additive is usually Portland cement or hydrated lime which is added to the mix to control its consistency, setting rate and mix segregation. It normally represents about 2% of the total aggregate. Where a rapid set is required to permit early trafficking or to guard against early damage from wet weather, a rapid setting anionic emulsion, class A4 or cationic emulsion class K3 are used. Where setting time is not critical a slow setting anionic emulsion, class A4 can be used. Depending on the characteristics of the aggregate, between 180 and 250 l of emulsion/t of aggregate are required. The composition of the cured material is about 80% aggregate, 20% bitumen.

Slurry seals are normally applied by specialist continuous-flow machines which meter and deliver aggregate, emulsion, water and additives into a mixer which continuously feeds a spreaderbox towed behind the machine. These machines are capable of laying up to $8000 \, m^2$/day. Depending on the composition of the slurry, light rolling with a pneumatic tyred roller may be required. Rolling is applied as soon as the material has set sufficiently to avoid rutting.

Recent developments in binder technology using polymer modification are now rapidly changing the scope and use of slurry sealing and further development work is allowing larger aggregate and thicker layers. Aggregate grading up to 11 mm has been tried and thicknesses up to 20 mm have been successfully laid. The use of fibre reinforcement is also being developed and it seems likely that this technique will help to overcome some of the problems of heavily-cracked surfaces.

9.4. Thin and ultra-thin asphalt layers

Asphalt layers between 3 mm and 35 mm thickness are generally described as thin asphalt layers; and are used for both new road construction and maintenance. They have mainly been used to improve road profile by filling in longitudinal rutting and sealing surface cracks. More recently they are being used as thin wearing courses to overcome level problems in road maintenance i.e. bridges and edge detail especially in urban roads, and motorways. In comparison with surface dressings they can improve profile, considerably reduce the change of flying chippings, they are less noisy and can be laid during a wider range of climatic conditions throughout the year.

SMA

The material was developed in Germany about 25 years ago but not included in the national specification until 1984. It is now also extensively used in Scandinavia, the Netherlands and France. The material is typically made with

11 mm nominal size aggregate but in some cases 5 mm, 8 mm and 16 mm stone is used. The coarse aggregate used must also satisfy test limits for abrasion resistance, flakiness index and impact crushing strength. The filler fraction is in the range 8−13% of the aggregate weight and the sand fraction is in the range 12−17%. The binder used is generally 65−80 pen and cellulose or mineral fibres are often added during mixing to improve stability and prevent binder drain-off. Polymer modified binders have also been used and typical binder contents are in the range 6−7%.

SMA is usually designed with an air voids content of 3%. Compaction on site is needed primarily to orientate the aggregate exposed in the upper part of the pavement layer, and little actual densification occurs under the roller. Because the mastic receives relatively little compaction during construction it is rich in binder to ensure the low void content needed for satisfactory durability. The expected life of SMA surfacings on heavily trafficked roads is 10 to 15 years.

The stone skeleton of SMA must accommodate all the mastic while maintaining point to point contact of the large aggregate. Too much mastic will separate the coarse aggregate, leading to a pavement layer susceptible to deformation. Too little mastic will give unacceptably high voids leading to accelerated ageing and moisture damage. Very little latitude is therefore permissible in the production control of the aggregate gradation, binder content or fibre content if used.

It seems likely that the excellent reputation that this type of material now enjoys in many parts of Europe will lead to its future use within the UK, especially as the performance-based specifications currently being prepared are brought into use.

Hot-mix thin surfacing

Hot-mix thin surfacing is a relatively new technique that has been used in France since 1988 and has recently been introduced into the UK market by two large surfacing contracting companies. It consists of a layer of hot-coated aggregate applied over a sprayed layer of binder by a specially designed machine. Safepave is a proprietary hot-mix thin surfacing (Fig. 9.14).

The machine consists of a self propelled chassis designed to carry the following components

- (a) a hopper to receive the coated aggregate
- (b) an aggregate storage chamber
- (c) a binder storage tank
- (d) a spray bar sited immediately before the screed unit
- (e) a screed unit.

The binder is generally polymer-modified bitumen emulsion and the pre-coated aggregate contains 20−25% mastic type material. The grading of the aggregate is similar to a 10 mm porous asphalt grading and with a binder content of 5.2% ±0.2%, it is a workable and durable mix. The screed reorientates the aggregate and the resultant surface is much quieter than a normal surface dressing.

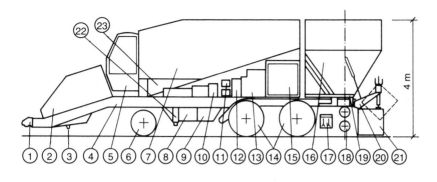

1	Towing hook	9	Fuel tank	17	Spray bar
2	Hopper	10	Electronic control panel	18	Auger
3	Tractor coupling	11	Hydraulic pump	19	Mix gate
4	Chassis	12	Gear box	20	Screed arm
5	Driver's cab	13	Engine	21	Screed
6	Front axle	14	Rear axles	22	Screw conveyor
7	Emulsion tanks	15	Engine radiator	23	Hydraulic lifting
8	Hydraulic tank	16	Holding hopper		ram

Fig. 9.14. Diagram of a machine for laying 'Safepave'. Diagram courtesy of Associated Asphalt Ltd

9.5. Asphalt pavement recycling

The 1973 oil crisis increased the costs of energy, bitumen and aggregate dramatically and made the recycling process a potentially attractive, indeed necessary, proposition. Significant research and development work produced a considerable number of technically viable techniques. These used between 10% and 100% reclaimed materials in the final product which could be used for pavement maintenance and rehabilitation. During the 1980s the economic drive was reduced as the price of oil products dropped towards their pre-crisis values. Although in real terms the price of both bitumen and aggregate have remained fairly constant over the last decade increasing ecological forces have renewed the interest in recycling.

It is now becoming clear that the over-riding factors in the economies of recycled versus virgin material are the penalties imposed by the road authorities on the disposal of old material coupled with any incentives that may be given for the re-use of the asphalt pavement. It would appear that without these penalties or incentives the use of virgin material is more economic in most countries, certainly those with reasonable or good access to stone which is certainly the case in the UK. Environmental considerations such as concern about landscape affected by opening new quarries and the effects of global warming will probably lead in the longer term to incentives for both hot and cold recycling.

There are few technical reasons against recycling as the performance of recycled materials can, with correct design and operation, virtually equal that of virgin material. Recent evidence suggests that within the recycling process

Fig. 9.15. The 'Repave' process

the emphasis is shifting from the recycling of the binder to the recycling of the aggregate. During the recycling process the opportunity can be taken to introduce an upgraded or modified binder which in turn can enhance the performance of the recycled material.

Recycling of bituminous materials can be broadly divided into two categories.

9.5.1. In situ recycling

In the UK in situ recycling can be further subdivided into three processes known as repaving, remixing and retread. Repaving and remixing are hot in situ recycling processes, whereas the retread treatment is a cold process. The repave process (Fig. 9.15), which was introduced into the UK in 1975, involves heating and scarifying the existing surface to a depth of about 20 mm. Approximately 20 mm of new asphalt is laid directly on the hot scarified material and the material compacted. These operations can be completed by a single pass of a purpose-built machine. As both layers are hot when compacted, a good bond is achieved between the recycled material and the new material.

In the remixing process a hot scarified material is mixed with new material in the pugmill mixer of a machine and the blended mix is paved and compacted on the scarified surface. A report on the surface characteristics of roads using the repave process has been published by the TRL,[18] and the Department of Transport issued a Standard and Advice Note on the repave process in 1982.[19,20]

A cold in situ procedure known as retread has been established UK practice for over 30 years. This process consists of scarifying the existing road to a depth of approximately 75 mm, breaking down the scarified material to the required size and reshaping the road profile. The material is then oversprayed two or three times with a bitumen emulsion. After each application of emulsion the material is harrowed to distribute the emulsion. When harrowing is complete the material is compacted with an $8-10$ t dead weight roller and finally the surface is sealed with a surface dressing. This process has been successfully carried out on minor roads for many years.

9.5.2. Off-site or plant recycling

In this process material removed from the surface of an existing road is transported to a hot mix plant where it may be stockpiled for future use or processed immediately. Both batch and continuous plants have been successfully converted to produce recycled mixes, using a range of methods which heat the reclaimed material before mixing with correctly designed plant. Recycling has been carried out without excessive fuming and blue smoke emissions during manufacture, and performance tests on asphalt containing a proportion of recycled material has shown identical properties to mixes using virgin components.

Recycling in batch plants is achieved by superheating the virgin aggregate and then adding the material to be recycled either immediately after the drier or directly into the pugmill. Heat transfer takes place during the mixing cycle. Although this method successfully minimizes the problem of blue smoke, it entails keeping aggregate in the heating drum for longer and consequently the maximum output from the plant is reduced. As it is not possible to obtain adequate heat transfer with high percentages of recycled material, it is widely accepted that the maximum quantity of recycled material that can be added is between 25% and 40%.

Other disadvantages using batch plants for recycling are high heating costs and accelerated wear and tear on the plant and dust collectors due to the higher manufacturing temperatures. Recycling using continuous mixers involves introduction of the reclaimed material into the drum itself. During early trials using this type of plant, environmentally unacceptable blue smoke was produced. Blue smoke is produced when vapourized bitumen condenses. The condensate takes the form of particles too small to be removed by conventional emission control equipment and so escapes through the stack.

Bitumen vapourizes at about 450°C and as gas temperatures in the drum of a continuous mixer can reach 2000°C the introduction of reclaimed material has to be carefully controlled. By various modifications continuous mixers have been designed which can take up to 60% recycled material, without exceeding statutory pollution standards. The entry point for the recycled material is approximately half-way down the drum and the flights in the drum are modified to produce a homogeneous mix of virgin and recycled material before the addition of the bitumen and filler. The re-designed flights also shield the recycled material from the intense radiant heat from the heating unit.

Recycling in the UK is still in its infancy and although plant developments have enabled significant quantities of reclaimed material to be successfully recycled and full-scale road trials and laboratory evaluations of these materials are encouraging, the long-term performance of recycled surfacings has yet to be established.

9.6. Mastic asphalt

Mastic asphalt consists of a mortar of bitumen and fine aggregate and a proportion of coarse aggregate. Full details of the basic raw materials are given in BS 1446[21] and BS 1447.[22] The majority of the mastic asphalt laid in the UK is manufactured to BS 1447; material to BS 1446 is seldom used because of the need to import natural rock asphalt. To comply with BS 1447 the fine aggregate has to be naturally occurring limestone rock with between 40–55% passing the 75 μm sieve and no more than 3% retained on the 2.36 μm sieve. The percentage of coarse aggregate added varies depending on the application. Normally the aggregate is mixed with either a 25 pen or 15 pen bitumen.

The manufacture of mastic asphalt is complex. It can be carried out in a single process only when the material is to be used immediately; in such cases all the aggregate is combined with the bitumen in a large mixer equipped with slow-moving blades. Otherwise, only the fine aggregate is mixed with the bitumen, producing a mortar which is cast into 25 kg blocks and left to cool. These blocks can be stored and supplied to sites when required. They are then re-melted in a special mastic mixer, together with the desired proportion of coarse aggregate.

Mastic asphalt is normally hand-laid by skilled and experienced asphalters who use wooden floats to work the material at temperatures of between 175°C and 230°C until it has stopped flowing. Thicknesses may vary between 20 mm and 50 mm, according to the intended end use. Its impermeability is assured by the fact that the void content of the laid material is generally below 1%. These stringent requirements — and the relatively small areas normally laid at any one time — have in the past ruled out machine-laying. Large areas, such as on the Humber Bridge, have been recent exceptions (the routine machine-laying of mastic asphalt on German roads is not strictly comparable since the German *gussasphalt* mix differs from mastic asphalt to BS 1446[21] and BS 1447.[22]

The high percentage of fines in mastic asphalt gives a smooth surface with poor skid-resistance properties. To increase the latter, a variety of treatments may be applied. Rolled-in pre-coated chippings are used for busy road surfaces in the same way as with hot-rolled asphalt: the chippings are applied while the material is still plastic enough to allow partial but secure embedment and are rolled in using either hand or light power-rollers. On footways and other lightly-trafficked surfaces, the desired finish may be obtained by applying a special sand, using the floats. Another method, particularly used for multi-storey car park surfaces and footpaths, is to crimp the surface with an indentation roller while it is still hot.

9.7. Coloured surfacings

The appearance of traditional bituminous surfacings, especially when finished with the normal contrasting white lines, is generally pleasing to the eye. In some locations a specific colour is desired and there are a number of ways that this can be achieved

(a) incorporating coloured pigments into the mix during manufacture
(b) application of suitably coloured chippings to the surfacing during laying

(c) application of a coloured surface treatment after laying
(d) using a conventional bitumen with a coloured aggregate
(e) using a suitably coloured aggregate with a translucent binder

9.7.1. Incorporating coloured pigments

The majority of coloured mixes are manufactured by adding pigment, usually iron oxide, during the mixing process. Appropriately-coloured aggregates are used to ensure that, when aggregate is exposed after trafficking, the overall appearance of the materials is maintained. The main drawbacks to colouring mixes using conventional bitumens are that the only acceptable colour that can be achieved is red and the quantity of iron oxide required to achieve an acceptable red is fairly high, increasing the cost of the mix substantially.

To enable mixes to be pigmented in colours other than red, binders have been developed specifically for pigmenting. These are synthetic binders which, because they contain no asphaltenes are readily pigmented in virtually any colour. They possess rheological and mechanical properties similar to conventional bitumen. Coloured mixes manufactured using these binders require $1-2\%$ of pigment to achieve a satisfactory colour, whereas only red can be achieved with conventional bitumen, using typically $5-6\%$ iron oxide addition.

The Department of Transport recently introduced a novel application for coloured mixes in Clause 2003 of the standard specification.[17] This is the application of red-pigmented mixes, usually sand carpet, as a protective layer over waterproof membranes laid on bridge-decks. When the wearing course on the bridge-deck needs replacing, contractors planing off the old wearing course will only plane down to the red sand carpet, so avoiding damage to the expensive waterproof membrane.

9.7.2. Application of coloured chippings

Hot-rolled asphalt and fine-graded macadam (previously termed fine cold asphalt) can have decorative coloured chippings rolled into the surface during compaction. Hot-rolled asphalt is suitable for most traffic conditions whereas fine-graded macadam is only appropriate for lightly trafficked areas, private drives and pedestrian areas.

To provide a decorative finish to these types of surfacing, pigmented bitumen-coated or clean resin-coated chippings can be applied during laying. The bitumen or resin coating promotes adhesion to the surfacing and it is not recommended that uncoated chippings are used. However, in the case of fine-graded macadam laid on areas subjected to little traffic, a light application of uncoated chippings e.g. white spar, produces an attractive finish if they are uniformly applied and well embedded. Decorative chippings cannot be successfully rolled into the surface of high stone content mixes such as dense macadams or high stone content rolled asphalts.

9.7.3. Application of a coloured surface treatment

Three surface treatments can be applied to a bituminous wearing course to provide a decorative finish. These are

(a) pigmented slurry seals

(b)   surface dressing
(c)   application of a coloured paint.

Pigmented slurry seals are available as proprietary products in a range of colours. They are applied in a very thin seal (about 3 mm) and are only suitable for pedestrian and lightly trafficked areas. Surface dressings are suitable for most categories of road application but are less suitable for pedestrian areas. The final colour of the surface dressing will be that of the aggregate used. Proprietary coloured paints are available for over-painting conventional black wearing courses. These are only suitable for pedestrian and games areas e.g. tennis courts.

9.7.4. Coloured aggregate bound by a conventional bitumen

When a conventional bitumen is used, the depth of colour achieved depends on the colour of the aggregate itself, the thickness of the binder film on the aggregate and the rate at which the binder exposed on the road surface is eroded by traffic. In medium and heavily trafficked situations, the natural aggregate colour will start showing through fairly quickly, but in lightly trafficked situations, where coloured surfacings are usually specified, the appearance of the aggregate colour may take considerable time.

9.7.5. Coloured aggregate bound by a translucent binder

Several proprietary macadams, in which the binder is a clear resin rather than a bitumen, are available. By selecting appropriately coloured aggregates a range of colour surfacings can be manufactured. The major advantage of this type of system is that the colour is obtained immediately.

References

1. DEPARTMENT OF TRANSPORT. *Design guide for road surface dressing.* HMSO, London, 1992, Road Note 39.
2. ROAD SURFACE DRESSING ASSOCIATION. *Code of practice for surface dressing.* Road Surface Dressing Association, Leeds, 1990.
3. SALT G. F. and SZATKOWSKI W. S. *A guide to levels of skidding resistance for roads.* Transport and Road Research Laboratory, Crowthorne, 1973, LR 510.
4. DEPARTMENT OF TRANSPORT. *Skidding resistance of in-service trunk roads.* DTp, London, Dec., 1987, HD 15/87.
5. DEPARTMENT OF TRANSPORT. *Skidding resistance of in-service trunk roads.* DTp, London, 1987, HA 36/87.
6. ASSOCIATION OF COUNTY COUNCILS on behalf of the LOCAL AUTHORITIES ASSOCIATION. *Highway maintenance. A code of good practice.*
7. BRITISH STANDARDS INSTITUTION. *Method for determination of the polished stone value.* BIS, London, 1989, BS 812: Part 114.
8. CATT C. A. An alternative view of TRRL research into skidding resistance. *J. Inst. Asphalt Techology*, London, July 1983, No. 33.
9. DEPARTMENT OF TRANSPORT. *Specification requirements for aggregate properties and texture depth for bituminous surfacings to new roads.* DTp, London, 1976, H 16/76.
10. BRITISH STANDARDS INSTITUTION. *Sampling and examination of bituminous mixtures for roads and other paved areas. Methods of test for the determination of texture depth.* BSI, London, 1990, BS 598: Part 105.

11. BRITISH STANDARDS INSTITUTION. *Specification for single-sized aggregate for surface dressing.* BSI, London, 1987, BS 63: Part 2.
12. BRITISH ROAD FEDERATION. *Basic road statistics.* BRF, London, 1989.
13. HOAD and HOBAN T. W. S. A cost performance relationship for surface dressing. *Road Surface Dressing Association symposium*, 1982.
14. SOUTHERN D. Premium surface dressing systems. *Shell Bitumen Review*, March 1983, No. 60.
15. INSTITUTE OF PETROLEUM. *Bitumen safety code. Model code of safe practice.* Inst. Petroleum, London, July 1990, Part 11, 3rd edn.
16. BRITISH STANDARDS INSTITUTION. *Bitumen road emulsions (anionic and cationic). Code of practice for use of bitumen road emulsions.* BSI, London, 1984, BS 434: Part 2.
17. DEPARTMENT OF TRANSPORT. *Specification for highway works.* HMSO, London, 1992, **1**.
18. COOPER D. R. C. and YOUNG J. C. *Surface characteristics of roads surfaced using the repave process.* Transport Research Laboratory, Crowthorne, 1982, Suppl. R744.
19. DEPARTMENT OF TRANSPORT. *In situ recycling — the repave process.* DTp, London, Dec., 1982, HD 7/82.
20. DEPARTMENT OF TRANSPORT. *In situ recycling: the repave process.* DTp, London, 1982, Advice Note HA 14/82.
21. BRITISH STANDARDS INSTITUTION. *Specification for mastic asphalt (natural rock asphalt and fine aggregate) for roads and footways.* BSI, London, 1973, BS 1446.
22. BRITISH STANDARDS INSTITUTION. *Specification for mastic asphalt (limestone fine aggregate) for roads, footways and pavings in buildings.* BSI, London, 1988, BS 1447.

10. Problems, failures and their causes

10.1. Introduction

All pavements ultimately fail but application of the appropriate treatment at the correct time can significantly prolong their life. One definition of engineering failure is that a characteristic falls below some threshold of acceptability.[1] An example would be a wearing course which is regarded as having failed when its skid resistance, as defined by SFC and measured by SCRIM, falls below the critical value for that part of the network. It may well be the case that other characteristics such as the structural strength or the degree of cracking or rutting have not reached critical levels. The current design methods in the UK are based on the assumption that the structural layers on a pavement will have a life of 40 years; with enhancement by overlay after 20 years. The wearing course will last at most 20 years before rutting excessively but will probably need a surface treatment such as surface dressing to restore skid resistance, and seal cracks during those 20 years.

This chapter considers symptoms which warrant consideration for maintenance treatment and some possible causes.

10.2. Problems during construction

10.2.1. General

Problems can occur with bituminous mixtures during the mixing, laying and compaction operations. Care should also be exercised during preparation of the specification to avoid requirements which may prove difficult to execute and may jeopardize the integrity of the pavement for dubious benefit, e.g. excessive texture depth in urban situations and roundabouts; too-high stability wearing course particularly in winter; traffic management requirements. The client's over-riding criterion must be that the pavement, after construction, will provide the required service life.

It is likely that the European initiative on standardization (as discussed in chapter 11) and the American SHRP will require achievement of end-product performance criteria for the layers in a pavement.

Current methods, largely based on recipes for the material and compliance with laying instructions, can blur the responsibility for the achievement of a satisfactory pavement.

It is generally assumed that if the contractor has produced material using any of the permitted ingredients; mixed them in the correct proportions within specified tolerances and the permitted temperatures; delivered them to site;

laid and compacted them, and finally opened the carriageway to traffic when they have cooled; all as described in the contract, then the performance of the layer will be acceptable to the client.

There have been situations where additional requirements have been placed on the laying contractor which have meant that the material supplier did not have all the normal tolerances available to him. Examples of this are where there are particular requirements related to air voids or binder content or there are stipulations about the grading envelope or proportions of constituents.

Contractors may also be expected to have some knowledge of the possible poor performance, under certain circumstances, of local aggregate sources. Where, for example, the local materials are flint gravels or acidic granites, the contractor may have to take steps to adjust mix designs or include additives to ensure a satisfactory performance of the material. The extent to which contractors should do this unprompted may lead to dispute where premature failure of a pavement layer occurs.

To assist suppliers, the Department of Transport is to introduce a mix design method based on air voids, which should ensure that the possibility of unstable macadam mixtures is avoided whilst maintaining adequate binder for durability.

This part of the chapter identifies a number of site problems, how these problems show themselves and the timing i.e. whether before laying, during laying or during compaction and suitable remedies. The reason for the problem is frequently an interaction of factors because, generally, bituminous mixtures are surprisingly tolerant of errors in site practice. It is assumed that bituminous mixtures are laid by a paver. Hand-laying cannot achieve the same level tolerances and degree of compaction. A major source of urban pavement failure is due to the intervention of public utilities which use hand-laying because of the limited scale of their operation. *The specification for the reinstatement of openings in highways*[2] should help reduce this problem.

10.2.2. Poor levels and mat shape
Observed defect: undulating mat surface across and/or along the mat

If the final product is not deemed to be satisfactory, there could be many causes e.g. the constituent parts of the mix could be unsuitable, the base could be poorly shaped, parts of the paver could be out of adjustment or the compaction may be at fault.

Causes before paver

Each layer in a pavement has a smaller level tolerance than the layer upon which it is laid. A paver is capable of removing about one half of the variation in the shape of the underlying layer so it is essential that all materials are laid within the specified tolerances. Roadbase is often laid in two layers and a contractor would be wise to monitor levels at the top of each intermediate layer even though there is no specified tolerance on levels. Any patches or potholes should be filled and compacted in layers and before paving begins.

The layer thickness should be one and a half to five times the nominal stone size for the aggregate in the mixture if good compaction is to be achieved. A regulating course is frequently used to ensure that the basecourse or wearing course has a uniform layer thickness. Where an overlay is required, regulation should take place as low in the pavement as possible to minimize the stresses on the regulation material. Where new works are under construction, regulation should not be necessary and is undesirable. In laying a regulation course the requirements for stone size above must be observed. If this is not possible, the existing layer may have to be planed so that a uniform thickness of material can be laid.

Generally, the smaller the stone size, the less stability a mixture has. Thin layers do not compact well and may not bond well to the underlying mat. The use of a tack coat i.e. spraying the layer with emulsion before laying, may provide the necessary bond but if done to excess, may lead to the introduction of a slip layer.

Specifying excessively thick lifts can cause difficulties in achieving surface tolerance after rolling.

Material temperatures must be reasonably constant because the paver screed will react to materials depending on their viscosity. A cold load followed by a hot load will inevitably produce different levels. Such materials, although technically to specification, can exhibit substantially different compaction characteristics. This can quickly be seen by the condition in which it arrives on site. For example, one truck load when unsheeted will appear perfectly flat, the next would be convex in appearance. The result will be different mat profiles when laid.

The introduction of Quality Assurance schemes at the plant, which frequently demand recording of the load temperature taken from the plant computer or taken manually close to the weighbridge and recorded on the delivery ticket, has resulted in more consistent production.

The mixing plant, the paver and the roller form a continuous process which acts upon the bituminous mixture; consistency of mixture in all its aspects will produce a consistent final product.

Causes at paver

Material delays cause inconsistency in density which, after rolling, will result in poor level consistency. As far as possible, the paver speed should match the truck deliveries. Delivery frequencies should be matched to the optimum laying speed of the paving operation. If the paver has to stop, the paver screed has to be lifted and a transverse joint formed by cutting back. Even the best surfacing crews will occasionally run into joint problems and these form the most frequent causes of poor levels and also of other faults as described later. Transverse joints are discussed in section 10.2.1.

As material is delivered, the sides and top of the loads chill; this effect is reduced by using insulated trucks and double sheeting. The sides of the loads invariably go into the sides of the hopper and are subject to further chilling. Most paver manufacturers offer insulated hoppers and this option is a small but worthwhile investment. The material in the hopper should be brought into the centre of the paver regularly by lifting the hopper wings. It will then mix with material at high temperature. The existence of cold material at the start or end of a run or as a result of delivery delays is a significant factor in poor ride and mat shape at joints.

The paver should move at constant speed without stopping. Maintenance of a constant pressure in front of the screed is important in maintaining a constant speed. Equipment is now available which provides a buffer in front of the paver. It holds approximately 50 t of material which is fed into the paver hopper between truck loads.

Poor levels resulting from paver malfunctions may result from many causes and a number of remedies are available. Poor levels result from more factors than any other type of failure.

The most likely cause of an undulating surface results from variation in the head of material, as shown in Fig. 5.6. This expression refers to the material in and around the augers, immediately in front of the screed. Asphalt pavers incorporate a free floating screed principle i.e. the screed is not held in a rigid position relative to the tractor part of the paver. Therefore, pressures bearing on the front of the screed are critical. If the pressure increases as a result of a rise or fall in the head of material against the screed, the complete screed will react accordingly and the results will be evident in the mat in the form of level variation.

Screed controls can be faulty. Electrical problems can cause hydraulic malfunctions; again poor levels will result. On most pavers, the hydraulic valves can be manually operated in an emergency. If the fault cannot be repaired quickly, the hydraulically operated tow point can be manually operated.

Automatic level control mechanisms can cause level problems. Care must be taken to ensure that the setting is not over-sensitive which may cause hunting i.e. frequent and erratic movement of the tow point. The sensor itself may be faulty and should be disabled if suspect.

Excessive lateral joint overlap causes the screed to be rigid on one side. The overlap should not exceed 25 mm to 50 mm. If it does, the screed width should be reduced by removing screed extensions accordingly. Most modern machines have hydraulically extending screeds the width of which are adjusted by the operator.

Bolts used to retain the screed on the sidearms in a positive rigid condition, will result in poor levels if loose as the screed will tend to rock in sympathy with the forward movement of the tractor unit. The bolt holes may become elongated. If this occurs, a workshop repair should be undertaken to correct it.

Bumping is a common operational fault and is caused by lorries reversing onto the paver and pushing it backwards. The progress of the paver will be impeded if the lorry brakes are not released. When laying up steep inclines or on geotextile fabric stress-absorbing membrane, traction is difficult to maintain and it may be necessary for the truck driver to drive his vehicle forward at a speed which matches that of the paver thus minimizing the required traction force.

Causes after paver

As the material leaves the paver as a planar uniformly-compacted mat, an inexperienced roller driver can destroy the transverse and longitudinal profiles. In general, the roller should work as close to the paver as possible where the mix is most workable, provided the mat does not shove or crack, and a smooth finish will be achieved with the minimum number of passes.

Vibrating rollers should have the vibrating mechanism switched off when the roller is stationary, moving off or changing direction. Where fitted, automatic speed control mechanisms should be used. At the end of a pass, a slight change in direction will minimize the possibility of roller marks. Steering direction should only change on compacted material. Constant roller speeds should be maintained. On sensitive mixes including polymer-modified mixtures, one or two static passes before engaging vibration improves level control. Each pass should finish closer to the paver than the adjacent one and there should be an overlap of 100–200 mm.

The adoption of rolling patterns ensures that each area across and along the mat is given the same number of passes. Rolling round sharp bends such as roundabouts requires a skilful combination of vibration, steering and roller patterns to maintain a good transverse profile.

10.2.3. Material insufficiently compacted

Observed defects: density meter readings too low, core densities too low, voided material observed in the cores, excessive texture depth. Long term effects: considerably reduced material performance and pavement life.
Causes before paver

The temperature of the mixture is of primary importance in determining its compactability. Hot bitumen has a low viscosity and thus coats the aggregate during mixing. It also acts as a lubricant during placing. As a bituminous mixture passes through the paver, it loses about 20°C in temperature; the rate of subsequent cooling is determined by the layer thickness, the surface and air temperatures, wind speed, the temperature and rate of application of pre-coated chippings (if any), the penetration of bitumen and the quantity of air voids in the mixture; open textured mixtures cool faster than dense mixtures. The relative influence of each of these factors is discussed in detail in chapter 7.

The amount of coarse aggregate, its particle shape and surface roughness have a significant effect on the compactability (workability) of a mixture; this is most noticeable in DBMs with crushed rock fine aggregate. Adequate, but not excessive, lubricant, i.e. hot bitumen, is required. Sand and gravel mixtures with their rounded shape and smooth surface and mixtures with limestone aggregate, can be compacted significantly more easily than crushed rock material, probably even with deadweight rollers. The structural strength (stiffness modulus) and deformation resistance of sand and gravel mixtures is less. Modern vibrating rollers can achieve high densities with all types of aggregate.

Engineers designing roads where winter working is planned should, if possible, ensure the maximum layer thicknesses possible to maximize heat retention. Wearing courses 45 mm or 50 mm thick rather than 40 mm thick are recommended for this reason. This is not a panacea as a skin of relatively cold stiff material can rest on a mobile mass which may result in surface cracking during rolling. Maximum layer thicknesses of 200 mm in roadbases are recommended for this reason. End product compaction can be specified in two ways, PRD[3] and air voids.[4] The value of the former is not materials-dependent, although achievement is. The latter depends upon the grading curve of the material, for example a well-compacted close-graded DBM will have

2%−5% voids while a well-compacted porous asphalt will have 18%−24% voids.

The PRD measures the density of a 150 mm core taken from the mat, compared with its density after reheating and compacting to refusal in a mould by a vibrating hammer. To achieve a uniform compaction of 95% PRD, precision exercises showed that an absolute minimum of 93% PRD is required and this is specified in the Department of Transport's *Specification for highway works*.[5]

One of the main drawbacks to PRD is that a full PRD cycle takes three to four days with a staff time in excess of seven hours for six specimens. This can be overcome on site by the use of nuclear density gauges.[6] In order to be accurate, each gauge must be calibrated for a particular material type. Clients and contractors should seek evidence of this, particularly when independent test houses are being used.

Air voids are an important factor in the structural performance and durability of dense materials, as discussed in chapter 2. They can be determined directly from cores taken from layer in accordance with ASTM D2041.[7] This method is known as the Rice method. The mixture is warmed and broken up into loose particles so that trapped air can escape. The mass of coated loose particles is then measured in air and water to determine the effective voidless density. The voids in the total mixture can then be obtained. Provided that the aggregates are well coated to prevent water absorption, or their water absorption is low, the test is simple and speedy. The test takes a day to carry out.

For contractual purposes, the air voids can vary with mix composition as well as compaction practices, which can lead to disputes between the parties to the contract where the requirements are not met. Where bituminous macadams to BS 4987[8] or porous asphalt to BS 594[9] are used, the supplier may need to carry out pre-contract trials to ensure any air voids requirements will be met.

Air voids can also be determined from the specific gravities of the constituents.[10] These are difficult to determine precisely, particularly for fine aggregates. The different aggregate fractions may come from different sources requiring the specific gravity for each source and for each fraction to be determined. This permits the specific gravity of the total mix to be determined. Any allowance for the absorption characteristic of the aggregate may not have the same value for bitumen as water. As a result, a considerable amount of pre-contract work is necessary.

Air voids of mixtures may be carried out on cores after analysis but this may take substantial amounts of time. Alternatively, but less accurately, the client and supplier may agree, before work starts, the maximum theoretical density for the mixture, at, say, mid-point grading and binder content. Achievement of this can be monitored by using a calibrated nuclear gauge. This method is speedy, removes the agreement of density from the area of dispute, sets a clear target for roller drivers and has the advantage to the client that the supplier will improve his chances of specification compliance by making the mixture more closely graded which will improve aggregate interlock, density and overall structural performance. Typical specifications require 2%−7% for DBMs and 2−5% for hot-rolled asphalts. A minimum level of voids is required to prevent instability during compaction and allow space for bitumen flow during long term traffic consolidation.

Causes at paver

Paver speed has a direct effect on the initial compaction imparted by the paver. Sufficient time must be given for the action of the tampers or vibrators (or a combination of both) to have an effect on the material. If the paver speed is too high, the tampers/vibrators will not make an appreciable difference. In the USA, paving speeds are two or three times as fast as European practice allows so vibratory screeds are the norm because mixes are generally easier to compact, maintenance costs are lower and they can produce satisfactory results under such conditions.

Pavers have either a tamping screed, a vibrating screed or a combination tamping/vibrating screed. It is the screed which imparts initial compaction to the hot material. The degree of compaction achieved is influenced by the material temperature. Where there are variations in the temperature of the material there will be variations in the level of compaction imparted.

Initial compaction is from the screed, so it is essential that the tamping/vibrating elements are maintained in good working order and properly adjusted. The tamper speed must be matched to the paver forward speed. The amount of protrusion of the tamper below the screed plate must be in accordance with the manufacturer's recommendations. The tamper extensions must be aligned with the main screed and not in opposition to them. Tamper edges must not be badly worn and tamper frames must be securely bolted together.

When vibrating screeds are used, optimum balance conditions must be maintained between frequency and amplitude. If either is too great, the screed will tend to bounce. If either is too slow, under-compaction and a poor surface will result. Vibrators must not be in phase.

The auger box i.e. the space between the screed and the tractor unit, must not be too large otherwise the material will cool and poor compaction will result. Poor compaction will also result if the screed is too wide; the overlap between adjacent rips should not exceed 25−50 mm.

Causes after paver

Modern pavers with heavy duty vibrating/tamping screeds can achieve almost all the compaction a thin layer (up to 50 mm thick) requires. The rollers smooth the mat, ensure good longitudinal joint bonding and embed pre-coated chippings in hot-rolled asphalt wearing course.

In thicker mats, the requirements for the rollers are more demanding and the size/performance of the roller must be matched to weather, layer thickness and output. Roller manufacturers produce charts for the performance of their rollers on commonly experienced materials as a result of field trials or feedback from users. An 8−10 t deadweight vibrating roller is able to compact 120 mm of hot dense macadam to low air voids in six passes. Over recent years, vibratory rollers, normally tandem rollers, have almost completely overtaken deadweight rollers for compaction operations. However, for finishing mats and thus obtaining good rideability characteristics, three-point rollers are unbeatable.

On airfield pavements and some other specialist uses, pneumatic tyred rollers or combinations of pneumatic tyred rollers and vibrating drum rollers are often used. Tandem vibrating rollers have operating weights in the range 1—12 t with heavier rollers having a choice of ampitudes for vibration. High amplitude settings up to 1 mm are used for thick lifts whilst low amplitude settings, 0.35—0.6 mm, are used for thinner lifts to avoid rippling, shoving and aggregate damage.

All rollers carry a reservoir of water to prevent the drums from sticking to the hot material but excessive water should not be used as it will cool the mat. The number of rollers provided should be adequate to ensure compaction can continue while one of the rollers is being refilled with water.

Again, the use of good rolling patterns, having 100—200 mm overlap, is essential to ensure that the required density level is achieved uniformly. When the required density is achieved rolling should cease to avoid the possibility of cracking the mat.

10.2.4. Inconsistency of density and texture
Observed defect: variations in density meter readings, core densities and texture readings, some voided material observed in the cores. Long term effects: local variations in material performance and pavement life.
Causes before paver

Poor control at the plant leads to variations in the temperature of mixed materials. A supplier who is manufacturing a wide range of materials will have to change mixing temperatures. This cannot be done instantaneously and a variable product will be produced. Hot storage capacity at mixing plants can assist in reducing fluctuations.

It is important that consistent sources of aggregate supply to the mixing plant are maintained as differences in achieved densities will occur because of variations in aggregate gradation and composition.

The contractor and client may wish to audit the Quality Assurance system of the available plants before accepting a particular supplier, particularly on larger contracts.

Causes at the paver

It is important that the head of material in the auger box is kept at a constant level. The optimum level is just above the auger shaft, with the top of the auger flights (or segments), visible above the material level, as shown in Fig. 10.1(a) and (b). The set of auger flights should be complete and unbroken otherwise distribution of the material in front of the screed is likely to be uneven. Modern pavers have automatic feed control mechanisms which sense the head of material and adjust the amount of material passed from the hopper.

Material will stick to the screed plates if they are insufficiently heated causing poor texture. Problems can also be caused if the tamper edge protrusion is too low, resulting in a build up of material behind the tampers. Excessive waiting between loads and material will inevitably result in loss of heat, inconsistent densities and produce variable textures in the finished mat.

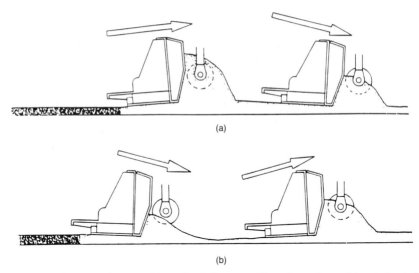

(a)

(b)

Fig. 10.1. Loading of augers: (a) overloaded augers will produce a hump (b) underloaded augers will produce a dip

Causes after paver

Consistency in operation of the rollers will lead to a consistent product. The roller driver must be sensitive to variations in temperature and compaction characteristics of the material. The subtle use of vibration and/or additional passes locally can offset these difficulties. Some rollers are provided with on-board compaction measuring devices based either upon the measured reaction of the layer to the vibrating force from the drum or using an on-board nuclear density meter. The former suffers from the fact that the roller response is dependent on density and temperature, the latter from the fact that the device cannot touch the top of the layer. This calls into question the accuracy of these devices but may provide additional guidance to the roller driver. They are most valuable on large schemes where some calibration of the devices may be possible.

Often there is variation in the achieved level of compaction across the width of a lane, as depicted in Fig. 7.7, with significantly lower values at the unsupported edge. Tandem rollers give more uniform performance across the lane width than deadweight rollers. The effect may be reduced by skilled rolling technique. On thick lifts the first pass should be approximately 300–400 mm from the edge to prevent the edge breaking away. This creates a platform so the remaining strip can be rolled at the next pass. Undue delay will lead to the edge cooling to an undesirable degree.

The specification normally allows a number of test results to fall to a certain level below the minimum provided that the average is within the specified limits. This allows gauge or core density determinations close to layer edges to take account of this effect. A value of 2% additional voidage is typically allowed since this area is outside the wheel path. However where the wheel path crosses the edge of the rip e.g. at motorway slip roads or on roundabouts, cutting away and removal of any low density material found at the unsupported edge may be necessary.

Inconsistent density across the mat can also occur at the supported edge if inadequate surcharge is left by the paver so that the roller is partly running on the adjacent rip. This effect is also apparent around manhole covers, etc. If ironwork is left in place while roadbase or basecourse materials are laid, special measures using plates, rammers or small rollers must be employed while the material is hot to achieve the specified density around the ironwork.

The level of texture achieved in chipped hot-rolled asphalt wearing courses depends on the quantity and shape of chippings applied, layer workability and temperature, and rolling technique. The UK Bristowes' chipper is capable of placing a uniform rate of spread of chippings when using free flowing chippings and in consistent site conditions i.e. an absence of kerb drops, changes in gradient and no stopping/starting. Back chipping by hand should be kept to a minimum.

Achieving texture is a balance between

(a)   placing the required chipping density
(b)   embedding chippings adequately
(c)   compacting the mortar round the chippings satisfactorily
(d)   achieving compaction of the mat as a whole.

If points (b) and (c) are met this, together with the compaction provided by the paver, will ensure the last criterion is also achieved except possibly at the edges of the mat where cooling is greater.

Harsher mixtures using manufactured sands, rather than natural sands, and higher stone content mixtures make consistent achievement of the correct texture depth more difficult. This can be alleviated to a degree by the use of EVA polymer and the use of a 1 t vibrating roller to embed chippings followed by three-point deadweight rollers.

10.2.5. Dragging of material
Observed defect: stones, congealed material or foreign objects are dragged along the surface of the mat either continuously or for varying distances until the object is freed.
Causes before paver
The road surface should be free of potentially loose material, bricks/blocks used for level purposes, sample bags, etc; and any sample trays should be fixed firmly in position. The material temperatures should be within the specified limits and any colder material at the side of trucks moved into the centre of the hopper by operating the hopper wings and shovelling. In extreme cases such material should be discarded. The thickness of the material to be laid should be at least one and a half the nominal aggregate size.

Causes at paver
If dragging is apparent across the whole width of the mat the paver speed, the screed temperature and the material temperature should be checked. The condition of the tampers is vital if dragging is to be avoided. The tamper speed must be correct and depends on paver speed and material. The protrusion below

the screed plate must be in accordance with the manufacturers instructions and the screed extensions in line with the main screed. Again, the head of material must be maintained at the correct level.

The screed heating system must be working correctly, and the material evenly spread across the width of the screed. The screed heating system cannot reheat material to make it workable; it is designed to act as an iron, producing a smooth and even surface. Modern pavers are equipped with automatic temperature ignition systems which are designed to conserve fuel and prevent the screed from over-heating which in the long term would cause the screed to crack.

If dragging occurs in the centre of the mat the crown setting must be adjusted i.e. the tail to lead crown differential. Dragging may occur at the edge of the mat caused by cold material building up at the ends of the augers. The screed end plates, the plates fitted at right angles to the main screed to retain the material in the auger box, must be perpendicular to the main screed and free to rise and fall in the end brackets. Problems will also occur if the mat thickness is less than one and a half times the nominal aggregate size in the mix. It is essential to maintain the paver in a good clean condition; a dirty screed will cause dragging.

10.2.6. Hairline cracking of laid material
Observed defect: narrow fissures across the mat.
Causes before paver

High mix temperatures are the most likely cause of this difficulty. Hairline cracking may also be the result of low workability. Where, for example in hot-rolled asphalt wearing course, the supplier is able to increase the binder content this problem may be eliminated. If the problem becomes apparent in roadbase or basecourse mixtures with certain sands and fine aggregates, adjustment of mixture constituents, particularly low quality natural sand, and proportions may be necessary. The macadam mix design method (see 10.2.1) may prove valuable. Once apparent, close liaison between the supplier and laying contractor is necessary. The problem may also occur where the specifier has asked for a too-large stone size for the layer thickness.

Causes at paver

The forward paving speed must be checked. If it is too fast, cracking may occur. Paving downhill tends to open up the mat introducing hairline cracks. A shower of rain will chill the surface which, after rolling, may reveal cracks.

Causes after paver

Where paving downhill, vibratory compaction can introduce tensile forces in the mat causing hairline cracking. Rolling when the material is too hot or over-rolling by continuing after the mat has cooled, particularly on thin layers, can also produce this. An excess of tack coat can lead to the formation of the slip plane and cracks forming in thin overlays. The roller can generate a bow wave of material ahead of the drum by premature use of heavyweight rollers or the non-driven drum facing the paver.

Excessive use of sprinkler water can cause a skin on the surface which cracks as it is under-supported by the hot material beneath. It has been suggested that subsequent cracking of pavements, particularly reflective cracking above Portland cement concrete roadbases, is accelerated by the existence of hairline cracking in the laid mat. One roller manufacturer has produced a roller which vibrates horizontally and it may be that this reduces the propensity of hot bituminous mixtures to exhibit hairline cracking. The use of pneumatic tyred rollers or combination rollers is an effective remedy.

10.2.7. Auger shadows
Observed defect: localized areas with a high sheen on the surface, radiating from the mat edges towards the crown.
Causes at paver
The situation is usually caused by missing, broken or worn auger flights. The material must flow evenly across the screed. The head of material must also be monitored to prevent an excess of material in the auger box.

10.2.8. Holes
Observed defect: holes or gaps in the finished mat.
Causes at the paver
An occasional fault caused by insufficient material in the auger box. Ensure that the operator is vigilant and that the automatic feed control is functioning correctly.

10.2.9. Ripples in the mat
Observed defect: ripples, mostly at the outer edges but occasionally across the width of the finished mat. The distance between ripples can vary from 20–800 mm.
Causes at paver
Ripples are usually caused by some form of hesitation in the forward movement of the tractor or screed units. Fluctuating paver speeds can cause the problem but it is more likely to be caused by the screed itself. It is essential that the head of material in the auger box is maintained at a constant height. The most common cause is looseness in the bolts which fix the screed to the side arms. Loose screed plates may also contribute to the problem. Looseness in these areas causes the angle of attack to change which directly affects the action of the screed causing ripples. A ripple can also be caused by a careless truck driver reversing on to the paver causing a jolt in the forward movement.

Causes after paver
Vibrating rollers normally vibrate at 50 Hz, so, if the forward speed is around 5 km/h, the pavement receives an impact every 30 mm. Rolling at a speed which is too high or improper use of vibration can lead to perceptible ripple on the surface which is more marked in thin layers, such as wearing courses, as they cool more rapidly and may be mobile only on the first pass. Good practice encourages drivers to travel faster on thin layers. Smoothing by deadweight rollers or only the most careful and selective use of vibration may eliminate the problem. Using vibrating rollers in deadweight mode only on the wearing course or using only light vibrating rollers is safer. The Bristowes' chipping machine may introduce an apparent ripple by placing chippings in regular rows

across the rip or mat; this should not be apparent if a uniform rate of spread of chippings is achieved.

10.2.10. Tearing
Observed defect: lateral tearing of the mat.
Causes at paver

The material may tear if the mat thickness is less than one and a half times the nominal aggregate size or the material is too cold, particularly at the auger ends. Tearing can also be caused by a failure in the screed heating system. If the screed plate is too cold, material will stick to the underside of the screed causing tearing. It is also essential that a constant flow of material is maintained across the front of the screed. Missing or broken auger flights and paddles may also cause this problem. Material which is spilled in front of the paver should be discarded.

10.2.11. Minimum mat thickness cannot be achieved
Observed defect: when the laid thickness is checked the thickness is less than that specified.
Causes before paver

The thickness of the laid material should be compatible with the nominal size of aggregate otherwise excessive forces will be generated under the paver screed. The prepared surface should always be checked during the dipping operation to ensure there are no high spots which may occur between the measuring locations specified in the contract.

Causes at paver

If the tow points at the forward end of the screed arms are at their lowest or highest position and difficulty is experienced in attaining the correct depth this may be caused by loose bolts at the end of the sidearms enabling the screed to rock and thus altering the angle of attack of the screed. The tamping rate may be too great causing additional compaction under the tampers. High points in the existing base material may be greater than the mat thickness preventing the screed from falling to its lowest position. The screed must be completely free floating i.e. free to move up or down without support. A fault in the hydraulic system could pressurize the screed rams, preventing free movement and preventing the screed reacting to tow point adjustment.

10.2.12. P or longitudinal joints
Observed defect: poor compaction of the longitudinal joint leading to chipping loss and/or fretting of the laid material. Poor matching of levels at the longitudinal joint between the two rips. Long term effects: inadequate sealing of the joint may permit the entry of water into the pavement structure as water drains across the crossfall leading to subsequent pavement failure.
Causes before paver

Compaction at unsupported edges is difficult to achieve, low density material must be removed. Low density in wearing courses will lead to fretting failure. A thickness, normally equal to the laid thickness of the layer, should be cut cleanly by a roller-mounted wheel, circular saw or pneumatic breaker before adjacent paving takes place. A layer of hot penetration grade binder is applied to the cut face to fill any voids and ensure effective sealing. The use of cold applied emulsion materials may be satisfactory for basecourse application but rarely gives an adequately thick coating for wearing courses. A proprietary pre-formed bituminous tape is now available. This is particularly valuable when applied at ironwork and adjacent to concrete kerbs where some thermal movement occurs and where water-tightness is important.

Causes at paver
Insufficient compaction is the most common cause of joint deterioration. It also causes mat tearing. Bumps or depressions along the longitudinal joints are likely to occur if the screed is operated manually. Automatic level control is recommended. The sensor shoe should ride as close as possible to the previously laid mat approximately 150 mm from the edge. Excess overlap causes insufficient compaction along the joint. The length of screed which sits on the existing mat should be around 75 mm. The head of material must offer a constant pressure against the tamper shield. Use of automatic feed control should ensure that this is the case. A clean edge will only be produced if the screed end plates are perpendicular to the screed extension. The correct surcharge of material to allow for compaction should be set on the screed.

A longitudinal joint heater attached to the paver is an advantage in cold weather to assist roller matching of the joint. The best longitudinal joints are obtained by two or more pavers working simultaneously in parallel.

Causes after paver
Poor edge matching occurs if the roller driver does not adopt a good rolling technique. The preferred method is for the roller barrel to overlap the adjacent cold mat by 100–200 mm. The speed of the roller should be low and the adjacent mat should be clear of debris.

10.2.13. Poor transverse joints
Observed defect: joint breaking up, bumps or hollows evident leading to poor rideability, chipping loss adjacent to the joint.
This problem is the largest single cause of poor rideability on an asphalt pavement.

Causes before paver
The last few metres of previous work should be inspected for vertical alignment and, in the case of hot-rolled asphalt, chipping retention. Any unacceptable material should be removed. The cut face should be treated as for a longitudinal joint. Some clients prefer a 90° cut to reduce joint length and assist in joint matching; others prefer a 45° cut, accepting that joints cannot be matched perfectly, to give an improved ride as defects are, to an extent, masked by this practice. These preferences should be specified in the contract. Transverse joints should be reduced to a minimum by good planning.

Causes at paver

Inadequate attention to the following points can lead to a poor joint. At the start of surfacing, the paver should stand over the joint with the screed heaters working to pre-heat the joint which assists with matching materials. A suitable block should be placed under the screed to ensure the correct surcharge and the machine correctly set at the tow point.

The auger box should be properly filled with hot material to produce the correct head of material. The first load from a plant should be checked to ensure temperature compliance as it is this load which is most likely to be outside the specification for temperature if the plant does not operate rigorous quality control procedures or a Quality Assurance system.

Any material on the adjacent rip should be cleaned of extraneous matter. The joint should be checked quickly after rolling and, only if strictly necessary, the area scarified and additional material added or excess removed by hand, before adding pre-coated chippings (if any) and final compaction. A dip in the mat can also occur some distance after the joint if the head of material is insufficient. If it is too great a bump will occur. If there have been delays then material at the back of lorry may be cold and should be discarded.

Causes after paver

As soon as the paver has moved away the joint should be rolled transversely so that levels are matched. The edge of the roller wheel should be predominantly on the cold material with about 100 mm width on the new material. If necessary, start with static passes followed by vibratory rolling or use a small tandem roller. Compaction should take place at the earliest possible moment and should continue along the length of the mat so that there is no differential compaction which will result in dips or bumps in the mat.

The chipper should check the rate of spread on the old surface before starting on the new material. The achieved rate of spread can then be checked and adjusted, as necessary. Although hand raking and level adustment may be visible on the completed surface such steps may be necesary to ensure a matched joint. Excessive rolling should be avoided to prevent dips in the surface.

It is likely that the last piece of the laid mat will be unsatisfactory in terms of compaction, texture, chip retention and levels as the cold material left in the paver hopper is passed through the paver and as the head in the paver box falls. A good transverse joint is dependent on this material being laid on sand or other debonding material to facilitate cutting back and removing all the material to a straight line.

10.2.14. Segregation of material
Observed defect: separation of the constituent parts of the mix.
Causes before paver

Some mixtures are more prone to segregation than others. DBM 40 mm nominal size to BS 4987[8] is particularly susceptible to segregation. Certain aggregate sources also seem more prone to this problem. Some plants and some hot storage methods introduce segregation into the mixture. Pavers have a limited re-mixing ability. Clients or contractors, when choosing or accepting plants for bituminous mixture production on particular contracts should be aware of any history of segregation problems so that they can be addressed and avoided.

Causes at paver

Segregation can occur as the material is unloaded from the truck into the paver hopper. One paver manufacturer now offers a re-mix unit which is fitted to the bulkhead of the paver hopper and can be engaged when a load is received. The re-mixed material is then presented to the augers. Excessive auger speed can also induce segregation within the auger box. The hopper should be emptied between loads to prevent excessive build up of coarser materials at the sides.

One of the most important factors at the paver is the crown adjustment shown in Fig. 10.2. Excess lead crown i.e. the centre screed adjustment turnbuckle nearest to the augers, can cause a bright streak at the centre of the mat, while insufficient 'lead' crown can cause bright streaks at the edges, with loose material at the centre. The lead crown affects the texture and the tail crown influences the shape of the mat.

Incorrect setting of the height of the auger with respect to the screed for the layer thickness and stone size for the material being laid may lead to transverse and vertical segregation. It is not always the case that the coarser fractions of the bituminous mixture fall to the edge of the box. Occasionally, the coarser stone falls to the bottom of the layer or, if the mixture contains some combinations of coarse aggregate and sand, the vibratory/tamping screed may bring finer material to the top of the layer. This can be detected by coring or the presence of unusually fatty i.e. bitumen-rich material at the surface. This problem rarely occurs with hot-rolled asphalt roadbase.

Apparent segregation may be due to defective sampling techniques. Correct sampling of materials is essential if spurious results which suggest segregation are to be avoided. Segregation may be signalled by poor density measurements and is usually most apparent in cores. Where it is suspected, samples should

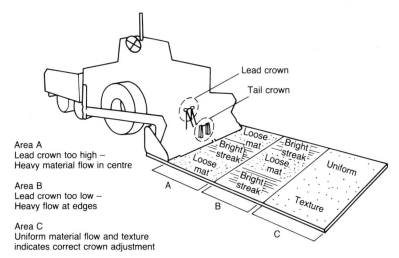

Fig. 10.2. Screed crown setting

be taken from across the finished layer before compaction and analysed individually.

10.2.15. Premature rutting
Observed defect: the layer deforms soon after the road is opened to traffic. This is most frequently observed where paving is taking place on contra-flow or signalled traffic management regimes.
Causes before paver

When wearing course mixtures are designed, the design must take into account the construction phase when heavy vehicles may be channelled and travelling slowly through the works. A two-fold multiplier is recommended for such conditions and a speed factor introduced into the wheel tracking rate formula (chapter 3) and appropriate values specified. Base mixtures, usually containing natural sand and gravel aggregates, at certain binder contents, can readily become unstable. These should be known to suppliers and avoided or the macadam mix design method used. Alternatively creep testing[11] may be carried out before work begins to ensure adequate deformation resistance at field density. The presence of air voids, in the basecourse and roadbase mixtures manufactured in the plant and compacted to the PRD,[3] within the range 1−3%, may be sufficient to prevent premature rutting.

Causes after paver

Over-compaction of mixtures can lead to instability especially in binder-rich materials. Opening the road to traffic prematurely is probably the major cause of this defect. The requirement of BS 594[12] for the wearing course to have reached ambient temperature is particularly relevant in hot weather (where air temperatures exceed 25°C in the UK) as the surface temperature can reach 55°C by solar gain. A recommendation for roadbases or basecourses not to be overlaid until their temperature has fallen to the softening point of the bitumen minus 15°C has been tried by some clients to prevent heat build-up on a pavement when thick lifts and high tonnages are being laid. Temperature criteria can lead to problems since the surfacing squad wishes to remove the traffic management measures at the earliest time. There may also be a financial incentive on lane rental contracts.

10.2.16. Poor texture depth in chipped hot-rolled asphalt wearing courses
Observed defect: inadequate texture measured by laser or sand patch methods.

Three reasons are possible: inadequate chipping, chip loss or excessive embedment of chippings. Some clients have accepted surface dressing as a remedy for the first two but only complete replacement of the mat is truly effective. Where pre-coated chippings are embedded excessively, the correct level of texture depth may be obtained through the action of traffic removing the matrix around the chippings. If it does occur, it should happen within the twelve month maintenance period. Significantly low texture depth may be a safety hazard especially on high speed roads. Physical removal of the bituminous matrix by hot air blasting or the like may be necessary but this may affect rideability.

Causes before paver

A specification requirement for high texture depth (1.5 mm sand patch; 1.03 mm laser texture depth) in areas of high lateral forces, particularly at roundabouts and junctions is unnecessary. It will almost certainly lead to chipping loss to a degree that the resultant mat has inadequate microtexture for skid resistance.

The specification of a high stone content mixture or high Marshall stability mixture which will be harsh and difficult to work with will lead to problems with embedment of chippings, especially when the material is laid in windy conditions. The use of 18/150 EVA (VA content/MFI) or styrene butadiene polymers, which 'skin' at about 90°C, will prevent the asphalt matrix moulding round the chippings and holding them in place. The use of 33/45 EVA helps to avoid this problem in windy conditions but the material may remain excessively workable in summer as its skin temperature is about 60°C.

The mix from the plant should be consistent in both composition and temperature. Paving should not begin if wind chill effects will mar success. Where high texture depths are required, the chippings should have a cubic shape and the percentage of single sized material should exceed 80% with not more than 15% undersize. Chippings with a soaked 10% fines value below approximately 240 kN are liable to crush under the roller, leading to reduced effective rates of spread. However, this is a chracteristic of high PSV gritstones, as is the presence of elongated or flaky particles which reduce the chance of achieving high texture. Restrictions on strength, elongation and flakiness are frequently specified. Obtaining the correct texture and avoiding chip loss is particularly difficult at transverse joints because of the time required for their construction. These should be kept to a minimum by good planning and continuous deliveries.

Causes at paver/chipper

The chipping machine should be in a good state of repair and hand chipping should be restricted to small areas. The hopper should be of adequate size. Chips should be clean, bright, free flowing and ice-free. The chipper should be the correct size for the width of mat being laid. As the gradient changes, the chipping machine will need adjustment. A minimum rate of spread of 11 kg/m^2 is recommended for good skid resistance; high texture demands at least 70% of the surface to be covered with chippings. Some additional chippings may have to be added to fill holes left by the chipper but this should be done sparingly to prevent over-chipping and subsequent chipping loss.

Causes after chipping machine

Obtaining the correct texture depth is a complex interaction of embedding an adequate quantity of chippings to the correct depth without pushing them too far into the material. It requires consistent temperature of material after placement, correct rolling pattern and controlled use of vibration time (depending on the response of the mat to the first pass). If polymer-modified binder is used, the roller driver should be informed. When the required texture is achieved rolling should cease and the rollers parked on a cold part of the surfacing until required.

Table 10.1. *How bituminous surfacings fail: types and causes of failure*

Type of failure	Causes of failure		Distinguishing features	
	Primary	Contributing	Specific	General
Disintegration	Low cohesion	Soft bitumen Poor aggregate grading Low density	Tenderness Over-sanded Under-compacted	Scuffing Marking Indenting Gouging
	High abrasion	Insufficient bitumen Brittle bitumen Soft aggregate Chains and studs	Dryness Brittleness Crushed aggregate Grooving	Pitting Ravelling Washboarding Potholing
	Debonding of bitumen/aggregate	Hydrophilic aggregate Displacement of bitumen by water Clay in aggregate Displacement of bituminous mix by jet blast	Stripping Fuel spillage Blast erosion	
Instability	Low interparticle friction	Excess bitumen Smooth, polished aggregate Clay/water present Rounded aggregate	Bleeding Fatty surface Ball bearings	Flowing Pushing Shoving Rippling Rutting
	Low mass stiffness	Rounded aggregate Soft binder Poor aggregate grading (insufficient coarse material) Insufficient fines	Ball bearings Undue workability	Corrugating Fretting Rutting
	Low density	Poor mix design Insufficient compaction Cold compaction Improper compaction Incorrect layer thickness	Porous Poor durability	

Fracture (or cracking)			
Changing foundation support	Differential settlement Differential expansion Frost heave Trenching	Local depressions Dishing Mounding Frost boil Trench settlement	Longitudinal waves Transverse waves Porpoising Cracking Potholing
Shrinkage	Absorption by aggregate Ageing bitumen Temperature fluctuations Volume change of coatings (paint, joint seals, epoxy dressing) or underlying material (cement-bound)	Right-angle cracking Random cracking Curl cracking	Transverse cracking Longitudinal cracking Diagonal cracking Load-associated Non-load-associated
Brittleness	Bitumen embrittlement Burned bitumen Brittle base e.g. cement-treated Low temperature exposure	Pattern cracking Block cracking Ladder cracking Low temperature cracking	
Fatigue	Resilient or 'springy' base Inadequate paving material stiffness Channellized traffic Poor drainage of pavement section	Alligatoring or chicken-wire cracking Wheel track cracking	
Slippage between layers	Insufficient bond Low tensile strength of overlay Thin overlay No lateral support e.g. shoulders Over-rolling	Vee-cracking Crescent cracking Micro cracking	
Reflection	Tensile stresses/strains Shear stresses/strains Bending stresses/strains (overlay to cement-bound base)	Reflective cracking	Transverse cracking Longitudinal cracking
Settlement and heave	Deep fill settlement Expansive soils Tree roots	Settlement cracking Spread cracks Heave cracks	

Table 10.2. Classification of the condition of the road surface

Classification	Visible evidence
Sound	No cracking, rutting under a 2 m straight edge less than 5 mm No cracking, rutting from 5−9 mm
Critical	No cracking, rutting from 10−19 mm Cracking confined to a single crack or extending over less than half of the width of the wheel path, rutting 19 mm or less
Failed	Inter-connected multiple cracking extending over the greater part of the width of the wheel path, rutting 19 mm or less No cracking, rutting above 19 mm Cracking confined to a single crack or extending over less than half of the width of the wheel path, rutting above 19 mm Interconnected multiple cracking extending over the greater part of the width of the wheel path, rutting above 19 mm

10.3. Failure in service

Failure in service is said to have occurred when the road pavement shows signs of defects before the design life is reached. Table 10.1, based on the paper by Vallerga,[13] depicts the main reasons for premature deterioration in durability. Additionally, a surface can suffer skid resistance problems if the PSV is inadequate, insufficient chippings have been applied or have been lost through inadequate embedment, or the level of achieved texture depth does not meet the specification. A road pavement deteriorates at an ever-increasing rate, as shown in Fig. 3.8, using the concept of serviceability index.

Table 10.1 relates to the totality of the performance of the pavement i.e. rideability, the loss of this is an indicator of loss of structural strength; rutting and cracking, rarely seen together during the early life of the pavement as they tend to cancel each other out, and potholing and fretting.

The pavement should not be allowed to fail if the investment in the underlying layer is to be realized. Appropriate maintenance measures should be carried out when intervention levels are reached. These are specified for the UK in LR 833[1] and are shown in Table 10.2. Current UK design information is based on intervention at the critical time and is discussed in chapter 3. Appropriate maintenance treatments must be undertaken to ensure that the pavement structure remains serviceable until major structural works are required. Such measures are shown in Table 10.3.

References

1. KENNEDY C.K. and LISTER N.W. *Prediction of pavement performance and the design of overlays.* Transport and Research Laboratory, Crowthorne, 1978. LR 833.
2. HMSO. *New roads and streetworks Act 1001, HAUC specification for the reinstatement of openings in highways, a code of practice.* HMSO, London, 1992.
3. BRITISH STANDARDS INSTITUTION. *Sampling and examination of bituminous mixtures for roads and other paved areas. Methods of test for the determination of density and compaction.* BSI, London, 1989, BS 598: Part 104.

Table 10.3. Appropriate maintenance treatments for premature deterioration

Defect	Surface treatment
1. Transverse cracking (a) Hairline, < 0.5 m	Hot air blast clean and fill with hot poured sealant to BS 2499.
(b) Significant, single or multiple	Rout out affected material, min 50 mm wide to depth of wearing course and fill with hot poured sealant to BS 2499 or proprietary joint sealing system. (Not to be used for longitudinal cracks exceeding 50 mm wide).
(c) Frequent hairline, multiple, random (especially if accompanied by loss of skid resistance)	Double surface dressing using a bitumen impregnated non-woven geotextile fabric as first layer to keep pavement structure dry. (Indicates that structural strengthening is imminently required).
(d) Reflective cracking	Plane out a width of 300 mm min. to a depth of the wearing course — reseal with a proprietary rubberized/polymer modified bitumen sealant (these will not have significant texture depth and may rut/shove under heavy traffic). Local overbanding at top of cement-bound layer or road surface are relatively ineffective long-term solutions. Crack and reseat concrete. Include geotextile (not geogrid) stress absorbing interlayers and use polymer-modified materials in overlay.
(e) Random 'fatigue' or brittle cracking	Plane out affected material and replace macadam or hardened hot-rolled asphalt with new hot-rolled asphalt material, investigate support and replace as necessary.
2. Longitudinal cracking	If in wheel track this is probable evidence of structural failure. Full reconstruction/overlay strengthening is recommended. the crack must be kept sealed to prevent water ingress to the pavement structure and subgrade (1(a) or 1(b)).
3. Inadequate skid resistance	Surface dressing with appropriate binder and chipping combinations (see Road Note 39).[14] If other defects present plane out/overlay with new wearing course with aggregate of adequate PSV for site.
4. Surface wear	Check the abrasion resistance of the aggregate. Surface dress or replace the wearing course. In extreme cases concrete block paving may have to be substituted.
5. Rutting/deformation	Replace wearing surface with one adequate for the site (including during construction), check roadbase for deformation and replace as necessary.
6. Potholing/fretting	Patch potholes and cut out fretted areas. Over banding with a proprietary system (temporary).
7. Scuffing/marking	Use of inappropriate material, e.g. soft macadam on footways, soft asphalt/macadam on loading areas. In extreme cases concrete block paving may need to be substituted.
8. Settlement	Particularly of statutory undertakers trenches. Complete excavation down to the service and reinstatement with good quality materials may be necessary.

4. AMERICAN SOCIETY FOR TESTING AND MATERIALS. *Standard test method for per cent air voids in compacted dense and open bituminous paving mixtures.* ASTM, Philadelphia, 1989, D3203−88.

5. DEPARTMENT OF TRANSPORT. *Specification for highway works.* HMSO, London, 1992, **1**.

6. TRANSPORT RESEARCH LABORATORY. *Nuclear gauges for measuring the density of roadbase macadam.* TRL, Crowthorne, 1982, SR 754.

7. AMERICAN SOCIETY FOR TESTING AND MATERIALS. Standard test method for theoretical maximum specific gravity of bituminous paving mixtures. *Annual book of ASTM standards.* ASTM, Philadelphia, 1990, **04.03**, D2041−90.

8. BRITISH STANDARDS INSTITUTION. *Coated macadam for roads and other paved areas. Specification for constituent materials and for mixtures.* BSI, London, 1993, BS 4987: Part 1.

9. BRITISH STANDARDS INSTITUTION. *Hot-rolled asphalt for roads and other paved areas. Specification for constituent materials and asphalt mixtures.* BSI, London, 1992, BS 594: Part 1.

10. BRITISH STANDARDS INSTITUTION. *Methods for determination of physical properties.* BSI, London, 1975, BS 812: Part 2.

11. BRITISH STANDARDS INSTITUTION. *Method for determination of creep stiffness of bitumen aggregate mixtures subject to unconfined uniaxial loading.* BSI, London, 1990, DD 185.

12. BRITISH STANDARDS INSTITUTION. *Hot-rolled asphalt for roads and other paved areas. Specification for the transport, laying and compaction of rolled asphalt.* BSI, London, 1992, BS 594: Part 2.

13. VALLERGA B. A. Pavement deficiencies relating to asphalt durability. *Proc. Ass. Asphalt Paving Technologists.* 1980, 50.

14. DEPARTMENT OF TRANSPORT. *Design guide for road surface dressing.* HMSO, London, 1992, Road Note 39.

11. Some European practices and the harmonization of bituminous mixture specifications

11.1. Introduction

Bituminous materials comprise no more than mixtures of aggregate and binder, usually bitumen, but there is an almost infinite variety of products which material technologists currently produce in the various member states of the European Community (EC) and European Free Trade Area (EFTA).

The concept of the EC is that it should form a European free trade zone which, in purchasing power, will be the largest single market in the world. This requires the removal of barriers to trade.

One step towards achieving this aim was the issue in December 1988 of the Construction Products Directive (CPD) which requires construction products used in member states to be fit for their intended use, satisfying certain essential requirements whenever required by regulations covering the works. The essential requirements are

- (*a*)   mechanical resistance and stability
- (*b*)   safety in case of fire
- (*c*)   hygiene, health and environment
- (*d*)   safety in use
- (*e*)   protection against noise
- (*f*)   energy, economy and heat retention.

At first sight, these may appear bewildering requirements for construction products but the list is the result of work in the 1980s to rationalize the basic safety requirements for all products of member states so that these did not form barriers to trade.

The essential requirements have to be taken into account in the drafting of European standards, or norms, referred to subsequently as ENs. They have already been embodied in UK Statute Law through the Construction Products Regulations which came into force on 27 December 1991, and will become enforceable when the necessary harmonized ENs have been produced. Their significance will be seen later in this chapter when the extent to which they are likely to re-cast the responsibilities of producers' and clients' technical inputs to material quality control work is discussed.

The task of producing ENs is entrusted to the Comité Européen de Normalisation (CEN) and in turn to CEN technical committees. TC 227 has been formed specifically to produce ENs for materials used in the construction and maintenance of roads, airfields and other paved areas. Materials such as proprietary products, and therefore not covered in ENs, may, instead, be subject to EOTA, a form of CEN agreement board approval. In this chapter, attention will be focused on some of the many non-proprietary European flexible pavement materials and their use, laying practices and also on moves towards harmonized standards.

11.2. Constituent materials
11.2.1. Aggregates
The haulage of road aggregates requires energy and money. As a direct result, a common factor in road construction from pre-Roman times has been the use of locally available materials. In the twentieth century, as material specifications have formalized the customs and practices within local industries, so highway engineers everywhere have followed the same economic dictate of using local materials wherever they have proved to be suitable. It is important to realize that uniformity in quality is as important as quality itself in road pavement works. Airfield pavement engineers do not hesitate to specify high quality aggregates even if transport costs appear unduly excessive. This is because they are primarily concerned with durability and an absence of fretting in their relatively large paved areas of airfield and taxiways since fretting can result in extremely expensive foreign object damage to both planes and jet turbines. The use of airfield quality aggregates in road pavements would not only be expensive, but also unnecessary.

This chapter is, therefore, concerned with road pavement bituminous mixtures. In general, and certainly in Western Europe, whenever a material fails in practice due to normal wear and tear, it is usually due to oxidation, fatigue of the binder, or less commonly now, the stripping of the binder from the aggregate. Low-grade aggregates with poor durability have long since been identified and physical tests developed to eliminate their use in relatively expensive bitumen-coated pavement materials.

In the UK the vehicle driver has grown accustomed to road surfaces which not only enable him to travel smoothly and quickly, but more importantly, enable him to stop quickly. While the Department of Transport has only recently formalized skidding resistance requirements, and other UK highway authorities have yet to follow, highway engineers have striven for many years to provide good skid resistance within the limits of their budgetary constraints.

The provision of non-slip road surfaces started in the mid nineteenth century when civil engineers involved in building dock access roads realized that granite sett paved roads soon polished and resulted in many injuries to draught horses. In this country there is a long tradition of using relatively high PSV aggregates in wearing courses which has been possible because the UK has a wide diversity of natural rocks which crush to provide such aggregates.

Not all European countries have high PSV rocks and even where they do an absence of demand for particular road properties, such as skidding resistance, has sometimes resulted in a lack of their commercial exploitation. For instance, all member states which border the Mediterranean have plentiful supplies of limestone. Their relatively low annual rainfall and absence of frost at the low altitudes where the bulk of the populations reside, has resulted in them using limestone aggregates in all their pavement mixtures, even wearing course layers. Naturally enough, their road traffic accident statistics bear witness to polished road surfaces but, to date, the indigenous populations have accepted these driving risks as part of normal life. Other more northerly member states also make free use of limestone and harder low PSV rocks in wearing course materials and their road users face similar skidding risks.

At the other extreme, the Netherlands has no dry land aggregates sources and aggregates for road pavement materials are either gravels dredged from the relatively shallow estuaries of the rivers Rhine and Maas or imported rock. Environmental pressures have resulted in the control of even these major dredged sources and from 2001, materials will have to be imported from adjacent member states, or deep-water sea-dredged which will necessitate washing to eliminate salt and other mineral contaminants.

The north German plain comprises virtually all natural sand, and this has resulted in the development of a peculiarly German wearing course material, *Gussasphalt*, similar to mastic asphalt but machine-laid and used for surfacing all classes of road from autobahns to city streets. The surface is finished with a crimping roller while still hot to produce the effect shown in Fig. 11.1.

On autobahns, in addition to the crimping, the hot surface is sprinkled with

Fig. 11.1. Gussasphalt, a German wearing course material

a grit to improve surface texture. Needless to say, there are no national skid resistance standards within Germany at present.

Further north again in Scandinavia, the longer winters with consequent increased periods of hazardous driving conditions have resulted in the widespread use of winter tyres with steel road-stud inserts. These give vehicle wheels more grip on ice-covered roads but result in wearing courses being seriously abraded in the wheel tracks. The prime aim of highway engineers in Scandinavia is therefore not the construction of carriageways with wearing courses having high skid resistance but the provision of roads which will last for several winters before failing as a result of being abraded away.

A significant departure from the use of natural aggregates occured in Denmark which, despite its relatively small size and population, has a road building industry which is second-to-none and a highly respected state highway authority which has introduced some concepts which are novel in the UK. For wearing courses in the state controlled parts of the highway network, the highway authority always specifies a degree of light reflectivity to enhance night driving visibility. This reflectivity has been achieved either by the use of an artificial aggregate, Synopal, a glass-like ceramic material significantly more expensive than natural aggregates, or a white crushed rock aggregate imported from Norway, or a combination of both. Synopal is now too expensive and its use has been discontinued.

Within EC and EFTA member states' current bituminous mixture specifications, aggregates are described in terms of

(a)   type of rock, gravel etc
(b)   resistance to abrasion
(c)   resistance to polishing
(d)   strength
(e)   shape
(f)   flakiness
(g)   angularity
(h)   soundness
(i)   compatibility with bitumen
(j)   cleanliness.

These parameters are measured using a bewildering array of test equipment, and each piece of equipment is used within a contractual framework in its parent member state despite its frequently abysmal precision. This is as true in the UK as elsewhere.

Fortunately for TC 227, CEN Technical Committee TC 154 is responsible for producing the EN for aggregate and has carried out an enormous amount of work in producing its first draft specification in November 1992.

Aggregate will not be divided into classes for specific uses, as has previously happened in the UK with BS 882.[1] Instead, it is seen as a feedstock for many other products which may need differing aggregate properties. The draft specification emphasizes that customers should only specify those characteristics which are necessary for their particular production processes and not the full range of specification parameters.

From a plethora of sieve size sets used in the different states it has been

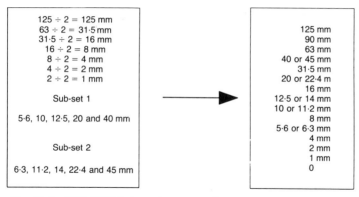

Fig. 11.2. CEN TC154 sieve size proposals

established that there will be one prime set, based around the 63 mm sieve, with two sub-sets. Specifiers will be able to use the main sieve set plus one or other, but not both of the sub-sets as shown in Fig. 11.2.

11.2.2. Fillers

All states appear to use crushed limestone fillers although, frequently, different specifications permit Portland cement or hydrated lime. The Netherlands have a highly developed filler technology with three grades of filler, all of which are imported. The three grades are defined in terms of their stiffening effects on the binder in the mix and represent soft, medium and hard stiffening qualities. Across Europe, the pre-CEN range of specified filler requirements include

(a) type
(b) maximum sieve size (63, 75 and 90 μm all being used)
(c) specific gravity
(d) shape
(e) absorption
(f) swelling
(g) voids according to Rigden
(h) solubility in water
(i) loss on heating (for pulverized fuel ash fillers)
(j) calcium hydroxide content
(k) bitumen number.

Hydrated lime filler is specified for use in all porous asphalts and in some premium, high performance mixtures.

11.2.3. Binders

Binders are vitually all penetration grade bitumens, with a relatively small tonnage being polymer modified in recent years for high performance uses. Tar is not widely used on the Continent, and in at least one member state tar-bound materials excavated from old road pavements are regarded as hazardous wastes, despite public relations exercises by the few surviving tar producers.

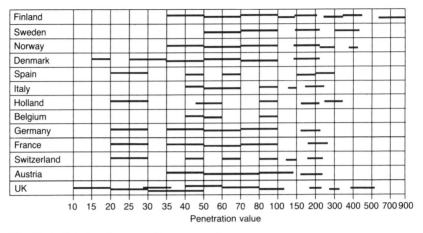

Fig. 11.3. Existing European bitumen grades

However, TC 227 is charged with producing any EN needed for tar binders.

Fig. 11.3 shows both the grades of bitumen currently used within Europe and the proposed rationalization of grades which has already been proposed within CEN/TC 19 by its sub-committee SC1 which is charged with producing ENs for bituminous binders. Finland uses 800 pen bitumen in soft asphalt which is understood to withstand its extremely cold winters and the retention of the 40−60 pen grade binder traditionally used in UK hot-rolled asphalt but only as a result of pressure on SC1 by UK.

A prime objective of TC 19 has been to minimize any disruption in member states caused by the rationalization process. The enormity of this task can be judged from the properties included in specifications of the various member states and listed in Table 11.1.

After six years work, some measure of success has been achieved as can be seen from Tables 11.2 and 11.3 which list the properties for use in binder ENs.

11.3. Recycling

Despite numerous trials and even more learned tracts on the subject, recycling remains a relatively minor activity in most countries, ranging from the addition of reclaimed asphalt planings to virgin materials in asphalt plants, to in situ recycling of upper pavement layers with the addition of new binders and sometimes new aggregates.

Significant in situ recycling is carried out in Italy but the Danish Government has decreed that a specific percentage of the road maintenance programme should be carried out using in situ recycling. It recognizes the technique is not yet more economical than work using virgin materials, but argues that unless it encourages such work, it never will be more cost-effective since contractors will have no incentive to develop and improve techniques. Fig. 11.4 shows a rural Danish road being recycled to a depth of 100 mm using an Italian-built mobile plant. This processes the existing pavement, adds supplementary virgin

Table 11.1. Existing specifications for bitumen properties used in Europe

Bitumen property	Original bitumen	Thin film oven test residue*
Penetration	Yes	Yes
Pen — ratio	Yes	No
Softening point	Yes	Yes
Viscosity 60°C	Yes	Yes
130°C	Yes	No
135°C	Yes	Yes
Penetration index	Yes	Yes
Fraas point	Yes	Yes
Ductility 7°C	No	Yes
10°C	No	Yes
13°C	Yes	Yes
25°C	Yes	Yes
Flash point COC	Yes	No
PM	Yes	No
Loss on heating test	Yes	No
Thin film oven test	Yes	No
Rolling thin film oven test	Yes	No
Rotating flask test (German)	Yes	No
Density	Yes	No
Solubility (6 solvents)	Yes	No
Ash content	Yes	No
Paraffin (wax) content	Yes	No
Permittivity	Yes	No
'Judgement Macroscopique' (microscopic examination, French)	Yes	No

* Or other hardening test

aggregate and new bitumen to the heated mix and re-lays the material, in a single pass, at the rate of 120 t/h. This satisfies the mix-design grading requirements of the recycled mix.

Within TC 227 Working Group 1 (WG1) which is specifically charged with producing ENs for bituminous mixtures, it has been agreed that specifications for recycling will not be produced until the work on hot and cold mixtures has been completed, although it is acknowledged that increasing environmental pressures will inevitably result in wider use of the range of recycling processes.

Table 11.2. CEN TC19/SC1 initial proposals for bitumen properties, 1

Bitumen property	Standard grades
Penetration	Yes
Softening point	Yes
Viscosity at 135°C	Yes
Rolling thin film oven test — weight loss — penetration on residue	Yes Yes
Solubility	Yes
Flash point*	Yes
Density*	Yes

* Flash point and density are to be reported but not specified. Their use is within commercial supply contracts.

Table 11.3. CEN TC19/SC1 initial proposals for bitumen properties, 2

Bitumen property	Grades, very cold areas	Grades, special safety areas
Penetration	Yes	Yes
Softening point	Yes	Yes
Viscosity at 135°C	Yes	Yes

Rolling thin film oven test
   — weight loss
   — penetration on residue
	Yes
	Yes
		Yes
		Yes

Solubility	Yes	Yes
Ductility at (X)°C† or maximum stiffness at critical low temperatures	Yes Yes	Yes No
Permittivity	No	Yes
Flash point*	Yes	Yes
Density*	Yes	Yes

* Flash point and density are to be reported but not specified and are to be used for commercial supply contracts and not end-user's specification
† X is a figure to be inserted in the specification

Fig. 11.4. A Danish road being prepared for in situ recycling

Fig. 11.5 shows the Marini recycling machine combining mixer and paver. It has a bitumen tanker and the initial breakdown rolling of the new mat is by pneumatic tyred roller.

11.4. Bituminous mixtures used in Europe

11.4.1. Asphaltic concrete

This is the basic hot-mix material in the majority of European countries. In terms of standard UK bituminous mixtures, asphaltic concrete is a form of DBM in which the binder content has been optimized by design processes. These usually result in lower binder contents than are used in the UK.

The aggregates are continuously-graded and this type of material is used in all pavement layers. For wearing courses, there is a compromise to be struck between achieving maximum durability, which requires a high binder content to completely fill almost all the voids in a layer and reduce its permeability to water and oxygen, and achieving resistance to deformation under traffic which demands good aggregate interlock, with stone-to-stone contact and thus some voids.

If the binder content is too high, and fills all the voids in the mineral aggregate structure, the layer deforms readily under traffic, particularly as ambient

Fig. 11.5. Marini recycling machine combining mixer and paver

pavement temperatures in some member states commonly approach the softening point of the binder. This is because the stone-to-stone interlock is destroyed. It is therefore vital that wearing course mixtures are produced with sufficient voids to ensure that the layer does not deform readily under loading, but also with sufficient bitumen to ensure durability. Mix design procedures to optimize binder contents are, therefore, a necessary part of asphaltic concrete production.

Tables 11.4–11.6 show gradings for typical French and German wearing course asphaltic concretes, and a UK grading for close-graded wearing course macadam with a BS 4987 wearing course macadam.[2] The French and German materials perform well in general use but would be judged by UK standards to have too low a surface texture for high speed roads. Fig. 11.6 shows the grading envelopes for all three materials.

For a mix to be durable in the pavement it needs a minimum of voids so that the bitumen binder is oxidized as slowly as possible. On lightly trafficked roads, mixes with high binder contents give trouble-free lives, but with increased traffic loadings, mixes need resistance to wheel track rutting because of their high mechanical stability, and this is frequently achieved by using an

Table 11.4. Gradings for French 'semi-grained' asphaltic concrete wearing course

Property	BB 0/10	BB 0/14
Passing sieve 14 mm	–	94 – 100%
10 mm	94 – 100%	72 – 84%
6.3 mm	65 – 75%	50 – 66%
4 mm	45 – 60%	40 – 54%
2 mm	30 – 45%	28 – 40%
0.08 mm	7 – 10%	7 – 10%
Binder content by mass	5.8 – 6.1%	5.5 – 5.8%

Table 11.5. Gradings for German asphaltic concrete wearing courses

Property	0/11	0/16
Retained on sieve 16 mm	–	< 10%
11.2 mm	< 10%	> 15%
8 mm	> 15%	25 – 40%
5 mm	–	–
2 mm	40 – 60%	55 – 65%
0.09 mm	7 – 13%	6 – 10%
Crushed/Natural sand	> 1:1	> 1:1
Binder content by mass	6.2 – 7.5%	5.2 – 6.5%

Table 11.6. Gradings for UK close-graded wearing course macadam

Property	10 mm	14 mm
Passing sieve 20 mm	–	100%
14 mm	100%	95 – 100%
10 mm	95 – 100%	70 – 90%
6.3 mm	55 – 75%	45 – 65%
3.35 mm	30 – 45%	30 – 45%
1.18 mm	15 – 30%	15 – 30%
75 μ	3 – 8%	3 – 8%
Binder content by mass	5.3 ± 0.5%	5.1 ± 0.5%

absolute minimum of binder with subsequent adverse affects on durability.

As wearing course mixtures of asphaltic concrete have been refined to satisfy both those apparently conflicting needs, so the importance of obtaining optimum values of voids in the designed mixes has risen. UK highway engineers unfamiliar with the concepts of VMA, voids in the mineral aggregate, and, VM, voids in the mix, should read through the current PSA *Specification for airfield runway wearing course materials*.[3] Asphaltic concretes have to be designed to satisfy these conflicting needs, and for the most heavily trafficked pavements, conventional unmodified bitumens are approaching the limits of their performance. For critical situations therefore, there is an increasing need to use modified bitumens, as is now happening in the UK, but not to the same extent as in some other member states.

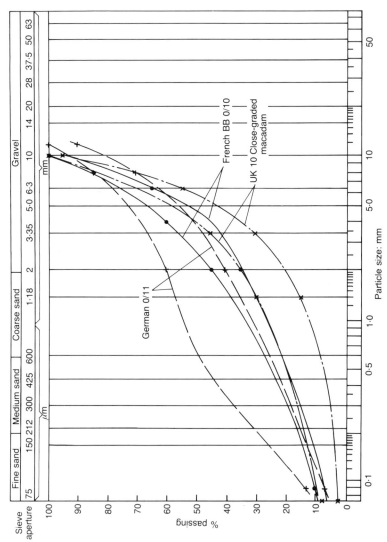

Fig. 11.6. Grading envelopes for French and German asphaltic concretes and UK bitumen macadam

Asphaltic concrete basecourses

Since the basecourse is protected from oxidation by the surface course, the highway engineer can concentrate on providing resistance to deformation under traffic loading and this may be achieved by using a relatively open-textured grading. The void content is higher, and since durability of the binder is not the dominant factor, the binder film thickness can be reduced by using an absolute minimum of bitumen in the mix. The binder content must still be sufficient to give the mixture the required degree of workability at the laying stage and adequate tensile strength in the finished layer. For optimum resistance to deformation, crushed rock sands are used or a blend of crushed rock and natural sands with harder binders.

Asphaltic concrete roadbases

This type of asphaltic concrete is used in far greater tonnages than all other types combined since it provides the main load-bearing strength in a flexible road pavement. Economy of production and the dynamic stiffness of the compacted material are both paramount. Since there is an even lower possibility of bitumen oxidation than in basecourse material, binder contents are reduced to the lowest practical minima and maximum aggregate sizes are increased to take advantage of the greater thicknesses of constructed layers.

Gussasphalt

There is little superficial difference between *gussasphalt*, widely used in all types of carriageway wearing courses in Germany and mastic asphalt sometimes used on railway station platforms and other very heavily trafficked pedestrian areas in the UK. Typical grading envelopes for these two materials are shown in Fig. 11.7 and it is immediately apparent that both require very high binder contents when compared to asphaltic concretes or to any other bituminous mixture.

In the case of *gussasphalt*, the high binder contents and resulting low air voids give the material enhanced durability and working lives well in excess of twenty years are not unusual. Unlike UK mastic asphalt, which is used on relatively small scale works and mixed on site in mobile, heated mixers or pots, *gussasphalt* is mixed in large scale plants and delivered to sites in insulated tanker-type road vehicles. It is then laid with paving machines modified to deal with the hot, low viscosity material, followed immediately by either gritters which lightly spread 2−5 mm chippings on the hot surface and rollers which embed the chippings, or by rollers which texture the surface as shown in Fig. 11.1.

Although typically produced with 30−50 pen grade binder, the material will deform under heavy loadings, and particularly under stationary loads. Its relatively high cost results in *gussasphalt* being laid in thin layers, usually between 15 mm and 30 mm thick, which also limits wheel track rutting.

Despite its durability, the rapid increase in the price of bitumen in the early 1970s and the need for enhanced resistance to wheel track rutting as traffic loadings increased have both resulted in the development of a new type of mix, stone mastic asphalt, which is becoming widely used in Germany, Austria and Switzerland, and also to a lesser extent in the Netherlands and Belgium.

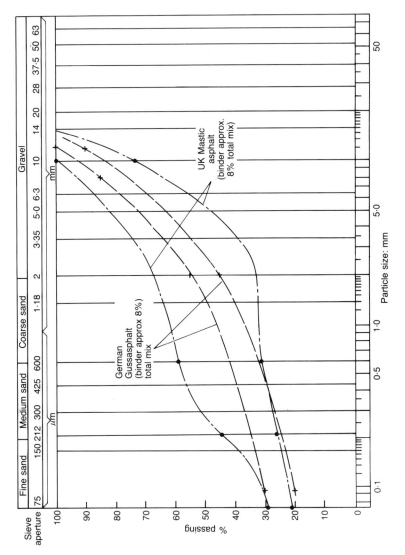

Fig. 11.7. Gradings for German Gussasphalt and UK mastic asphalt

Stone mastic asphalt

Extracts form the German Asphalt Pavement Association specification[4] for stone mastic asphalt (SMA), also known as split mastic asphalt are shown in Table 11.8. These differ slightly from the Federal German specification but apparently produce a more durable mixture.

The material consists of a stable mix of coarse aggregate in which the voids are filled with a mixture of sand, filler and binder. These combine to form the mastic asphalt element of the material. The coarse aggregate/sand ratios are selected to ensure that the compacted material has a low air void content, but minor variations in the 'ideal' mix ratios can result in rapid changes in the performance of the surface layer under traffic. For instance, the Danes report that with some of their aggregates, the percentage passing 2 mm is critical. Less than 22% gives a mix which ruts easily under traffic. Nevertheless it is a very durable mix by virtue of its relatively high binder content, but there is more to achieving a high binder content in a mixture than simply adding more bitumen.

Few UK engineers realise that a mix can only carry an amount of binder related to its total surface area, and higher binder contents simply result in binder draining within the mix, or in very extreme cases, off the delivery lorry. In SMA therefore, the surface area of the mix constituents is increased significantly without a corresponding increase in the mass by adding cellulose fibres or rock-wool fibres to the aggregates.

The effect of these fibres on the mix can be understood by thinking of one stone particle being coated with binder in the mixer, and then being rolled

Table 11.7. *Specifications for German split mastic asphalts*

Property	0/11	0/8	0/5
% > 11.2 mm sieve	≤ 10%	—	—
% > 8 mm	≥ 25%	≤ 10%	—
% > 5 mm	50−70%	45−70%	≤ 10%
% > 2 mm	70−80%	70−80%	60−70%
% > 0.09 mm	8−13%	8−13%	8−13%
Crushed sand, natural sand	≥ 1:1	≥ 1:1	≥ 1:1
Binder penetration	65 pen	65 pen/80 pen	80 pen
Binder by mass	6.5−7.5%	6.5%−7.5%	6.5%−7.5%
Stabilizing additive* by mass	0.3−1.5%	0.3−1.5%	0.3−1.5%
Marshall test temperature	135 ± 5°C	135 ± 5°C	135 ± 5°C
Voids in Marshall specimen	2−4%	2−4%	2−4%
Layer thicknesses	25−50 mm	20−40 mm	15−30 mm

* Cellulose fibres or rock-wool fibres

Table 11.8. Specification for UK high-stone-content medium-temperature asphalt

Property	Limits
Passing 14 mm	100%
10 mm	90–100%
6.3 mm	35–70%
2.36 mm	35–47%
600 μ	25–47%
212 μ	5–30%
75 μ	4–8%
Max. passing 2.36 mm and retained on 600 μ	9%
Binder by mass	5.5% min.
Nominal thickness of layer	40 mm

in a mass of small feathers, a number of which will adhere to the bitumen on the particle without adding noticeably to its weight. Each feather will then carry more binder by virtue of its very large surface area, and in consequence, the feather-coated aggregate particle is able to carry a considerably enhanced volume of binder. Seen under a microscope some of the fibres added to SMA are not unlike feathers and they perform in the same way. Fibres can also be used to increase the binder capacity of porous asphalt.

SMA is cheaper and performs better under heavy traffic loading than *gussasphalt* wearing course. It is more expensive but more durable than asphaltic concrete wearing course mixtures, and this greater durability gives it a greater cost-effectiveness in whole-life-costing terms. It is this which explains its increasing use in the German-speaking countries and in those northern EC member states which share common borders with Germany.

The initial surface texture is extremely low, due to the high binder content but this problem is immediately resolved to some degree by gritting with either coated or uncoated 2–5 mm stone chippings which are rolled into the surface of the material while it still has a temperature of at least 90°C. As the surface binder oxidizes, the larger aggregates are exposed and surface texture is further enhanced.

Table 11.8 shows the composition of the most similar UK mixture, the high stone content medium temperature asphalt introduced in Table 3 column 4 of the 1992 edition of BS 594: Part 1[5] and the similarity between the two materials can be readily judged from Fig. 11.8.

Porous asphalt

Porous asphalt is the new CEN description for what has previously been known in the UK roads industry as pervious macadam and in the UK airfield industry as friction course. In the USA, similar material is commonly referred to as popcorn mix, which gives an image of the compacted material in which the aggregate grading is formulated for a mix in the pavement layer with 15–20% air voids even after compaction. Voids are the result of the grading and not a lack of rolling. Surface water percolates through the porous asphalt

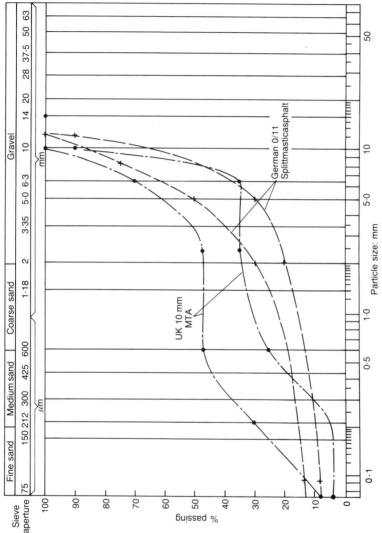

Fig. 11.8. Comparison of gradings for German split mastic asphalt and UK high stone-content medium temperature asphalt

wearing course to drain by way of the crossfall on the dense underlying basecourse and it provides the motorist with excellent visibility in wet weather, unimpeded by spray from the wheels of other vehicles.

Ironically this material was developed in the UK for airfield use where it provides a safeguard against aircraft wheels aquaplaning at high landing speeds but it has been hardly used on roads despite numerous TRL controlled development and proving trials. In Europe, however, it is widely used and has been for at least a decade, particularly for its property of significantly reducing the generation of road-traffic noise. In otherwise identical UK conditions of traffic loading and environment, perceived noise levels even in dry weather are up to one third lower where porous asphalt wearing course is used to replace hot-rolled asphalt and it is this relatively extensive sound-reducing use of the material in other member states which gives it a place in this chapter on European practice.

With the same concept, only the composition differs between member states and trials have shown conclusively that material using an aggregate grading from 20 mm down is considerably more resistant to void clogging than material using an aggregate grading from 10 mm down. Most mainland European states use 14 mm, 10 mm or smaller material.

Table 11.9 shows the areas laid to 1991 in member states, and it is difficult to understand the past reluctance of UK engineers to use the material other than on experimental sites.

A significant tonnage of porous asphalt has been used in urban areas in Europe and UK engineers have been astounded to see it laid even across bridge decks in urban Paris. There is a perceived difficulty in the UK of draining surface water from the top of the impermeable basecourse at the low edge of the carriageway in such situations. In rural areas this is simple enough; it discharges into a stone-filled drain trench at the edge of the carriageway, a french-drain.

On kerbed roads, the porous asphalt may be laid to within, say, 300 mm from the face of the kerb and water draining across the top of the basecourse emerges into this exposed channel and discharges into road gullies set at

Table 11.9. Use of porous asphalt in Europe in 1991

Country	Area laid: million m²
Austria	> 5
Belgium	> 1.5
France	> 3.5
Germany	Test sections only
Italy	> 5
Netherlands	> 2
Sweden	> 2.6
UK	< 0.1

basecourse level. The nominal 40−50 mm difference in level can be a hazard to pedal cyclists and this practice is therefore only used on rural high speed roads with traffic volumes which are so high as to deter even the boldest cyclist.

The problem has been tackled successfully in mainland Europe by the types of drainage details shown in Figs 11.9−11.11. It is acknowledged that while the traffic on high speed roads keeps the drainage voids relatively clear in the wheel tracks, on urban sites it tends to clog-up with discarded paper, etc. which is pulped by vehicle wheels.

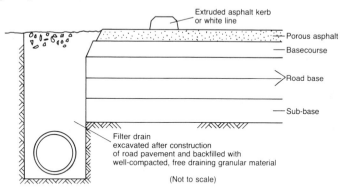

Fig. 11.9. Detail for non-kerbed carriageway

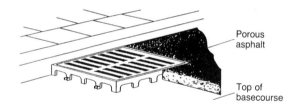

Fig. 11.10. Gully grating used to drain porous asphalt on kerbed roads in Holland

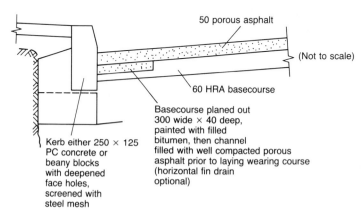

Fig. 11.11. Detail for kerbed carriageway

Fig. 11.12. High pressure jetter/vacuum sweeper as used to maintain voids in French porous asphalts

This problem has been solved in France by the simple expedient of high pressure jetting at least once a year using a mobile combined jetter and suction cleaner. Fig. 11.12 shows a high pressure jetter/vacuum sweeper, operating at a water pressure of 90 atmospheres, as used to maintain voids in French porous asphalts. The machine is cleaning a sett-paved road in a steel works.

Other member states have developed thin, high-texture materials which are used for noise reduction in urban areas, and French contractors have also developed similar materials which do not need pressure jetting to the same extent as porous asphalt. Some of these are now being laid on UK roads by UK contractors who have formed commercial links with their French counterparts. Two such materials are Safepave and ULM (Ultra Mince). These are patented materials which initially fall outside the scope of TC 227 ENs.

Grave-emulsion

This is a peculiarly French material, comprising an aggregate mixture coated with one of several proprietary cold emulsions which are formulated to begin

Table 11.10. Grave-emulsion performance results

Property	Limits
180/220 bitumen	> 2 GPa
80/100 bitumen	> 3 GPa
40/50 bitumen	> 4 GPa
Duriez result*	> 0.55
PCG compaction*	> 85%

* Duriez result and PCG compaction are explained later in this chapter

breaking as the material is laid. The breaking process continues for six to nine months within the compacted pavement which consequently does not reach its optimum strength until this process has been completed. Voids after laying are 8−12% for the 0/14 mix and the emulsifier evaporates through these voids.

Since the material is used in less heavily trafficked road pavements, this is not a serious disadvantage. The ability to haul the material over significant distances without it losing workability is considered to be one of its prime advantages. This is particularly so for work in those areas which are remote from mixing plants and where the scale of the work is insufficient to justify the expense of establishing a mobile asphalt plant.

The material is used in several gradings, 0/14, 0/20 and 0/31.5, the latter only being used for roadbase layers. The emulsions are cationic and the binder content such that the bitumen content of a 0/14 mixture on analysis in the laboratory must be 4%. It is not a high performance material but is required to have a range of compressive strengths which depend on the stiffness of the base bitumen as shown in Table 11.10.

French high modulus roadbases and basecourses

These are mixtures which have unusually high stiffness. There are two types, those having a high binder content, and those with a more normal binder content. Gradings are usually 0/14 and 0/20. The French consider 20 mm to be a maximum acceptable aggregate size since the use of larger aggregates results in both segregation of laid materials and lesser qualities of ride. Recommended nominal thicknesses are up to 300 mm for 14 mm material and up to 400 mm for the 20 mm size. This is the maximum stone size multiplied by a factor of 20 whereas in UK and elsewhere, macadam layers are usually laid at up to six or seven times the maximum stone size.

The stiffness of these materials is the result of their manufacture with extremely hard bitumens, and acceptable penetration values are within the range 5−35 pen, with a ring and ball softening point value of at least 65. Binder contents and other specification requirements are shown in Table 11.11.

The mixtures are laid at paver-out temperatures of at least 140°C and initially compacted using pneumatic-tyred rollers before finishing with conventional deadweight rollers. The binders are usually modified and the mixtures marketed as proprietary materials.

Table 11.11. Specification summary of French high modulus materials

Property	High binder content		Normal binder content	
	0/14	0/20	0/14	0/20
Binder content	6.0	5.8	4.8	4.7
PCG compaction at 100 revs	< 5	–	< 10	< 10
PCG compaction at 200 revs	–	< 5	–	–
Duriez result	> 0.75	> 0.75	> 0.70	> 0.70
Rut depth 30,000 Hz−100 mm thick slab	< 8	< 8	< 8	< 8
Complex modulus at 10 Hz in GPa	> 14	> 14	> 14	> 14

Hot-rolled asphalt

Hot-rolled asphalt, as widely used in the UK, is included in this chapter not only because small tonnages are used in several member states, but also to compare it with some of the materials previously described.

Unlike *gussasphalt*, it can be produced either to have a very high stability, with minimal binder content, in order to withstand wheel track rutting, or to have a high degree of flexural strength by virtue of an enhanced binder content. It is relatively easily produced with only sand, filler, binder and a nominal single size of coarse aggregate going into the mixer and is relatively insensitive to minor changes in composition.

These factors make it a strong competitor to SMA, particularly in the recently introduced 55% stone varieties, which are not subsequently finished by the application of high PSV pre-coated chippings. Its skid resistance and resistance to deformation make it a superior material to *gussasphalt*. It is more expensive than asphaltic concrete but has far superior skid resistance, and its unconnected voids lessen the tendency for its constituent binder to be oxidized. It also has superior durability. Where low texture depth values prevail immediately after laying, it can be remedied by the German practice of rolling grit into the layer whilst it still retains heat although this has yet to be adopted in the UK.

11.5. Mix design methods

Since 1979, most hot-rolled asphalt on heavily trafficked roads in the UK has been subject to a mix design process described fully in BS 598: Part 107.[6] The process involves the optimization of the binder content in a mix where the constituent coarse and fine aggregate gradings are derived from within empirically-produced grading limits derived from experience.

In the UK, the basic BS 594[5] hot-rolled asphalt aggregate grading was produced in the early twentieth century by a North American asphalt technologist using his experience of failure analyses of contemporaneous rolled asphalts. In all member states, their recipe grading envelopes for asphaltic concretes and other materials have been produced as a result of two or three generations of experience of bituminous mixtures.

The BS 598: Part 107[6] mix design method involves the use of the Marshall type apparatus based on that which was initially developed by another North American, Bruce Marshall, to optimize the design of asphaltic concretes for use by the Mississippi State Highways Department. Before his work, roads were either rutting under increasingly heavy loadings because the binder contents were too high, or cracking because they were too brittle, as the binder contents were too low. His work was carried out more than 50 years ago and followed earlier work carried out by another North American, Francis Hveem in the 1930s for the Californian State Highways Department.

Marshall's work has been refined since, but, for this chapter, BS 598 practice can be summarized as the carefully-recorded production of cylindrical test samples with identical aggregate compositions, representing those of the aggregates to be used in full scale production, but with differing binder contents. The binder is increased in increments of 0.5% and the small samples of mixture moulded and compacted to produce the test speciments. These are then conditioned in a heated water tank to a test temperature of 60°C before being placed in the curved plattens of the test machine and subject to a controlled strain of 50 mm/min until they fail. The two essential elements of Marshall test equipment are the mechanically-driven hammer used to compact the test specimens and the test loading apparatus, as shown in Figs 11.13 and 11.14.

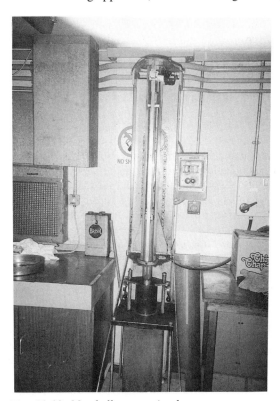

Fig. 11.13. Marshall compaction hammer

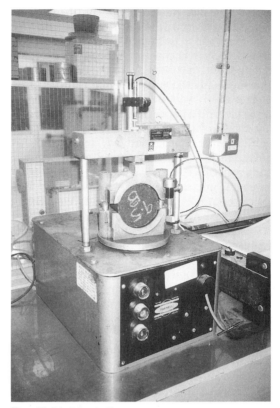

Fig. 11.14. Marshall test apparatus

The load at which a sample fails to carry increased loading is referred to as the stability and the deformation of the specimen under this loading is termed the flow. In the UK BS 598 method, the technician plots the test results in terms of binder content against stability, compacted aggregate density and mix density. The mean value of these three binder contents, the 'design' binder content, is increased by 0.7% for 30% stone mixes to give the target binder content used in the mixing plant for the production of the mixture. Many member states also use mix design methods derived from Marshall's work but with some additional steps, generally including the determination of air voids at each binder content.

It has already been explained that hot-rolled asphalt is a very tolerant material so far as a producer is concerned, whereas both asphaltic concretes and stone mastic asphalts can be much more sensitive to minor compositional changes. Mix designs in most member states using these materials, and also for UK airfields where asphaltic concrete is specified, have to take careful account of the problems of voids being over filled with binder.

Typical specifications for asphaltic concretes in most EC states require that contractors not only record their stability and flow, but also the ratio of stability divided by flow, the so-called Marshall quotient, the voids in the mix, and the voids in the mineral aggregate filled with bitumen. Limits are established for all these variables and the skilled technologist aims to produce mixes which satisfy the specifier at a cost which gives his own company a competitive edge in a tender.

Danish engineers also use Marshall, but once again, their specifiers are in the vanguard of purchasers who give a contractor the greatest possible freedom in producing satisfactory cost-effective materials. Instead of therefore specifying values of stability and flow, some clients give contractors the design parameters for a layer in terms of finished thickness, skid resistance, resistance to wheel track rutting and the traffic volumes and axle weights to be carried by the layer and leave it to the contractor to produce a mix with a specified life of at least five years. Sometimes, light reflectivity is also specified.

To encourage contractors, part of the payment is withheld until the material has been clearly seen not only to have survived, but to have performed as required for this period. If the contractor wishes to use a modified bitumen in the mix, then he has to be aware that conventional Marshall testing will not give reliable results and samples of the trial mix should undergo testing in a laboratory wheel tracking test rig. Transverse and longitudinal profiles are covered by a six month guarantee period.

The German technician however, works within an established tradition of using crushed aggregates and recipe grading envelopes and producing his Marshall test samples with differing binder contents but not loading them to determine their stabilities. Instead the Germans are confident that their asphaltic concretes will perform satisfactorily in the pavement as long as the air voids in this mixtures are controlled. They too wish to see the binder contents maximized to enhance durability but not to the extent where the voids are over-filled and the mixtures lose their aggregate inter-lock. This is particularly important and, in the UK in a recent development, the Department of Transport is encouraging producers of heavy duty macadams to ensure that their materials have a low void content after compaction to total refusal in the PRD test.

The Belgians have devoted a great deal of effort to ensuring that the correct volumetric combinations of coarse and fine aggregates and filler are used in asphaltic concretes and have written a *Code of Good Practice*[7] to assist producers in this task.

In Scandinavia, where the abrading of wearing course mixes by spiked vehicle tyres is a matter for serious concern, the largest possible aggregate sizes are used in mixtures and since the Marshall test mould, with 102 mm dia. and 64 mm depth cannot usefully handle an aggregate larger than 20 mm, great reliance has been placed on traditional recipe mixtures. Volumes of traffic are generally low compared to the UK, but where they are concentrated, asphaltic concrete is designed using conventional Marshall techniques to achieve the balance between durability and resistance to wheel track rutting in their warm summers. However, high stability mixtures are not necessarily abrasion-resistant.

Austria, Italy, Spain, Portugal and Switzerland all use asphaltic concrete in their airfield pavements and more heavily trafficked roads, designed in the conventional Marshall manner. France is a notable exception.

The French have an extremely well-organized national roads administration and an innovative national road research organization, the Laboratoire Central des Ponts et Chaussées (LCPC).

Some years ago, the French decided that the characteristics of Marshall design mixtures bore little relationship to the manner in which the mixtures behaved either at the laying stage or in service. They therefore developed an entirely new mix design philosophy involving fatigue tests on trial mixtures wherever a new aggregate source or similar fundamental change is involved, but not carried out on a routine basis, and a combination of a laboratory compaction test, wheel tracking test and test indicating the sensitivity to water of the mixture, all of which are performed frequently. Since they are relatively unknown within the UK, they are described in detail.

As elsewhere, contractors select an aggregate composition from within pre-defined grading limits seen by the French as initial recommendations. A formula is used to determine the trial range of binder contents based on the surface area of the aggregates. These trial mixtures are then tested in a gyratory compactor which kneads the mix and simulates the compaction action of a pneumatic-tyred roller. Its action is shown in Fig. 11.15.

The specification states the degree of compaction to be achieved at pre-defined mix temperatures and a stated number of machine revolutions. For a typical 10 mm wearing course material to be laid 60 mm thick the voids remaining after 60 gyrations must be between 5% and 8%. The gyratory compactor or PCG apparatus is illustrated in Fig. 11.16. It is a substantial machine and its relatively high cost means that there are only about 50 in various testing laboratories throughout France. Some are owned by contractors and other by specifiers. A less robust machine could not test the high strength French mixes.

This initial stage enables a technician to identify a promising trial mixture which is then used to produce compacted cylinders for the second stage, in the French mix design process which involves the Duriez test.

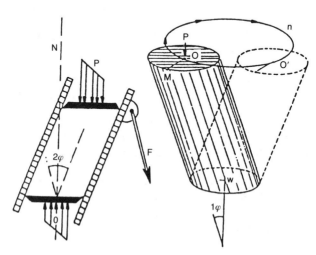

Fig. 11.15. Diagram of gyratory shearing press (PCG) used in France

Fig. 11.16. Gyratory compactor used in French mix design process

For this, the sample size will be determined by the maximum aggregate size, but for a stone size not exceeeding 14 mm the cylinder will be 120 mm dia. by 270 mm high. The trial specimen is steadily loaded with 180 kN over 60 seconds and the load maintained for a further 300−305 s, at temperatures which depend on the grade of bitumen used in the mixture. Thus, for instance, with 20−30 pen it is 180°C and with 80−100 pen it is 140°C, within limits of ±5°C.

All samples are removed from their moulds after 24 hours and two are then retained for air void determinations. A further four are stored in air for another seven days and another four have their voids filled with water under a pressure equivalent to 350 mm of mercury for two hours and are then totally immersed and stored in water at a temperature of 18°C.

These latter four are removed from the water, and after being wiped surface dry, are weighed in air at three, five and eight days after their initial manufacture, being re-immersed on days three and five.

On the eighth day, all eight specimens are then subjected to unconfined compression at a strain rate of 1 mm/s until failure, and the average failure loads calculated for both the air-stored samples R and the water-stored samples, r.

Then, depending on where the material is to be used in the pavement, it must have the following minimum values of r/R

For wearing course	− 0.75
For porous asphalt	− 0.80

The Duriez test is carried out for one of three temperature conditions, 0°C, 18°C or 50°C. For the 0°C test the air-cured specimens are kept at 0−1°C for seven days; for the 18°C test, both air and water stored specimens are kept at 18°C for six days and 50°C for one day.

Following satisfactory completion of the Duriez test, the selected mixture is then used to produce a test slab within the laboratory. The test slab is tested for skid resistance and wheel tracking in the LCPC wheel tracking machine which produces its own micro-climate without the need to heat a laboratory room, and uses a pneumatic tyre with no tread but of average car tyre dimensions. The rut depth is measured after 30 000 wheel passes at 60°C and the trial mix approved if the deformation is less than 10% of the initial thickness. The wheel tracking machine showing the treadless tyre is in Fig. 11.17. The machine is located within its own heated casing (shown here with the lid open) to achieve the 60°C test temperature at minimum expense. The mix is then approved for use on site with a maximum air void control of 90% of the voids found in the two initial samples kept for air void determination.

This is the routine French approach to mix-design. For a decade, it was run alongside the Marshall method while the industry developed enough confidence to adopt the new system. Where the contractor intends to use a new aggregate source, or a high performance mixture incorporating a modified

Fig. 11.17. French wheel tracking machine

binder, a fatigue test is carried out on the optimized mix. Trapezoidal specimens are sawn from laboratory compacted slabs and then mounted using epoxy resin in a test machine originally developed in the Shell Oil Company's laboratories. Fig. 11.18 shows specimens which are being tested for fatigue over one million cycles. The trapezoidal section enables the same stress levels to be developed throughout the height of the test specimen. The test is usually used only when a completely new mixture formulation is being developed. It is not a routine control test.

Dynamic stiffness testing is carried out more frequently and, typically, an asphaltic concrete wearing course with a nominal maximum aggregate size of 10 mm, produced with 40−50 pen grade bitumen will be required to have a dynamic stiffness of 8 GPa at 10°C with a 0.02 s loading time and 3 GPa at a temperature of 0°C and a loading time of 300 s.

The Belgians place great emphasis on constructing an aggregate structure for a mix which is derived from their *Code of Good Practice*.[7] This handy booklet deserves the wider readership which a translation into English would immediately give it and contains a three stage mix design method. The stages are

(*a*)   volumetric composition guidance, from which is derived the mass of the various mix constituents

(*b*)   an analysis of the suggested mix which examines the characteristics of those particular constituents which are available for the mix and assesses their effect on its functional characteristics

(*c*)   a verification stage which is based on a carefully detailed series of Marshall tests.

If the physical tests do not produce exactly what is required the process is repeated to include either a different aggregate or a modified composition.

Fig. 11.18. French fatigue testing apparatus

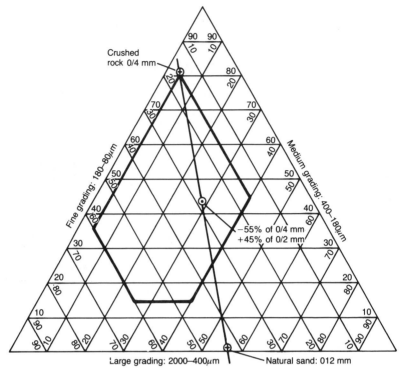

Fig. 11.19. Belgian diagram used to assist volumetric formulation of mix components

The Belgian Road Research Centre, who publish the code, have also produced computer software to facilitate stages (a) and (b). Fig. 11.19 taken from the guide shows an example of mixtures being classified according to their mineral skeleton.

11.6. German practice

While UK British Standards have never included a requirement that satisfactory adhesion should be achieved between layers, which sometimes appears to be easier said than done, the German Transport Ministry do not hesitate to specify it. Nevertheless, as in the UK, this bond is not always evident when trial cores are removed and discussions revolve around the question of the possibility that a bond was achieved but has been disrupted by the coring operation. Producers using hot bin storage have to ensure that the mix in the silo is not harmfully altered and this seems a reasonable request.

Also in Germany, no material except bituminous roadbase is allowed to be laid if an unbroken film of water covers the area being covered and this automatically covers the problem of what to do if rain starts to fall while laying is in progress. Cracks and joints in pavement areas which are to be overlaid are first filled with stone chippings or gravel of suitable size, and then impregnated with bitumen emulsion. This is probably more effective than over-

Fig. 11.20. Fusible bitumen sealing bands developed in Germany for use on vertical joint faces

Fig. 11.21. Patch completed using bitumen sealing bands

band sealing or filling joints with hot poured sealants, which frequently fail because of lack of care, but no guidance is given on how long should be allowed for the emulsion to break.

Transverse joints in mats are bevelled over their full thickness and painted or sprayed with binder before work continues. The Germans have made extensive use of the fusible bitumen joint-sealing bands for many years, as shown in Figs 11.20 and 11.21, and at last these are now being marketed in the UK.

The usual UK practice of paying for laid material on an area and thickness basis with only regulating course being paid for by weight, is also widespread in member states. Some realistically pay by weight on small jobs, typically less that 6000 square metres. Any tendency for material to be laid over-thick is countered by simply not paying for it, at least in theory. If a basecourse is laid too low and an over-thick wearing course is required, the Germans pay for the additional materal at 5% of the contract rate.

In Germany, Sweden and Austria contracts provide for work which does not fully comply with the specification to be accepted, subject to a reduction in payments made to the contractor. The reductions are calculated using formulae included in the contracts which cover shortfalls in thickness, binder content, compaction and poor riding quality.

11.7. Effect of EC pressures on current UK practices

These are only brief insights into black top practice within other member states but as UK firms increasingly become involved in work on mainland Europe, so translated documents will become more widely available. It is even possible that understanding of other languages will improve. It is likely that the book method and practice vary as they do in the UK but engineers are reluctant to acknowledge this.

Whereas hot-rolled asphalt has long been seen as the automatic choice of wearing course material on an increasing proportion of UK highways, EC pressures have resulted in some innovative UK laying contractors forming liaisons with mainland European firms. A number of French developed materials are already being laid in place of surface dressing or hot-rolled asphalt, particularly the French ultra-thin wearing course mixtures.

The French have experienced difficulties with wheel track rutting in their asphaltic concrete wearing courses as have the rest of Western Europe. Their reaction has been to develop stiffer basecourse and roadbase materials, and very thin wearing courses, only 15−25 mm. Even if these rutted by 50% of their initial thickness, the rut depths would only be 7−12 mm deep.

The mixtures are virtually single size macadams. The thinnest of all, marketed in UK as Safepave, is normally laid about 15 mm thick using a purpose-built, wheeled machine which carries several tonnes of both the coated hot mix and the tack coat. Both use normal binders but the process is unique in that the cationic tack-coat is applied to the old road surface immediately in front of the paving augers i.e. within the wheelbase of the specialist machine. The hot mix is distributed by the augers immediately in front of the compacting screed, as in a normal paver but the heat of the mix causes an immediate breaking of the emulsified tack coat. After two or three passes of the attendant rollers, the road can be trafficked immediately.

High PSV aggregate and a suitably large maximum aggregate size means the material is suitable for high speed roads with the 1.5 mm UK texture depth requirements. Its macadam-type structure with multiple surface interstices results in lower levels of road traffic noise than hot-rolled asphalt, even in dry weather, and so is also suitable for use in environmentally sensitive urban areas where French experience has indicated that porous asphalts became quite quickly clogged.

A similar material is marketed as ULM but this is a true macadam and is laid to a nominal thickness of 20 mm, using a conventional paver. ULM can be laid wherever a paver can operate whereas the Safepave machine, with its longer wheelbase, cannot easily negotiate sharp radii. The ULM is produced using a patented modified binder and, in experiments on the LCPC test track at Nantes, has been demonstrated to contribute to pavement strength despite its thinness. ULM can be produced in a variety of nominal sizes to achieve either high speed texture depth with the same noise reduction as Safepave or, with a small aggregate, give a lower degree of texture but apparently absorb noise from the higher frequencies in the spectrum. So overall noise reduction is much the same as it competitor but its lower frequency is considered by some to be more comfortable for residents of roadside properties.

Both mixtures are not only considerably quieter than hot-rolled asphalt but significantly quieter than surface dressing of equivalent aggregate size. Another advantage is that the old material does not need to be removed to accept either as an overlay since the reduction in kerb upstand is minimal. This can be particularly useful if resurfacing needs to be carried out on bridge or viaduct decks constructed with sub-standard protection to the bridge-deck waterproofing. The fact that both processes can be trafficked within minutes of rolling being completed is also useful on heavily congested routes.

There is some UK interest in trials of German split-mastic asphalt and it is only a matter of time therefore before this very durable material is being laid by UK contractors, particularly if it is gritted while being rolled to eliminate the otherwise initial, binder-rich, dangerous-looking surface.

Fig. 11.22. Safepave being laid

Fig. 11.23. ULM being laid

Figs 11.22–11.24 show Safepave being laid, ULM being laid and the texture of a Danish split mastic asphalt after trafficking has abraded the initial binder film. Light-coloured aggregate is used to enhance reflectivity in darkness.

11.8. Preparation of ENs for bituminous mixtures

The EC Construction Products Directive is hastening the preparation of standardized ENs for our national materials production. Bearing in mind that the purpose of ENs is to remove barriers to trade, in the black top field, the three classic barriers to trade would be

(a) different mix specifications in member states
(b) different test methods used in member states
(c) different material approval procedures.

While this work has so far been in progress for two years, it is all carried out by people working in either their own or their employer's time, with no cost contribution from the EC. In the UK there is a central government scheme for assisted travel costs but for delegation leaders only and on a very limited scale.

CEN Technical Committee 227 was formed specifically to deal with road and airfield construction and maintenance materials. TC 227 determines policy and delegates while the real work of producing specifications is the responsibility of the aptly named working groups (WGs) (Fig. 11.25). Each WG in turn then delegates specific tasks to task groups (TGs) and it is the members of TGs who burn considerable quanitites of midnight oil.

Bituminous mixtures, widely used throughout the EC are the responsibility of WG1, supported by four TGs

Fig. 11.24. Texture of split mastic asphalt

(*a*)   terminology
(*b*)   test methods
(*c*)   specifications
(*d*)   quality and attestation of conformity

WG1's priorities are to initially produce ENs for hot mixtures, then cold mixtures and finally recycled mixtures. WG2 is dealing with surface dressing, slurry seals, tack coats, and any other bituminous product which emerges from a nozzle. Its work and those of the other TC 227 WGs is outside the scope of this chapter.

The EC is determined that the CPD should acquire teeth within member states as soon as possible, despite the relatively late establishment of TC 227, and these teeth will be the ENs. There is, therefore, considerable pressure to produce standards quickly. This is a laudable aim, but it demonstrates a total lack of understanding of the problems likely to arise in future contracts if they are based on poorly drafted specifications.

In response to this CEN pressure, WG1 has had to move forward on several areas of work simultaneously although it might have been more logical to have drafted specifications within TG3 and in doing so, identified the need for test

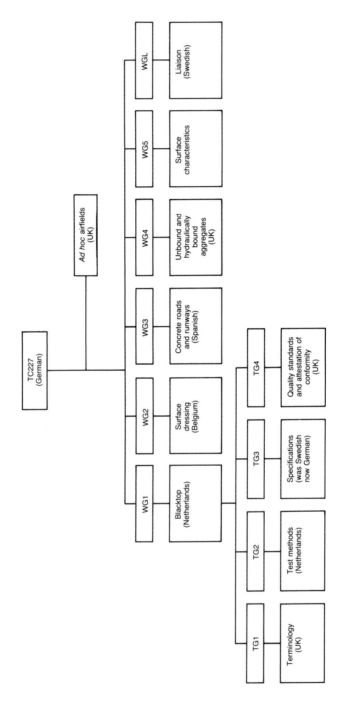

Fig. 11.25. CEN Technical Committee 227 working groups

methods which would then have been dealt with by TG2. Any terminology problems encountered could be referred to TG1 and eventually the work of all three groups would have been edited and printed.

Instead, all three TGs have had to plunge into the uncharted waters of producing draft standards without any substantial guidance from CEN and while 1992 was the oft-mentioned year of European unity, in free-trade zone terms at least, there were very few ENs by then.

In 1992, almost two years after the initial meeting of TC 227, CEN issued draft mandates for a number of products used in construction works and gave the first significant guidelines on how it expected ENs to be produced and work.

Almost simultaneously the subject of quality assurance was withdrawn from the scope of the CEN TCs because of concern that it was not being handled in a satisfactory manner. Instead it has been stressed that the so-called attestation of conformity which is referred to in more detail later, but which can result in a product carrying a CE mark, is no more than an assurance that a product will satisfy one or more of the essential requirements referred to at the start of this chapter, and complies with the relevant CEN specification. The CE mark is not to be regarded as equivalent to the BSI Kitemark in the UK for instance. It is not a quality mark.

WG1 also decided to produce specifications to cover two contractual interfaces, namely between the asphalt producer and laying contractor, and between the laying contractor and the purchaser. The latter could be a public sector body, a small private road owner or even a toll route operator.

CEN requires that specifications are written, wherever possible, in terms of the end-product performance of the materials. Despite the express wish of many delegates with TC 227 that ENs should specify the in-pavement performance of mixtures measured using non-destructive testing of the pavement, its official terms of reference do not allow this.

Instead, ENs will be mainly about materials since the aim of the CPD is to remove barriers to trade by facilitating the sale of materials across boundaries between member states. Each of WG1's task groups is working to this end.

11.8.1. TG1 — terminology

All WG1 plenary session work is conducted in English, with written work in English, French and German. The need for a common understanding of terminology is paramount. The oft-quoted PIARC *Dictionary of highway engineering terms*[8] simply translates English into French without explanations and is therefore inadequate. So far as the UK is concerned, the main results of TG1's work will be the changes from wearing course to surface course, basecourse to binder-course, pervious macadam to porous asphalt and the blurring of our traditional understanding of asphalts as gap-graded mixtures and macadams as continuously-graded. This is not to under-rate the work of TG1. Its paper on terminology definitions is included in the Appendix to this chapter.

11.8.2. TG2 — test methods

It is fundamental to cross-border trading that those involved in the industry on both sides of any border should understand the methods used to judge compliance with specifications. TC 227 recognizes this and all WGs are working to produce one test method only for any single specification parameter. Where several test methods exist in member states it is intended that the one having the highest precision should be selected as the European standard test method. This is the theory but the principle will have to withstand such pressures as the costs of re-equipping test laboratories in those member states whose national standard test methods are not selected as the TC 227 standard.

Test methods being discussed include

(a)   sampling
(b)   sample preparation
(c)   binder content determination
(d)   binder properties, rotary evaporator
(e)   binder properties, fractioning column
(f)   mix temperature
(g)   moisture content of mix
(h)   texture depth, sand patch methods
(i)   grading density of mix using solvents
(j)   density of mix using water
(k)   density of sample, hydrostatic
(l)   density of sample, measurement
(m)   density of sample, gamma ray
(n)   void content
(o)   compaction, degree achieved
(p)   compactability, resistance of the mix to compaction
(q)   sensitivity to water, uncompacted mix
(r)   sensitivity to water, compacted mix
(s)   abrasion by studded tyres
(t)   abrasion of porous asphalt
(u)   permeability of porous asphalt.

These represent the so-called basic tests and to the UK users of recipe mixtures, the array must seem bewildering. Inevitably the need for some of them will be questioned. There is increasing environmental pressure within the EC to dispense with tests requiring the use of solvents.

The implications of moving away from established test methods used in a recipe mix regime within a member state include both a period of uncertainty and cost. So far as uncertainty is concerned there will be a six month overlap period after the CEN Specification is formally published and before British Standards are withdrawn. Once draft ENs are available, it is expected that both sides of the industry will start to use them to gain familiarity and to reveal any practical problems. Direct costs will be those of simply purchasing new equipment, and for the routine sieve analysis and binder content tests, the costs of UK totally changing from its present systems has been estimated at £1–2 million. There is no likelihood of any UK Government or EC contribution toward these costs.

In the TG2 discussions it was accepted with some relief that the concept that these two test methods are described only in terms of their common principles but with specified performance characteristics. Tests themselves are used to assess compliance with the specification. While all member states make wide use of recipe mixes, with design methods used to optimize binder contents of essentially empirical gradings, most delegates see the CEN work as a splendid opportunity to eliminate the problems which arise when a material is found to be outside the specification by one or two per cent on one sieve size. So efforts are being made to specify mixtures in terms of their functional characteristics such as fatigue strength, dynamic stiffness, resistance to wheel track rutting and skid resistance.

This approach has many advantages. The foremost is that traditional national materials may be specified within a TC 227 produced catalogue of mixtures, not in recipe terms, as now, but in the terms referred to. For instance, hot-rolled asphalt wearing course could be described in the following terms, dispensing with fatigue requirement

(a)   stiffness — 3 point bending test, 5 Hz at 20°C, > 3.5 GPa.
(b)   wheel tracking < 2 mm/h at 45°C.
(c)   permeability < 10−5m/day.
(d)   surface texture — min. 1.5 mm (sand patch).

An innovative contractor could produce a totally new mix having these characteristics.

TG2 has therefore also identified more fundamental tests which are required for measuring the following properties

(a)   fatigue
(b)   permanent deformation
(c)   dynamic stiffness
(d)   low temperature cracking and
(e)   ageing of binder.

Fatigue is currently a problem because while there are many university-based test rigs of some complexity, none have the degree of contractual robustness to gain a place in a working standard. In the short term, fatigue is likely to be taken care of by ensuring the mixtures contain adequate quantities of binder.

At this stage it is not possible to pre-judge the outcome of TG2's work but members of a BSI sub-committee supporting the UK delegates to TG2 worked for 400 hours to produce 350 hours of word-processing which in turn resulted in 150 sheets of A4 submission to TG2 in a period of five months. All this work was on a voluntary basis.

The batches of draft standards for tests are circulated for comment within TG2 then recirculated after amendment before being considered by WG1 in plenary session. The drafts are then circulated for national comment within member states, with any proposals for change being dealt with by TG2. The final drafts will be approved by WG1 and published as provisional European Standard test methods. Provisional standards are labelled pre-ENs.

11.8.3. TG3 — specifications
Each member state has its own national specifications and, whatever their weaknesses, the people involved are familiar with the contractual framework within which they are used. Faced with the EC-driven requirement to move

from their familiar documents to something completely new, they are naturally apprehensive. Ideally, TC 227 would preserve the best of current standards and quietly allow the others to be forgotten. The TC 227 work certainly gives member states opportunity to bring a fresh approach to drafting specifications and many who are involved at first hand with national specification work do not wish to see this chance wasted.

CEN requires that specifications for materials are written wherever possible in terms of their performance but despite this, many within TG3 have been attempting simply to rationalize existing national recipe mix requirements and classify them in terms of traffic and climate. Another thrust has been to either recover cores or slabs of the materials from road or airfield pavements, after placing and compacting and measure their functional characteristics or even to measure such characteristics in situ. While surface texture can only be measured in situ, the in situ testing as-laid and compacted to judge the producer's compliance with the material production specification will not be permitted by CEN.

Instead that part of the specification for the material covering the producer/laying contractor interface will ideally specify the material in terms of its functional behaviour, given certain pre-determined requirements for both placing and compacting which are, of course, outside the control of the producer. In the UK the producer and the laying contractor are sometimes the same but not on mainland Europe.

Thus TG3 has recommended that at least the following fundamental properties should be provided for in WG1's ENs

- (a)   fatigue resistance, including any healing effects
- (b)   deformation resistance
- (c)   load spreading stiffness
- (d)   durability, covering (i) stripping of binder, (ii) ageing, (iii) low temperature cracking, (iv) reflective cracking.
- (e)   compactibility and voids after compaction
- (f)   surface texture
- (g)   permeability
- (h)   resistance against abrasion.

TG2 has therefore been asked to consider this list and produce tests to measure them. These must include

- (a)   a standardized wheel tracking test (there are currently at least three types of wheel tracking equipment, English, French and German)
- (b)   a test for permanent deformation other than a wheel tracking machine
- (c)   a test for dynamic stiffness
- (d)   a test for fatigue strength
- (e)   a test for workability or compactability.

TG3 also identified the need for a standard method of sample preparation and compaction, to provide test pieces for these tests but until the latter have been determined, it will be difficult for TG2 to know what size of sample is needed.

The difficulty in measuring these properties has been recognized and it is

Table 11.12. Possible functional specification for asphaltic concretes

Property	Test method	Sample preparation	Test criteria	Level	Remarks, test conditions
Mechanical properties					
Fatigue	bending tests, trapezoidal specimens	slab compactor	ε6 (10°C at 50 Hz)	100 — 130 — 160	
Resistance against deformation	wheel tracking test, 60°C	slab compactor 100 mm thick	30 000 cycles	< 10% < 5% test can be out to 100 000 cycles	
Healing	fatigue test with pauses	slab compactor			research test
(Dynamic) stiffness	flexural test	slab compactor	15°C at 10 Hz	5·4 GPa 8·0 GPa 10·0 GPa	
	complex modulus tensile test	Slab compactor	10°C at 0.02 s	6·0 GPa 9·0 GPa 12·0 GPa	
Durability					
Resistance against chemicals (thawing agents)					
Cracking					
Resistance against wear					
Ageing					
Stripping	Duriez test	compression	r (in water) R (in air)	0.80	
Workability					
Compactability	gyratory shear compactor		80 gyrations (0/14)	4 to 9 % voids	
Segregation	gyratory shear compactor		60 gyrations (0/10)	4 to 9 % voids	

* Asphalt concrete 0/10, 0/14, thickness > 50 mm

Table 11.13. Empirical specification of dense asphaltic concrete for surface layers

Property	8 mm	11 mm	16 mm
Retained on sieve 8 mm/11 mm/16 mm 2 mm 63 μ	< 5% 45–60% 90–94%	< 5% 50–60% 92–94%	< 5% 55–65% 92–94%
Binder content	6.5–7.5%	6.0–7.0%	5.5–6.5%

possible that, for instance, in the absence of an immediately available, contractually sound, robust fatigue test, with good precision, minimum binder contents for mixtures may have to suffice in the short term.

At present, efforts are being made to define mixtures in terms of both functional or fundamental criteria and empirical criteria. To some degree, the fundamental concept is pursuing a UK idea emanating from TRL which is if it is possible to specify hot-rolled asphalt wearing course in functional terms, relating to its use in the road, WG1 will have satisfied CEN. UK purchasers could then specify hot-rolled asphalt in these terms and know exactly what they will receive. Conversely, if an innovative contractor recognized an opportunity to develop a mixture to satisfy these same criteria, he would ideally be able to submit a bid for work on this basis.

So, at one attempt, WG1 would not only have satisifed CEN but also have encouraged the production of new and improved mixtures. This concept of the contractor innovating mixtures is still considered too bold to receive widespread acceptance but it may yet happen.

Extracts from a WG1/TG3 discussion paper concerning a possible functional specification for dense asphaltic concrete are illustrated in Table 11.12 and demonstrate current thinking. The particular materials used as examples are French, and only the French specify materials to this extent in performance terms and even they are not yet able to specify for all characteristics which TG3 see as important.

Alternatively an empirical specification for similar types of mixtures could be as shown in Table 11.13.

For its empirical specification work TG3 has stated that mix type, binder content and grade, Marshall properties, void content, layer thickness and aggregate type/properties must be included. It has identified that in considering test methods, TG2 may need to accept more than one fundamental test for a particular property since different tests may be more suitable for different aggregate sizes. Also, until fundamental testing is harmonized, the performance level of empirical mix types should be demonstrated by fundamental testing before equivalence i.e. equality of performance, could be accepted.

Acceptable quality levels

Given the wide disparity in standards of production and laying in some member states, many UK engineers will be concerned to ensure that the quality of UK materials is not diminished when British Standards are superseded by ENs. This aspect of standards has raised yet again the question of whether

(*a*)   a set of category B standards for asphalt materials
(*b*)   a set of category Bh standards for attestation
(*c*)   a set of category Bh standards for test methods
(*d*)   a category A standard for laying.

It also appears that category B product standards will operate under the regulations deriving from the directive and public sector purchasers will not be allowed to enhance these specifications.

Category A standards will operate under the contract law of a member state unless other CEN directives more specifically relate to the essential requirements of public completed works. The problem faced by all TC 227 WGs is that the CPD is written for building materials whereas TC 227 deals with roads and airfield pavements. It is the care and competency with which the materials are transported, laid and compacted which will determine the durability, safety in use and environmental well-being of completed pavements.

It will be vital to ensure that the laying standard does not prevent the use of a material considered suitable by the product standard. TG4 sees the WG1 standards fitting together in the following way.

Product standards — category B. These will need to contain the requirements for the raw materials, bitumen, aggregate, filler and any additives. Also, the requirements for mixed asphalt will have to be expressed in one of two formats, either fundamental or recipe/empirical.

An attestation standard — category Bh. This will set out the procedures which must be carried out to demonstrate conformity with the product standard to affix the CE Mark. It is envisaged that this will work in the following way.

I. Demonstration of mix properties. The first stage is a set of procedures to demonstrate that the mix will give the required properties. Some would call this design, others job standard mix; the CEN term is type testing.

Both the fundamental and recipe/empirical approaches will have sets of requirements linked to performance. The material producer must demonstrate that the proposed mix will meet those requirements. The aim of this exercise is to fix a mix composition which will then be subjected to factory production control routines which will be as simple as possible.

Some questions still remain such as what is the required frequency of this type approval procedure and how long does a mix design stay valid? Should the type testing procedure be used to determine the range of composition over which the required properties are achieved i.e. should compliance limits be set based on the sensitivity of individual mixes and should type testing be carried out on laboratory produced mixtures or on samples removed from the mixing plant? This is a hotly disputed matter.

II. Factory production control. Factory production control is a quality system or set of procedures which ensure consistent quality of the type tested mix. The elements covered in factory production control could be

(*a*)   quality of incoming raw materials
(*b*)   control of raw materials in depot
(*c*)   management of plant, including mixing temperature, mixing time, etc
(*d*)   type of plant

(e)   testing and inspection frequency
(f)   management of quality of testing and inspection
(g)   acceptable Quality Level (AQL) and
(h)   quality of transport/insulation/temperature control.

Compliance with the above requirements would be necessary before the CE mark could be stamped on the delivery tickets for materials produced at the plant. It is clearly important to establish the AQL for mixtures as this will have a major impact on limits and test frequencies.

III. Test method standards — category Bh. This is self-explanatory and covers the work produced by TG2.

IV. A laying standard — category A. WG1 must draft a standard which can be used to specify all those properties which are under the control of the laying contractor but which does not duplicate or conflict with the category B product standard. This will specify what the contractor must do with the CE marked material when it arrives on site. The elements which can be controlled by this standard would be

(a)   application of tack coats/surface preparation
(b)   joints
(c)   thickness
(d)   evenness
(e)   segregation
(f)   temperature during laying
(g)   temperature/time between mixing and laying
(h)   compaction degree
(i)   texture.

11.8.5. Overview of TG2, TG3 and TG4

While it is easy to define an interface between TG2 and TG3, the TG3/TG4 boundary is more difficult to specify. Initially, it had been planned to prepare draft specifications then add attestation of conformity clauses. Only later was it realized how dominant the AOC requirements could become. WG1 operates in plenary sessions and no TG has the authority to make decisions. Instead it recommends courses of action to the WG which accepts, rejects or modifies them.

An important matter concerning the TG3/TG4 work is the question of whether the material used to establish mix type approval should be mixed in the laboratory or in the mixing plant. Clearly, it is convenient to mix and test it in the laboratory, but the fact is that apparently identical mixtures prepared in the laboratory and taken from a real-life plant often behave quite differently. Mix-type approval is not mix design. It is the process by which the purchaser assures himself that the material to be laid in his pavement will have the fundamental characteristics he has specified. It is anticipated, therefore, that a producer will either carry out a new mix design for a job, or use a mix which, while new to the client, is familiar to the producer. The necessary mix type approval steps will then be demonstrated before defining that same mix in composition terms which will be checked in the mixing plant factory production control processes.

While the difficulties implicit in taking a representative sample from a mixing plant should not be under-estimated, it is equally necessary to appreciate the fundamental differences which can occur between laboratory and plant-mixed materials. These include

(a) the abrasion of the aggregate in the plant mixer thus producing extra filler with the subsequently increased binder demand

(b) the effectiveness of aggregate drying processes in the plant and hence the moisture content of the plant mix

(c) mixing temperatures which affect the hardening of the bitumen and will be generally much easier to control in the laboratory than on a plant working to capacity and producing sequential mixes with different binders

(d) mixing time, which is easy to control in both laboratories and plants but could be prolonged in plants because of e.g. efforts made to reduce the moisture content of aggregates kept in outside stockpiles.

In Denmark, where contractors are required to give a five year guarantee for work, producers test plant-mixed materials to ensure that they themselves are not at risk of laying poor materials. Where materials are not checked at post-production stage, therefore, the client takes the risk. Many consider this totally unacceptable and feel that since only the producer handles the materials, he should be the one who carries the risk. The debate continues.

11.9. Summary

It is clear that many mainland European black top practices could be used to advantage in the UK. Enterprising UK contractors are already establishing technical links with their counterparts on the European mainland and at least two French-developed ultra-thin wearing course mixtures are already being specified in UK. This practice can only grow and it is likely that split-mastic asphalt will also be laid here soon, so long as its surface texture can be sufficiently enhanced in its early life when the high binder content produces a very smooth surface texture. At best, ENs will not displace BSI documents until 1997 and it is anticipated they will tend to confirm what, by then, will be relatively common UK practice in the use of what are currently seen as novel mixtures.

Certainly, if the change introduces robust contractual test methods and the opportunity to use these in specifying and measuring fundamental characteristic requirements also allows the industry to have the freedom to use its virgin or reclaimed materials to best advantage then all concerned should benefit from CEN-related work. TC 227's ENs will certainly result in producers having to enhance their own technical strengths, whereas in the UK at least, WG1's intentions for CE mark procedures and the growth in non-CEN related contractor testing will certainly reduce the work-load of public sector test laboratories.

In an era of increasing financial pressure, this will be welcomed by all involved in the design, production and maintenance of black top pavements. Significant change will be needed in client culture here, not least because it is understood that clients will not be able to include their own particular requirements as supplements to ENs. Only time will show how dramatic the efforts made by the EC to reduce barriers to trade within the community will affect current UK black top practices, but this opportunity to make substantial improvements should not be lost.

11.10. Appendix 1

CEN TC227 WG1: bituminous mixtures
TG1: terminology: explanatory note to the third draft

In preparing this document, it has been assumed that the scope is defined by the following considerations

- (*a*)   the termonology for binders is the responsibility of TC19 SC1.
- (*b*)   the terminology for aggregates is the responsibility of TC154.
- (*c*)   the terminology for surface dressings, slurry seals, concrete and waste/marginal materials is the responsibility of the relevant WGs of TC227.

The TG has adopted the principle of attempting to define the essential character of the terms, without explanation. These proposals are shown in bold text, in parentheses. Explanatory text has been added where appropriate, in *italic* text. After the terminology has been agreed, this explanatory text could be deleted, but the TG proposes that some of it should be retained.

The terms included in this draft are those which are relevant to the work of TC227 WG1, in so far as they are apparent at present. Consequently, this draft does not contain terms for all the materials currently employed in all the CEN countries. There may be a need to add more terms, for materials in particular, after the work of TG3 has progressed.

Soft asphalt is included but there is no attempt to define it. This can only be done after TG3 has considered a proposal that is being prepared by the Nordic countries, which is presumably intended to define soft asphalt specifically.

TG1 suggests that WG1 should reconsider whether it is necessary or advantageous to harmonize specifications for soft asphalt, as their use is predominantly in the Nordic countries; they could therefore be covered by a non-CEN regional specification.

This third draft takes account of the comments received on the second draft, which were not extensive, and the discussions at the meeting of WG1 in December 1991.

Discussions at meetings of WG1, TG2 and TG3 have led to the conclusion that the retention or use of any root word (e.g. surface or base) in more than one term will inevitably result in misunderstanding. Each root word is therefore used only once in the definitions proposed in this draft.

The TG has also retained in the draft the principle of defining terms for courses and for materials separately and not combining them (e.g. surface course asphaltic concrete) on the grounds that the meaning of such composite terms is obvious from the definitions of its components.

Convenor
22 February 1992

CEN TC227 WG1: Bituminous mixtures
Terminology: Draft
Scope
All materials and applications under the jurisdiction of CEN TC227 WG1
are defined.
they are considered in two categories, i.e.:

1   Terms for *Materials*
2   Terms related to *Pavement Structures*

The definitions are shown in *bold* lettering. Supporting or explanatory text
is given in *italic* lettering.

A MATERIALS

1   *Asphalt*
        'A mixture of mineral aggregate and a bituminous binder.'
        This is a general *term, embracing all of the particular cases
        enumerated below.*
            *The term 'bituminous' has been adopted by TC19 SC1 to exclude
        tar and this has been accepted by TC227 WG1. The use of 'bituminous'
        i.e. an adjustive, instead of the noun 'bitumen', allows modified
        bitumens to be embraced by the definition proposed above.*

2   *Rock Asphalt*
        **'Asphalt composed of rock which has been naturally impregnated
        with bitumen.'**
        *Those natural asphalts, such as lake asphalt, which have the
        characteristics of a mastic, are considered to be* binders *and are
        therefore not included in this document.*

3   *Asphalt Types*
        3.1 *Asphalt Concrete*
        **'Asphalt in which the aggregate particles are essentially
        continuously-graded to form an interlocking structure'.**
        There are three sub-types of asphalt concrete, as follows:
            dense, medium and open.
        *These sub-types are not at present precisely defined, but they may
        be so in the future. The air voids content of these sub-types are*
        characteristically, *but* not precisely *3–6%, 6–12% and >12%
        respectively.*
            *In the UK these materials have been described previously as
        macadams.*
            *When asphalt concrete is used for heavy duty, it is normal practice
        to employ dense materials which have been formulated by a design
        procedure.*

3.2 *Porous asphalt*

'An asphalt formulated to have a very high content of interconnecting voids so as to facilitate the passage of water.'

Porous asphalt is the generally-accepted term for an asphalt of the gap-graded type which has been designed to have a very high voids content, i.e. of 20% by volume or more. It was previously known as pervious macadam in the UK.

3.3 *Hot-Rolled Asphalt*

'An asphalt composed of coarse aggregate dispersed in a mortar composed of sand, filler and bitumen.'

3.4 *Mastic Asphalt*

'An asphalt in which the volume of the combination of filler and binder substantially exceeds the volume of the voids in the skeleton of the remaining aggregate, they produce an asphalt which may be poured when hot and requires no compaction.'

Mastic asphalts contain either no coarse aggregate or a relatively small amount of such. They are normally hand-laid.

3.5 *Gussasphalt*

'A mastic asphalt containing a relatively hard grade of bitumen and a substantial amount of coarse aggregate and with characteristics which enable it to be laid mechanically or by hand.'

It is common practice to use Trinidad lake asphalt and/or a modified bitumen as part or all of the binder.

3.6 *Stone Mastic Asphalt*

'A gap-graded asphalt composed of a coarse crushed aggregate skeleton bound with a mastic mortar.'

These materials are not pourable.

It is common practice to use additives and/or modified binders in the manufacture of these materials, especially to allow the binder content to be raised and to reduce segregation between the coarse aggregate and the mortar.

3.7 *Soft Asphalt*

(Definition awaiting the decisions of TG3 on this matter.)

B PAVEMENT STRUCTURES

1 *Pavement*
   'A structure, composed of one or more courses, to assist the passage of wheeled traffic over terrain.'

2 *Layer*
   'A structural element of a pavement laid in a single operation.'
   This term does not include membranes, tack coats, etc.

3   *Course*
'A structural element of a pavement constructed with a single material.'
A course may be laid in one or more layers.
   Courses which may be employed in pavement construction are: surface course, binder course, regulating course, base (if constructed with a single material), upper base, and sub base.
These terms are defined below.

   3.1 *Surface Course*
'The upper layer of the pavement which is in contact with the pavement.'

   3.2 *Binder Course*
'The part of the pavement between the surface course and the base.'
A binder course is not always used.
   This course was previously known as a base-course in some countries. This latter term was also sometimes used in reference to the base.

   3.3 *Regulating Course*
'A course of variable thickness applied to an existing course or surface to provide the necessary profile for a further course of consistent thickness.'

4   *Base*
'The main structural element of a pavement.'
The base may be laid in one or more course, described as upper base, lower base, etc.

5   *Sub-base*
'A course of material placed immediately beneath the base.'
A sub-base may be composed of bound or unbound material and it may be laid in more than one layer.

6   *Formation*
'The surface of the subgrade in its final shape after the completion of any earthworks.'

7   *Subgrade*
'The material immediately below the formation upon which the pavement is constructed.'

8   *Capping Layer*
'A layer of material immediately below formation level.'

   8.1 *Improved Sub-grade*
'A Capping Layer formed treating the sub-grade material in-situ.'

9   *Full-depth asphalt construction*
'A pavement in which all courses above the formation are composed of asphalt'.
Asphalt is used here in the general sense. The courses will usually be composed of different asphalt mixes.

Convenor
22.2.92

References

1. BRITISH STANDARDS INSTITUTION. *Aggregates from natural sources for concrete.* BSI, London, 1983, BS 882.
2. BRITISH STANDARDS INSTITUTION. *Coated macadam for roads and other paved areas. Specification for constituent materials and for mixtures.* BSI, London, 1993, BS 4987: Part 1.
3. PROPERTY SERVICES AGENCY. *Standard specification clauses for airfield pavement works. Part 4 Bituminous Surfacing.* Airfields branch, PSA, Croydon, 1989.
4. GERMAN ASPHALT PAVEMENT ASSOCIATION. *Splittmastixasphalt.* DAV e.v., Bonn, 1992.
5. BRITISH STANDARDS INSTITUTION. *Hot-rolled asphalt for roads and other paved areas. Specification for constituent materials and asphalt mixtures.* BSI, London, 1992, BS 594: Part 1.
6. BRITISH STANDARDS INSTITUTION. *Sampling and examination of bituminous mixtures for roads and other paved areas.* BSI, London, 1990, BS 598: Part 107.
7. CENTRE DE RECHERCHE ROUTIERS. *Code of good practice for the formulation of dense bituminous mixtures.* CRR, Brussels, 1987, R61.
8. PIARC Dictionary of Highway Engineering Terms.
9. CEN Discussion paper on attestation of conformity — Luigi tree.

Index

Guildford College
Learning Resource Centre

Please return on or before the last date shown.
No further issues or renewals if any items are overdue.
"7 Day" loans are **NOT** renewable.

- 3 JUL 2009
25/09/09